T0321941

Security and Trust Issues
in Internet of Things

Internet of Everything (IoE): Security and Privacy Paradigm

Series Editors:
Vijender Kumar Solanki, Raghvendra Kumar, and Le Hoang Son

Privacy Vulnerabilities and Data Security Challenges in the IoT
Edited by Shivani Agarwal, Sandhya Makkar, and Tran Duc Tan

Handbook of IoT and Blockchain
Methods, Solutions, and Recent Advancements
Edited by Brojo Kishore Mishra, Sanjay Kumar Kuanar, Sheng-Lung Peng, and Daniel D. Dasig, Jr.

Blockchain Technology
Fundamentals, Applications, and Case Studies
Edited by E Golden Julie, J. Jesu Vedha Nayahi, and Noor Zaman Jhanjhi

Data Security in Internet of Things-Based RFID and WSN Systems Applications
Edited by Rohit Sharma, Rajendra Prasad Mahapatra, and Korhan Cengiz

Securing IoT and Big Data
Next-Generation Intelligence
Edited by Vijayalakshmi Saravanan, Anpalagan Alagan, T. Poongodi, and Firoz Khan

Distributed Artificial Intelligence
A Modern Approach
Edited by Satya Prakash Yadav, Dharmendra Prasad Mahato, and Nguyen Thi Dieu Linh

Security and Trust Issues in Internet of Things
Blockchain to the Rescue
Edited by Sudhir Kumar Sharma, Bharat Bhushan, and Bhuvan Unhelkar

For more information about this series, please visit: https://www.crcpress.com/ Internet-of-Everything-IoE-Security-and-Privacy-Paradigm/book-series/ CRCIOESPP

Security and Trust Issues in Internet of Things
Blockchain to the Rescue

Edited by
Sudhir Kumar Sharma, Bharat Bhushan, and Bhuvan Unhelkar

CRC Press
Taylor & Francis Group
Boca Raton London New York

CRC Press is an imprint of the
Taylor & Francis Group, an **informa** business

First edition published 2021
by CRC Press
6000 Broken Sound Parkway NW, Suite 300, Boca Raton, FL 33487-2742

and by CRC Press
2 Park Square, Milton Park, Abingdon, Oxon, OX14 4RN

© 2021 Taylor & Francis Group, LLC

CRC Press is an imprint of Taylor & Francis Group, LLC

Reasonable efforts have been made to publish reliable data and information, but the author and publisher cannot assume responsibility for the validity of all materials or the consequences of their use. The authors and publishers have attempted to trace the copyright holders of all material reproduced in this publication and apologize to copyright holders if permission to publish in this form has not been obtained. If any copyright material has not been acknowledged please write and let us know so we may rectify in any future reprint.

Except as permitted under U.S. Copyright Law, no part of this book may be reprinted, reproduced, transmitted, or utilized in any form by any electronic, mechanical, or other means, now known or hereafter invented, including photocopying, microfilming, and recording, or in any information storage or retrieval system, without written permission from the publishers.

For permission to photocopy or use material electronically from this work, access www.copyright.com or contact the Copyright Clearance Center, Inc. (CCC), 222 Rosewood Drive, Danvers, MA 01923, 978-750-8400. For works that are not available on CCC please contact mpkbookspermissions@tandf.co.uk

Trademark notice: Product or corporate names may be trademarks or registered trademarks, and are used only for identification and explanation without intent to infringe.

Library of Congress Cataloging-in-Publication Data
Names: Sharma, Sudhir Kumar, editor. | Bhushan, Bharat, 1949- editor. |
Unhelkar, Bhuvan, editor.
Title: Security and trust issues in Internet of Things : Blockchain to the
rescue / edited by Sudhir Kumar Sharma, Bharat Bhushan, Bhuvan Unhelkar.
Description: First edition. | Boca Raton : CRC Press, 2021. | Includes
bibliographical references and index.
Identifiers: LCCN 2020032471 (print) | LCCN 2020032472 (ebook) |
ISBN 9780367490652 (hardback) | ISBN 9781003121664 (ebook)
Subjects: LCSH: Internet of things—Security measures. | Blockchains
(Databases)
Classification: LCC TK5105.8857 .S427 2021 (print) | LCC TK5105.8857
(ebook) | DDC 005.8—dc23
LC record available at https://lccn.loc.gov/2020032471
LC ebook record available at https://lccn.loc.gov/2020032472a

ISBN: 978-0-367-49065-2 (hbk)
ISBN: 978-1-003-12166-4 (ebk)

Typeset in Times
by codeMantra

Contents

Preface

The recent advancements in the field of information and communication technology (ICT) have enabled the promotion of conventional computer-aided industry toward becoming smart industries that are powered by data-driven decision-making. In this paradigm shift, Internet of Things (IoT) plays a vital role in connecting the cyberspace of computing systems and the physical industrial environment. IoT aims to improve production throughput and operation efficiency and to mitigate the machine downtime. Decentralization of IoT systems, heterogeneity of IoT data, diversity of systems and IoT devices, and network complexity are the characteristic features of IoT. Owing to the massiveness of IoT and inadequate data security, the impact of security breaches may turn out to be humongous, leading to severe impacts. In addition to connecting devices, IoT connects people and other entities, thereby making every IoT component vulnerable to a large range of attacks.

Blockchain, a tamper-resistant and distributed ledger that maintains consistent data records at different locations, has the capability to address IoT security concerns. Owing to the fault-tolerance capabilities, decentralized architecture, and cryptographic security benefits such as authentication, data integrity, and pseudonymous identities, security analysts and researchers consider blockchain in resolving privacy and security issues of IoT. Apart from providing data security solutions to IoT, blockchain also encounters numerous inherent IoT challenges such as limited computing power, massive numbers of devices, error-prone radio links, and low communication bandwidth. In addition, blockchain is used in identity management, voting, healthcare, government records, and supply chain. Blockchain finds application in various economic sectors to build trust and has been applied successfully to digital cryptocurrencies. Hence, blockchain technology in today's world is becoming an emerging and disruptive future technology providing numerous opportunities to various industries.

The emerging and promising state-of-the-art of ubiquitous computing, future IoT and blockchain technology motivated us toward this book which focuses on the various aspects of IoT and blockchain systems such as trust management, identity management, security threats, access control, and privacy. The book aims to report advances in the field of IoT, ubiquitous computing, and blockchain, including applications, concepts, and theory. Moreover, the book aims to highlight the inherent blockchain possibilities for various economic sectors, as well as the additional value that blockchain can provide for the future of these sectors.

Editors

Sudhir Kumar Sharma is currently a Professor and Head of the Department - Computer Science, Institute of Information Technology & Management affiliated to GGSIPU, New Delhi, India. He has extensive experience of over 20 years in the field of Computer Science and Engineering. He obtained his Ph.D. in Information Technology from USICT, GGSIPU, New Delhi, India. Dr. Sharma obtained his M. Tech degree in Computer Science & Engineering in 1999 from Guru Jambheshwar University, Hisar, India and M.Sc. degree in Physics from the University of Roorkee (now IIT Roorkee), Roorkee in 1997. His research interests include machine learning, data mining, and security. He has published a number of research papers in various prestigious international journals and international conferences. He is a life member of CSI and IETE. Dr. Sharma is an Associate Editor of the International Journal of End-User Computing and Development (IJEUCD), IGI Global, U.S.A. He is a convener of ICETIT-2019.

Bharat Bhushan is an Assistant Professor in the Computer Science and Engineering (CSE) Department at School of Engineering and Technology, Sharda University, India. He is an alumnus as well as a Ph.D. scholar of Birla Institute of Technology, Mesra. He received his Undergraduate Degree (B-Tech in Computer Science and Engineering) with Distinction in 2012 and his Postgraduate Degree (M-Tech in Information Security) with Distinction in 2015 from Birla Institute of Technology, Mesra, India. He has earned numerous international certifications, such as Cisco Certified Network Associate (CCNA), Cisco Certified Entry Networking Technician (CCENT), Microsoft Certified Technology Specialist (MCTS), Microsoft Certified IT Professional (MCITP), and Cisco Certified Network Professional Trained (CCNP). In the last 3 years, he has published more than 50 research papers in various renowned international conferences and SCI-indexed journals, including *Wireless Networks* (Springer), *Wireless Personal Communications* (Springer), and *Emerging Transactions on Telecommunications* (Wiley). He has contributed several book chapters in various books and is currently in the process of editing seven books from the famed publishers such as Elsevier, IGI Global, and CRC Press. He has served as a Reviewer/Editorial Board Member for several reputed international journals, including *IEEE Access*, *IEEE Communication Surveys and Tutorials*, and *Wireless Personal Communication* (Springer). He has also served as a Speaker and Session Chair at more than 15 national and international conferences. In the past, he worked as a Network Engineer in HCL Infosystems Ltd., Noida and as assistant professor in the department of computer science and engineering, HMR Institute of Technology and Management, New Delhi, India. He has qualified GATE exams for successive years and gained the highest percentile of 98.48 in GATE 2013. He is also a member of numerous renowned bodies, including IEEE, IAENG, CSTA, SCIEI, IAE, and UACEE.

Bhuvan Unhelkar is an accomplished IT professional and Professor of IT at the University of South Florida, Sarasota–Manatee (Lead Faculty). He is also Founding Consultant at MethodScience and a Co-Founder/Director at PlatiFi. He has mastery in business analysis and requirements modeling, software engineering, big data strategies, agile processes, mobile business, and green IT. His domain experience is in banking, financial, insurance, government, and telecommunications. Bhuvan is a thought leader and a prolific author of 20 books, including *Big Data Strategies for Agile Business and The Art of Agile Practice* (Taylor & Francis/CRC Press, U.S.A.). Recent Cutter executive reports (Boston, U.S.A.) include psychology of Agile, business transformation, collaborative business and enterprise agility, and Agile in practice - a composite approach. He is a winner of the Computerworld Object Developer Award (1995), Consensus IT Professional Award (2006), and IT Writer Award (2010). He received his Doctorate in "Object Orientation" from the University of Technology, Sydney in 1997. Bhuvan is a Fellow of the Australian Computer Society, IEEE Senior Member, Professional Scrum Master, Life member of Computer Society of India and Baroda Management Association, Member of SDPS, Past President of Rotary Sarasota Sunrise (Florida) & St. Ives (Sydney), Paul Harris Fellow (+6), Discovery volunteer at NSW parks and wildlife, and a previous TiE Mentor. He also chaired the Business Analysis Specialism Group of the Australian Computer Society.

Contributors

K. Adalarasu
Department of Electronics and
    Instrumentation Engineering
SASTRA Deemed to be University
Thanjavur, India

Rangel Arthur
School of Technology (FT)
University of Campinas (UNICAMP)
Limeira, Brazil

Vidyadhar Aski
Computer and Communication
    Engineering
Manipal University Jaipur
Jaipur, India

R. Aswath Srimari
Department of Information
    Technology
PSG College of Technology
Coimbatore, India

Bharat Bhushan
School of Engineering and
    Technology
Sharda University
Greater Noida, India

A. Boulmakoul
FSTM, Hassan II University of
    Casablanca
Casablanca, Morocco

A. Chauhan
Delhi Technological University
New Delhi, India

Reinaldo Padilha França
School of Electrical and Computer
    Engineering (FEEC)
University of Campinas (UNICAMP)
Campinas, Brazil

M. Gębska
Warsaw University of Life Sciences
Warszawa, Poland

Sukriti Goyal
HMR Institute of Technology &
    Management
New Delhi, India

Moksh Grover
HMR Institute of Technology &
    Management
New Delhi, India

M. Hajder
University of Information Technology
    and Management
Rzeszow, Poland

P. Hajder
AGH University of Science and
    Technology
Kraków, Poland

U. Hariharan
Galgotia College of Engineering and
    Technology
Greater Noida, India

P. Harini
Department of Electronics and
    Instrumentation Engineering
SASTRA Deemed to be University
Thanjavur, India

Y. Hasija
Delhi Technological University
New Delhi, India

J. Horn Nord
Spears School of Business
Oklahoma State University
Stillwater, Oklahoma

Yuzo Iano
School of Electrical and Computer
    Engineering (FEEC)
University of Campinas (UNICAMP)
Campinas, Brazil

M. Jagannath
School of Electronics Engineering
Vellore Institute of Technology (VIT)
    Chennai
Chennai, India

L. Karim
National School of Applied Sciences,
    ENSA Berrechid
Hassan 1st University
Settat, Morocco

G. R. Karpagam
Department of Computer Science and
    Engineering
PSG College of Technology
Coimbatore, India

Ila Kaushik
Krishna Institute of Engineering &
    Technology
Ghaziabad, India

Aditya Khamparia
Lovely Professional University
Punjab, India

A. Koohang
Middle Georgia State University
Macon, Georgia

Abhijeet Kumar
HMR Institute of Technology &
    Management
New Delhi, India

Wu Linjing
Donghua University
Shanghai, China

M. Liput
University of Information Technology
    and Management
Rzeszow, Poland

M. Mądra-Sawicka
Warsaw University of Life Sciences
Warszawa, Poland

Ana Carolina Borges Monteiro
School of Electrical and Computer
    Engineering (FEEC)
University of Campinas (UNICAMP)
Campinas, Brazil

M. Nahri
FSTM, Hassan II University of
    Casablanca
Casablanca, Morocco

M. Nycz
Rzeszow University of Technology
Rzeszow, Poland

J. Paliszkiewicz
Warsaw University of Life Sciences
Warszawa, Poland

Anubha Parashar
Computer Science and Engineering
Manipal University Jaipur
Jaipur, India

Apoorva Parashar
Computer Science and Engineering
Maharshi Dayanand University
Rohtak, India

Nilotpal Pathak
Galgotia College of Engineering and
    Technology
Greater Noida, India

K. Rajkumar
Galgotia College of Engineering and
    Technology
Greater Noida, India

M. Saračević
Department of Computer Sciences
University of Novi Pazar
Novi Pazar, Serbia

A. Selimi
Faculty of Informatics
International Vision University
Gostivar, North Macedonia

Nikhil Sharma
Computer Science & Engineering
HMR Institute of Technology &
    Management
New Delhi, India

Rajveer Singh Shekhawat
Computer Science and Engineering
Manipal University Jaipur
Jaipur, India

Keith Sherringham
A.C.N. 629 733 633 Pty. Ltd.
Sydney, Australia

Shihu Shu
Donghua University & Shanghai
    National Engineering Research
    Center of Urban Water Resources
    Co. Ltd
Shanghai, China

S. Sridevi
Department of Computer Science and
    Engineering
PSG College of Technology
Coimbatore, India

B. Tharunika
Department of Electronics and
    Instrumentation Engineering
SASTRA Deemed to be University
Thanjavur, India

Bhuvan Unhelkar
College of Business
University of South Florida
    Sarasota–Manatee
Sarasota, Florida

B. Vinoth Kumar
Department of Information
    Technology
PSG College of Technology
Coimbatore, India

S. Vishnuvardhan
Department of Information
    Technology
PSG College of Technology
Coimbatore, India

Liu Xinyue
Shanghai National Engineering
    Research Center of Urban Water
    Resources Co. Ltd.
Shanghai, China

1 IoT Fundamentals and Challenges

U. Hariharan, K. Rajkumar,
and Nilotpal Pathak

Galgotias College of Engineering and Technology

CONTENTS

1.1 INTRODUCTION

Internet of Things (IoT) is a paradigm that links real items to the Internet, enabling issues to gather, progress, and talk informatively with no human intervention. The perspective of IoT is to create a better society for people where items (referring to actual physical things; the condition of objects, devices, entities, and elements are utilized interchangeably) around us can fully grasp our likeness and personal preferences to act appropriately with no explicit directions [1, 2]. The quick advancements

1

in inexpensive sensor production, correspondence protocols, lodged programs, actuators, and hardware miniaturization have contributed to the exponential development of the IoT. Physical items belonging in the real world are lodged with the solutions to produce them intelligently. The performance of intelligent products can be abstracted by a programmer, and IoT software can function by merging innovative products with solutions that help deal with issues of daily tasks. The various uses of IoT can be categorized in three different domains: society, business, and environment, as illustrated in Figure 1.1. When the overall culture is transitioning toward IoT, the number of sensors deployed close to the planet is growing at an immediate rate, and the receptors consistently generate a substantial amount of information [3].

Nevertheless, not all this information offers strategies that aid in creating procedures. Employing the unit search function supplied by an IoT program, the size and range of data gathered can be lowered. Therefore, a vital system is one that enables to look properly for smart devices according to the real-world attributes collected by the receptors. The benefits of the search engines and find functionalities in IoT have been emphasized. Nevertheless, because of a wide variety of free receptors to choose from and the source limits of an IoT program, developing the search engine program is tough [4].

1.1.1 MOTIVATION OF STUDY/MAJOR CONTRIBUTION

In the last 2 years, technological innovation has progressed significantly, particularly in terms of interconnections of computer systems, products, items, and

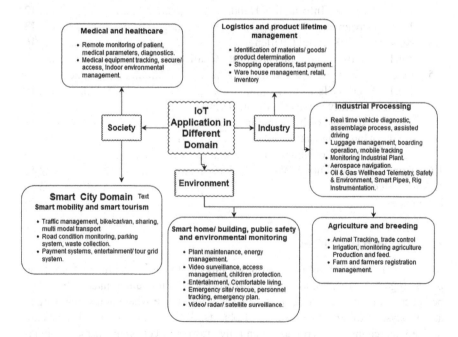

FIGURE 1.1 Utilization of IoT in various domains.

personals. In doing so, Internet has matured with substantial improvements. The very best instance of this progression is the creation of virtually the most amazing system, the Internet. Instead we look for those breakthroughs that will truly change how we live and work with computing group [5]. Some technologies are mailed out computing, power method estimations, cloud bases computing services, and ubiquitous computing. The improvisation and extemporization of technologies and computing strategies drive the planet toward another unexpected paradigm, IoT.

According to the International Telecommunication Union (ITU) [6], IoT is essential with respect to five projecting and noticeable investigative domains, including mobile computing, wireless sensor networks, pervasive computing, IoT, cyber-physical methods. IoT is an upcoming field that is moving upwards and surrounding the entire society under its umbrella. This particular effort describes the condition of IoT in the long-term situation. While heading through the creation of IoT, we now present a few definitions in the field of IoT.

1.1.2 DEFINITION OF IoT

Based on ITU [7], the IoT is "a worldwide infrastructure with the info culture which allows innovative services by interconnecting (virtual and physical) issues based on existing and evolving interoperable info and communication technologies." In regards to connectivity, ITU-T realized IoT as a system with anyplace and anytime connectivity for anything or anyone.

Based on ITU [8] report, "IoT is going to connect the world's items in an intelligent and sensory way by integrating technical advancements in product identification (tagging things), wireless sensor networks and sensors (feeling things), embedded methods (thinking things nanotechnology and shrinking things)".

Section 1.2 discusses the growth and origin of IoT, Section 1.3 presents the different technical challenges and a comprehensive survey of several issues along with the suggested answers within IoT. Section 1.4 deals with environmental and societal concerns. Future challenges of IoT are discussed in Section 1.5. An extensive conclusion of method domains of IoT discussed in this chapter is presented in Section 1.6.

1.2 ORIGIN AND GROWTH OF IOT

Retreat is observed but largely restricted to small communities or carried out for community information. These two phrases are not disjointed as they are glued collectively through the user interface and are fused collectively via the linkage of the Internet. This particular linkage is directed toward the genesis of its paradigm.

The development and evolution of IoT depend on the creation of new solutions and strengthening pre-existing ones. The solutions varied in terms of infrastructure, infrastructure less, ad-hoc networks, RFIDs, nanotechnology, ubiquitous computing, and correspondence protocols [9]. In the evolution of IoT, the focus is on producing a novel worldwide network; this paradigm utilizes the web as the main wedge. In a layered Internet in which a particular computing unit can communicate with an opposite computing unit with free energy generation, IoT evolves [5]. IoT backlinks every item and transforms them into an intelligent thing. This transformation of the original item into an intelligent item, for example, a home appliance into an intelligent home appliance, is attainable by embedding computing, transmitting, and sensing features into them.

The two outstanding, meaningful, and contradictory assertions supplied from 2006 to 2009 are [10] as follows:

- With the development of Internet (the 1990s) and movable Internet (the 2000s), we are currently proceeding on to the third even likely nearly all disruptive stages of the Internet revolution – the IoT.
- The IoT is capable of altering the planet as the Internet did when it evolved, and it might be a great deal more.

IoT is an unruly stage of engineering. IoT is capable of generating a conclusion PC handler capable of using vulnerable info. On the other side, competence and possibility of technological implementation is successful in all the cases [11].

Individual device can connect to digital networks and the Internet with such as smartphones and computers in order to share information. In 2003, the global population was projected to remain 6.3 billion, while the attached products were only 5 billion. By 2010, 6.8 billion people had 12.5 connected products. By 2015, the population grew to 7.2 billion, while the attached unit matter hit 25 billion. By 2020, the population is expected to reach 7.7 billion, and the connected products could reach 50 billion [10] (Figure 1.2).

Within this particular background, the evolution of IoT is considered. The development of IoT might be reflected as simple variations in profound

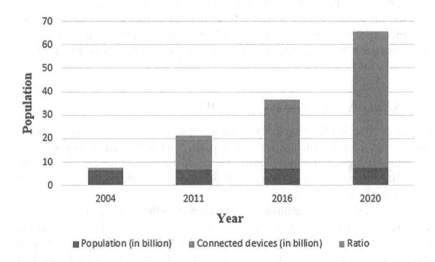

FIGURE 1.2   Comparison between connected devices and population.

technologies with time. At first, the computing community had merely stand-alone devices. After the arrival of social media, there was a need to achieve the demand to share and post information and materials. Interconnection of computing systems has led to the growth of nearly all attractions and the Internet, that is, prosperous networks. The web is made of intranets, organizations and unique computing systems, that offers limited infrastructure technologies to the users for communication. Creation of infrastructure-less communication in addition to the miniaturization of computing systems guides us toward portable systems [12].

Accessing the solutions together with electric gear as receptors and actuators led us to the current stage of technologies, in which each item is assumed to become sensed and possess several computing abilities, finally leading to the era of IoT. This development is presented pictorially in Figure 1.3.

The mass-scale deployment of homogenous IoT devices, the ability of some devices to automatically connect to other devices, the possibility to find out, identity, guide and the likelihood of fielding these devices in unsecure environments. In the future, the universe is going to appear as an interconnected net or perhaps

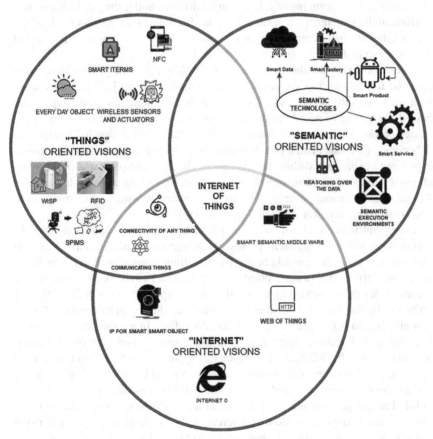

FIGURE 1.3 Evaluation of IoT.

linkage of each one of the pre-existing entities. The current initiatives are indulging computing power together with the potential to connect to just about everything [13]. IoT is not merely driven by the innovative developments in marketing communications and technologies, it has added features that are combined in a different way into daily everyday living. These issues might be intangible or tangible, invisible or visible, computing unit or perhaps a non-computing body [14]. Problems might belong to personal business or maybe a public organization. These issues might differ using things that are as basic as clothes, food material, and furniture to advanced units such as cellular cell phones, TV, and other electric home appliances. These items and entities are capable of efficiently communicating with one another, actively playing the job of sensors, which ultimately achieve the joint goal and carry out the typical job. Probably the most crucial element of IoT is its impact on the daily life of customers [15].

IoT will unquestionably enhance the quality of living by proposing brand new offerings to transform urban areas into smart towns and by strengthening the interaction between individuals and IoT devices/services. IoT has a remarkable effect within both domains of a house in addition to doing the job location. It is brilliant in the long term while supporting health, smart transportation, aided living, intelligent visitors, and creative community. The company situation cannot remain unblemished with its effects and effectiveness [16].

After examining all the choices, the U.S. National Intelligence Council has declared IoT among the six possible solutions, drawing in U.S. passions along with the way to 2025 [17]. Additionally, the case of interconnected gadgets surpassed the matter of people to drop to 2010, and it is expected that some interrelated products will reach 25 billion by 2020. This implies that IoT is expected to become one of the primary energy sources of crucial importance [18]. Such an exponential expansion within the vastness of the area and problems in the setup allows it to be a fascinating area of investigation. Nowadays, apps for IoT are also flipped to interpersonal everyday living uses, including smart grid, smart security, smart transportation, and intelligent homes [19].

The quantification of IoT is accomplished by learning and analyzing the accounts provided by different professionals and companies. Industry experts have predicted that there would be around 50 billion items related to IoT by 2020. Based on Gartner [10] (a conceptual investigation and advisory corporation), the matter of devices pre-existing within the IoT is expected to reach 30 billion by 2020. On the flip side, ABI Research predicted that the complete number of products attached to IoT wirelessly is over 30 billion [20]. In another survey done by Pew Research Internet Project, it is determined to have considerable and favorable significance by 2020 [21]. Therefore, it is apparent that IoT is going to consist of a large number of products getting connected via the Internet. In 2015 spending budget, the UK Government allocated £40,000,000 to research inside the domain of IoT. The UK government proclaimed that IoT might be the subsequent stage of the information revolution and then referenced the interconnectivity of most skin burns from the metropolitan transportation to medical-related products to home devices [22].

The European Research Cluster (IERC) [23] suggested that the IT paradigm is known as the IoT goals to create every technical entity that is in a place to communicate within a single package. Based on Miorandi et al. [24], IoT might be viewed as a technical paradigm pairing typical networks with networked entities. In fact, a system provider is achieved by the evolution of IoT. Academicians, scientists, and companies (private in addition to federal bodies) are anticipating building a functional, user-friendly, efficient, and convenient setting to live in and to operate. This revolution of moving ahead accessible design and practice of accessible development ensures both "direct access" (i.e. unassisted) and "indirect access" and media are lodging the computing merchandise figures, IoT such as a productive aspect of searching inside the approaching long term.

1.3　TECHNICAL CHALLENGES

From the viewpoint of a PC individual and IoT stakeholders, the difficulties and their solutions may differ. It is practical to believe that the present status of IoT might result in various obstacles. With this background, our objective is to examining and check several primary complications which should be concentrated to put IoT into action both productively and effectively.

Technical challenges are characterized into four categories: (1) Structure and privacy, (2) Security, (3) Efficient data handling, and (4) Resource management and heterogeneity.

1.3.1　HETEROGENEITY AND ARCHITECTURE

Numerous scientists have attempted to explain different architectural versions to put into action IoT; several of them are frequently placed on a specific program region. Castellani et al. [25] provided a structured layout, especially for an intelligent business office program. This particular program designed with the connection of WSN in addition to actuator links to the web is a neat program. These types of services include doorstep entree influence, granting authorization to an authenticated person, and, being an outcome, it takes a dependable networking and several identification solutions (e.g., RFID) [26]. It consists of three types of nodes, specifically base station node (BSN), mobile node (MN), and specialized node (SN). Every type seems to have character attributes of serotonin based around its mobility, its functionality assortment, and its specialization. A BSN is a fixed node; voice an IPv6 wireless router or perhaps IPv6 sink. Direct connectivity is provided by the node on the net. An MN is an external node, that is, a wireless dongle. It is based on compatibility problems toward neighborhood as well as neighborhood protocols configurations. Third, together with the main element of the entire framework, is an SN. It is a private node to confirm the particular service(s), claim local weather readings, provide regular water quantity readings, etc.

Beier et al. [27] provided the EPC (electronic product code) worldwide IoT structure. Its primary target is toward smart strategies and RFID network methods. This proposition is dependent on the idea of discovery expertise. Find products and provide a program which correlates the info about RFID-enabled goods, along with information exchange inside a source chain managing train. Find assistance is created by a well-defined data source with a set of total program interfaces. Some help is recommended by EPC global, particularly the object name service. It is viewed as being an example of a find program.

Additionally, they recommended to create an EPC quantity as a result of the correct URL, and then to access a history of EPCIS (Electronic Item Code Info Services) while using the current DNS. Whenever an organization utilizes the discovery service, it must become authorized by an official team. After becoming authorized, it is going to get a signed certification from an additional reliable team. Later, all of the transactions produced by the companies will likely be achieved through this specific certification. The five leading contents of the certification are:

(1) EPC quantity of the product, (2) Certification of the business foe which EPCIS submitted the record, (3) The web address on the EPCIS to show it has the custody of all of the items, (4) Timestamp was introduced, and (5) Visibility, a flag indicating in case the shoot may be discussed with anyone or otherwise.

To spread the example of IoT, a selection of layered architectures has been proposed by numerous scientists. Gronbaek [28] Dai and Wang [29] provided the OSI model as a structure. Tan and Wang [30] proposed that the structure will help support ubiquitous providers on an end-to-end schedule. A structure that incorporates four levels, specifically, things level, adaptation level, Internet application, and layered levels approach, was proposed by Dai and Wang [29].

Further, Tan and Wang [30] recommended a five-layered structure of the IoT. They suggested five levels that may stand for an IoT program, including application level, middleware level, coordination backbone level, and the fifth level comprising existing alone application system, access level, and edge technology level.

Ma [31] proposed that, before developing the structure of IoT, a person needs to think about the various viewpoints of its subscribers, designers, program suppliers, and system suppliers. In view of this, different interfaces, supporting protocols ,] and requirements are identified. Based on Ma [31], structure of IoT consists of four different layers, as shown in Figure 1.4, these are system level. The first fitness level, that is, object acknowledging amount is liable for seeing the points and collecting the required info; the second fitness level, that is, the information exchange level is in control of translucent communication that gathers information upwards; the third fitness level, that is, the information integration level is accountable for various tasks connected with the information acquired. This could inform follow-up decisions on information- gathering and other actions and ambiguous information gathered up non-networking and finally conditioned which failed information with important information. The final level, that is, the application program level is liable for giving the written content offerings to the conclusion.

Sarkar et al. [32] extended a three-layered construction to withstand the IoT structure. The following levels are used in their approach:

Virtual object layer (VOL) handles object virtualization, manages composite VOL, makes up the program, and then delivers. The service layer manages

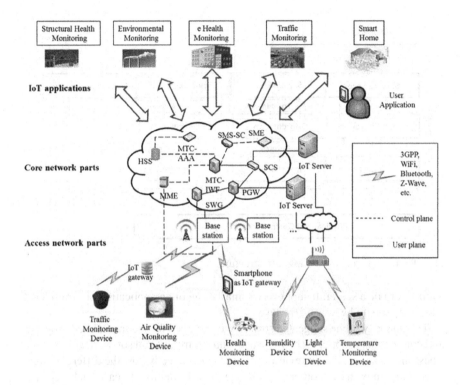

FIGURE 1.4   Four-layered IoT structure.

program development and control. Several other researchers have provided different directions toward the format. The improvement of IoT architecture have created the IoT architecture based upon human neural program and social organization framework (SOF)/three-layered IoT infrastructure, as shown in Figure 1.5. They have proposed two versions of IoT structure, namely,

   i. Device IoT receives the inspiration originating from a man-like anxious (MLN) design. It emphasizes to offer ways for various programs while using the MLN design. The following mechanisms are used:
     A.   Management and centralized data center
     B.   Distributed control nodes
     C.   Networks and sensors
  ii. The ubiquitous IoT receives the inspiration originating from an SOF. It can be categorized as industrial IoT or national or local IoT based on its assortment and connectivity boundaries.

Kovatsch et al. [33] have an available centralized structure to segregate the unit heterogeneity from program growth. They proposed the idea associated with a slim server. Other application program reason resides for this server, thin server is

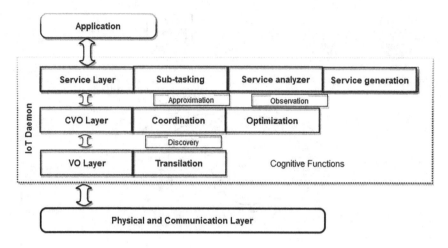

FIGURE 1.5   Three-layered IoT infrastructure.

a unit act to be a server. It supports the marketing of an application program level which appears to be a web-like level.

The majority of the suggested structure is intended for the specialized program and does not help support each industrial and environmental software application. These architectures do not review any issue about exactly how the different levels work and join with each other to exchange the information. In case each object is attached and items can swap information by themselves, the visitors and the storage space inside the system increases exceptionally quickly. Some development to manage any of these offered higher visitors of information.

Heterogeneity is among the primary critical characteristics of an IoT phone. An IoT product is made up of different device types, several community topologies, community setups, and many types of information representation. It is a complicated process to allow for heterogeneity within an effective manner to ensure that all of the equipment may be identified accurately. Based on Baraniuk et al. [34], conventional remedies have affected a range of practical wreckage which enable it to become very complicated fixes, while applying heterogeneity.

Because the IoT methods include the use of the Internet, they are much more vulnerable to a great variety of adversaries. Based on Hendricks [22], a generalized heterogeneous structure is necessary to allow for the different atmosphere comprising different kinds of products, their topologies, their configuration, underlying networks, etc.

The scalable contained structure is yet another problem managed by individual writers. Sinha et al. [35] pointed out that as IPv6 process offers an exceptionally huge standard address area, upcoming online programs will adjust it to deal with the incredibly wide variety of items. They distinctly indicate that the 128-bit IPv6 handling scheme is now being deployed in a lot of uses as it is in a position to cater to near trillion addresses. Thus, the scalability inside IoT is possible through the use of the IPv6 handling plan.

Chen et al. [36] likewise urged the thought associated with several architectures; most ought to adhere to openness. Transparency and adaptability are the two essential attributes of the structure. Whenever we talk about the framework of an IoT procedure, we cannot manage any one of ours to just one know-how, specific topology, just one form of computing free energy generation turbine, only one platform type, and same transmission velocity. Thus, the main worry here is with layout, not just one structure through a pair of architectures.

1.3.2   SOURCE MANAGEMENT

IoT is a technique that can be established by a number of mingling nodes, the nodes, and methods along with Internet resources. To be able to use the maximum capability and benefits of IoT, the supplies associated with IoT need to be managed in an efficient and well-planned manner. Several online resources, including human beings, smooth elements, sensible items, receptors, and actuators, will likely be hooked up and speak with one another to use the paradigm of IoT to its maximum.

According to Zhang and Sun [16], "in the future, people will probably be in the middle of trillions of devices which interconnect with one another and will meet up with or even comprehend the actual physical space." According to Sundmaeker et al. [37], "how to control, incorporate, and take advantage of these many, heterogeneous, or perhaps sent out methods is among the main difficulties for IoT." Several proposals are produced by scientists for the integration and to manage heterogeneous sources to send out, as well as the networked atmosphere.

Additionally, they opposed their suggested SHN type with a current semantic link network version, which is a semantic information design utilizing relational thought. They reported that the SHN design could stand for the associations between entities belonging in an all-natural manner, and help support the need for an IoT phone. Lopez et al. [38] projected various mechanisms for managing the possessions of the IoT. They are as follows:

(i) Grouping, (ii) User representatives, and (iii) Management approaches.

Arora et al. [39] recommended a specific structure for a particular IoT program. The usage of a cloud structure just for smart vehicular networks was proposed. On the foundation of different issues and different possibilities of cloud computing, they created a hierarchical structure to construct sensible vehicular networks. They depend on the idea of discussing methods. Computing systems, storage space systems, and the bandwidth are included with these materials. Source managing stands out as the primary target of their proposition. The suggested structure contains three levels, namely,

(i) The vehicular cloud, (ii) Roadside cloud, and (iii) Main cloud.

The mobile nodes in the system utilize materials and various cloud products which happen to be cars. The strategy used in deallocation and allocation was based on the inclined theoretic method. Lopez et al. [38] suggested that materials in IoT may be handled by utilizing strategies such as clustering, software CD representatives, and with the aid of synchronization strategies. Thus, information synchronization methods may also be helpful in keeping several duplicates of information relevant to items within an IoT phone.

According to Soni et al. [40], to help the interoperability along with numerous products and heterogeneous sources inside a dispersed natural environment, the IoT calls for a multitude of open architectures, supporting effective learning resource control. Thus, allocation and reallocation of Internet resources, in addition to arranging supply among several requesting methods, become an important situation of searching within an IoT phone. Yet another facet that costs much less than is the concurrent gain entry on to the supplies. At this stage within an IoT procedure, there may be the cases where two or perhaps a great deal more entities (it might be an end-user, a process, and assistance, offering user interface, or maybe a sensor itself) might attempt to get the same aid within a short time. Inside IoT, the materials (it might be a file, a free energy generator, a website, an information worth, software CD, hardware, or maybe a sensor again) are shareable online resources. Shareable energy is the materials that can be utilized by two or perhaps many more tasks.

The first class is the location where the useful resource could be seen and employed by two or maybe many more operations in the same period, and the second class does not enable concurrent entry. According to Varshney et al. [41], when someone thinks related to a sharable learning source, the cause is not handled and seen correctly, it may become corrupted, and maybe even could have a few sporadic statuses. Consequently, to avoid these inconsistencies, you have to provide the mutual exclusion of all the nodes while accessing the cause. It is becoming the most sought-after method to refers for implies when one node (thing/process/user) is using the cause, other node cannot utilize that specific source. In case the purpose is expelled, it can be used again by some other nodes.

1.3.3 EFFECTIVE INFORMATION HANDLING

In terms of little model disturbance, among the critical components is that it is useful. This information might be converted information, stored data, generated data, and information within the transit. In IoT, a significant struggle is information compilation. Numerous scientists are now talking about the value of information mining within IoT.

Goel et al. [42] highlighted the concept of information assortment. Although their primary focus is on information collecting, they talked about the authenticated version. Based on them, a group of information coming from different heterogeneous sources at the same time is a hard undertaking to achieve. Additionally, they proposed a unit for information compilation. According to them, because the assortment of every gain access to factor is extensive, it becomes harder to obtain information coming from them all at once. Thus, they proposed the cyclic way to gather information from a small number of units. According to Zhang and Sun [16], the significant problem in information controlling is information storage space as it indirectly or directly depends upon the storage space capacity of specific receptors. Thus, inside IoT, one cannot put into action conventional ways to manage information.

Tsai et al. [43] emphasized data mining concerning the information detection on records for IoT. They proposed three main aspects of interest:

i. Impartial (O): The first action would be to determine and establish the issue. It might have suggested assumptions, limits, and defined dimensions about the subject.
ii. Quality of information (D): Next and an inescapable part of the situation of IoT is information. Its attributes are similar to its sizing, its method of business presentation, its division.
iii. Pulling out procedure (A): Finally, a person needs to figure out what information mining algorithm needs to be applied after analyzing the abovementioned two needs.
   Based on Farooq et al. [44], inside IoT, the information getting gathered comes from different products, utilizing different prevailing solutions. This information is utilized to run among the equipment. Thus, a lot more attention is needed toward combating and mining this specific multidimensional information.

With this effort, the information associated with IoT is categorized directly into three types.

"Data were saying things:" This class comprises the information that describes the items and explains their characteristics. This feature could be their distinctive identification, actual physical place, express (busy or idle, readily available, or perhaps not), ownership in case needed, method of entry, and so on.

- "Data created by things:" This information includes the information created or even transferred by items.
- "Data assimilated by things:" This information includes the information sensed and taken by receptors.

Secure and efficient managing of this information formulates a significant problem of investigation within the area of IoT.

1.3.4 Privacy and Security

As IoT is made based on Internet, protection problems on the Internet will also appear in IoT. These might be broadly categorized as:

i. Verification strategies
ii. Information safety
iii. Confidentiality strategies

Jing et al. [45] recognized one of the many basic aspects: Privacy. They identified four distinct sizes of privacy:

  i. *Purpose*: It refers to the goal of the information getting utilized.
 ii. *Visibility*: It describes authorized customers to enter information.
iii. *Retention*: It refers to the period information in existence.
 iv. *Granularity*: It refers to the amount of information when it is sent.

Additional use of decision and discussion are raised by protection by realizing the information. Matta et al. [46] concentrated on the protection of information within an IoT phone. They proposed a generalized and multidimensional protection design to deal with the saved information a bit more appropriately.

Generally, computer users cope with the realizing information to be able to generate choices to come down with various program areas. According to Fazio et al. [47], accessing the parts of information inside a protected method is essential. Durable and Tim O'Reilly et al. [48] talked about IoT attempting to place all of the forms of products and items collectively on the web, which makes them capable of delivering various kinds of uses. Consequently, different methods supporting powerful protection are needed. Henceforth, a crucial problem within the IoT paradigm is protection programs and processes for movable gadgets. Information acquisition of IoT uses similarly yields significant protection issues. Three various risks could be realized concerning info acquisition: Data getting transferred from an individual IoT unit to numerous others, information being transferred from IoT unit to third bash, and the other way around. According to Lee et al. [49], one can get the quality finest protection inside a method, though subscribers' sense uneasiness when it concerns their privacy. For instance, in case of a healthcare program, people can really feel uneasy if their behavior, types of illness, and their reactions are disclosed.

According to Arseni et al. [50], suggested which the IT paradigm known as the Internet-of-Things (IoT) goals to create every technical entity that's in a position to communicate, within one package. Numerous scientists have recognized that many people involved in IoT arrived through a group of secrecy obstacles. These individuals might be investors, sponsors, application developers, and customers. Helsinger et al. [51] clarified that the different complications can be categorized in to the following:

  i. *Operator Consent*: User should have provided their consent to use the information.
 ii. *Independence of Choice*: This independence is designed for equally secret protocols and supporting criteria.
iii. *Anonymity*: The behavior of transmission and items of information must depend on the user's profile.

According to Steinberg [52], a study staff of the National Science Foundation, and the Faculty of Arkansas during the Little Rock learned that the secrecy of households using sensible household products might be putting them in danger by analyzing community visitors. Based on a selection of reasonable conditions, that is, web allowed products such as televisions, digital cameras, along with kitchen devices could now spy on individuals in their own homes. Several critical exploration concerns need to be tackled.

How to manage a home appliance in your home, and exactly how will the user's steps be protected to guarantee that absolutely no malicious program overtakes the settings without the user's knowledge?

How to examine residential settings, and which systems and policies can guarantee that the information given to a person is reliable and never provided by way of a malicious procedure?

Price may be a concern. Protection at an affordable price is an additional task or problem. There is a necessity to carry out the protection steps for dealing with secrecy and security, but in an inexpensive manner. Anytime protection is mentioned, it could be correlated on the distinctive identification of the unit or the person. To become part of the IoT process, each free energy generator, whether it's a person staying, or perhaps an item (virtual or physical), should have a few distinctive identifiers. Stankovic et al. [53] and Aggarwal and Das et al. [54] focused on the protection issues related to RFID system concerning IoT. Because of the fast development of IoT, cyber strikes are not virtual risks these days, but appear as a real-world risk. According to Vylegzhanina et al. [55], security methods for free tools, along with dynamic protection plans to prevent cross-procedure chance escalation attacks impacting intermediate PC and community user manipulations solutions, since they hook up to their environment and help make the vital issues in dangerous circumstance of an IoT.

As stated by Jing et al. [45], you will find three distinct levels in IoT; they pointed out that a person needs to guarantee safety measures found at all of the three levels. Sectional and segmental protection at every level is of severe value in addition to intact and entire protection that needs to be appropriately used. Individual scientists have centered on just one facet of protection, that is, authentication. Kawamoto et al. [14] have talked about the authentication of the terminology of location-based authentication. According to them, because the information gathered and refined is large and has been gathered using different products beneath mixed ownership, authentication should be attained in IoT. The verification strategy is shown in Figure 1.6.

This particular design mainly categorizes the user's demand into three groups, namely:

  i. Direct entry on the IoT services.
 ii. Access to the IoT products.
iii. Via Cloud provider.

Customers can also be characterized as experienced users in addition to fundamental users. Exchange of authentication assertions has been performed between customers and producers. For a reaction to boost worries regarding protection, the Internet of Things Security Foundation (IoTSF) was released on 23 September 2015.

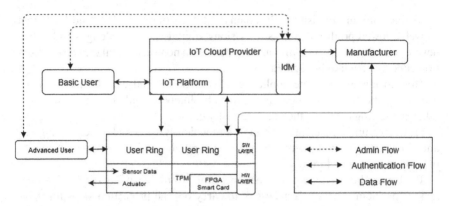

FIGURE 1.6   User authentication process in IoT cloud.

IoTSF aims to secure the IoT advertising by training and knowledge. IoTSF was founded by technology suppliers and telecommunication businesses such as BT, Vodafone, Imagination Technologies, and Pen Test Partners.

1.4   NONTECHNICAL/SOCIETAL CHALLENGES

After discussing the technical obstacles and a comprehensive literature survey, we now discuss another difficult task, that is, the societal problem. As we previously elaborated various uses of IoT, which includes many societal applications within Section 1.2, needless to point out, these difficulties will carry out in a place with the same enormity. The primary societal problem is skill breach. Quite simply, though the planet is relocated using a top rapidity toward IoT, the planet is deficient in the masses to apply IoT effectively. Mohanty et al. [56] analyzed the issue of analysts to go through large-scale information for effectiveness. Experts recommended the usage of big data analytics and algorithm-dependent approaches to face the circumstances.

An additional essential societal problem is disapproval on the novel paradigm by the masses. Although its growth continues to be hampered since its nonetheless not appropriate by individuals, utilizing IoT for different uses is actually among the top leadership aims of IoT. Asplund and Nadjm Tehrani [57] conducted a survey according to the concerns of various actors. The authors captured the shared risk and perception of masses on IoT. Secrecy problems, financial problems, and distress toward the adaptation of new technologies are the reasons powering the dissatisfaction of computer users.

For ensuring trust among computer users, providers must perform extremely difficult and extended tasks. Even though "efficient details handling" is a technical struggle, similar problem arises in business presentation. Unique details managing methods are utilized by different IoT platforms. The inspiration powering the usage of several information managing methods is the repeated feature for technologies, competitors within the marketplace, monetary problems, and brand name consciousness of customers. This leads to an increase in the difficulty and heterogeneity of interoperability, thus formulating data syndication is an essential societal struggle.

Businesses are steadily aware regarding the marketplace importance of IoT, and consequently require persuading actions toward the adoption of IoT paradigm, building the option of extensive and wide-ranging engineering. Nevertheless, this is not the sole arena, the possibility of wide-ranging for specific technologies would create issues. It is experienced by modern society to create the very best opportunities outside of the readily available alternatives together with the least chance adoption.

User-friendliness is an inescapable challenge. Last but not least, providers and IoT sponsors must work against the issues before virtually any IoT product may be built profitably for all of them. Mihovska and Sarkar [58] talked about the people-centric item contacts to facilitate their living. The authors argued the usage of information and correspondence methodologies. Bogdan et al. [59] proposed a cost-effective mathematical airier for complicated IoT methods. The suggested effectiveness of conditions of talking is an important information among products while processing the constant stream of metadata and information.

1.4.1 ENVIRONMENTAL SUSTAINABILITY CHALLENGES

The usefulness of technologies is exclusively connected to the enormity of relevant complications. As IoT is concerned, it is among the dominant paradigms, which creates the user's living simpler to follow. Development of internet, mobile computing and cloud computing, whose influences had been up-to-the information community. IoT technologies, directly or in-directly present architecture proposed to solve real-life problems with in the information community. When the Internet can try to make an information community significantly less hard to focus on, IoT creates the physical society environment more suitable to dwell in.

The primary environmentally friendly struggle with IoT is strength usage. As computing turns into an unavoidable component of every single element in our way of life, energy use turns into an inescapable problem. Thus, providers need to focus on energy usage that is lower in the smart gadgets used around the IoT setup. Mihovska and Sarkar [58] advised the use of typical operating systems to talk about information along with smart products for lower energy consumption. Their IoT platform uses robust solutions such as solar energy.

This particular dependency in the event of a technology disaster may result in bearable losses (within societal and domestic applications) to wild and unmanageable mishaps (in the case of healthcare and emergency applications). Jung et al. [60] proposed the usage of on-demand remote code execution strategy and the storage space of a lot fewer receptors in an ineffective manner. Experts have argued the setup of fixed put together and have predefined efficiency to improve the overall performance of the product. They have proposed an IoT framework in which one code is sent out to a camcorder obstructing essential updating.

One more obstacle is cost as well as an economic system. Numerous IoT products are now being utilized and will be placed in the near future. Lee and Lee [49] talked about the task of price and efficiency within the more considerable investments in unsure return shipping and IoT. Nevertheless, absolutely no substantive answer is available to reduce the price of IoT products. The cost of an IoT unit has to be incredibly low; thus, you must provide innovative solutions to the economic system problems before applying the science.

1.5 FUTURE CHALLENGES IN IOT

Crucial issues and difficulties need to be determined or addressed before further IoT device adoption occurs [61].

1.5.1 SECURITY ISSUES

When IoT becomes a vital component in the future along with its application for large-scale, partly mission-critical methods, loyalty as well as protection performs become essential. Innovative problems determined for secrecy, loyalty, and dependability are:

- Offering quality of information and trust in shared information clothes airers to allow reuse across numerous applications.
- Providing protected exchange of data between IoT equipment and customers regarding their information.
- Providing safeguard systems for insecure products.

1.5.2 USABILITY VERSUS COMPONENT COST

Technology is used by IoT to link actual physical items on the Internet. For IoT adoption to develop, the price of components required to help features including realizing, monitoring, and balancing systems need to become affordable in the near future.

1.5.3 EXCHANGE OF INFORMATION

In the conventional Internet, interoperability is the most essential primary value; the very first necessity of Internet connectivity is the fact that "connected" methods need to be in a position to "talk the very same language" of encodings and protocols. Distinct industries these days use standards that are different to allow their applications. With several energy sources of information in addition to heterogeneous products, the usage of regular interfaces between these several entities becomes crucial. This is particularly for uses that support cross-organizational and also different program borders. As a result, IoT methods have to deal with a higher level of interoperability.

1.5.4 MANAGING DATA

Information management is an important feature within IoT. When contemplating a planet of items interconnected and continuously swapping information, the amount of information generated and the procedures concerned within the managing of this information become important.

1.5.5 ENERGY CONSUMPTION

Among the many important difficulties in IoT is how you can interconnect "things" within an interoperable means while considering the power restrictions, as well as realizing the interaction would be the major power consuming task.

1.6 CONCLUSION AND FUTURE WORK

IoT offers interesting applications in healthcare, transportation, and agriculture. However, various factors such as security, privacy, and data storage need to be considered. It is also worth noting that things have been connected to networks for a long time without the guise of "IoT." However, heterogeneity exists in mining and collecting similar information, protecting information, improving the scalability of the device, and magnifying concurrency within the entire phone system. However, at this time, a lot more is still to be attained, with several efforts being taken by individual scientists. In the near future, to increase the livelihood, we need to pay attention to the regular format of the buyer, each small point, each facet on the consumer's way of life. In addition, to be "smart," we have to come up with much more fascinating, useful, reliable as well as novel methods for IoT paradigm. These solutions are intended to make our life "smart" life and therefore our world "smart" world. This chapter elaborated the treatments provided by notable scientists within the area of IoT in the present situation and offered an extensive guide for the scientists working in this field.

ACKNOWLEDGEMENT

We thank the International Telecommunication Union (ITU) for the definition of IoT, which we have used in this chapter.

REFERENCES

1. Sharma, T., Satija, S., & Bhushan, B. (2019). Unifying Blockchian and IoT: Security Requirements, Challenges, Applications and Future Trends. *2019 International Conference on Computing, Communication, and Intelligent Systems (ICCCIS)*. doi: 10.1109/icccis48478.2019.8974552
2. Aggarwal, S., Gulati, R., & Bhushan, B. (2019). Monitoring of Input and Output Water Quality in Treatment of Urban Waste Water Using IOT and Artificial Neural Network. *2019 2nd International Conference on Intelligent Computing, Instrumentation and Control Technologies (ICICICT)*. doi: 10.1109/icicict46008.2019.8993244

3. Saini, H., Bhushan, B., Arora, A., & Kaur, A. (2019). Security Vulnerabilities in Information Communication Technology: Blockchain to the Rescue (A Survey on Blockchain Technology). *2019 2nd International Conference on Intelligent Computing, Instrumentation and Control Technologies (ICICICT).* doi: 10.1109/icicict46008.2019.8993229

4. Perera, C., Ranjan, R., Wang, L., Khan, S.U., & Zomaya, A.Y. (2015). Privacy of big data in the Internet of Things era. *IEEE IT Professional Magazine*, 17, 32–39.

5. Tiwari, R., Sharma, N., Kaushik, I., Tiwari, A., & Bhushan, B. (2019). Evolution of IoT & Data Analytics using Deep Learning. *2019 International Conference on Computing, Communication, and Intelligent Systems (ICCCIS).* doi: 10.1109/icccis48478.2019.8974481

6. China Communication Standards Association. (2011). [YDB] Communication standard technical report. Retrieved March 20, 2017, from http://www.ccsa.org.cn/english/list_std.php?tbname=ydb_doc&keyword=& page_currentPage=4.

7. International Telecommunication Union (ITU). (2015). Internet of things global standards initiative. Retrieved January 8, 2017, from http://www.itu.int/en/ITU-T/gsi/iot/Pages/default.aspx.

8. International Telecommunication Union (ITU). (2012). Y.2060: Overview of IoT. Retrieved January 15, 2017, from https://www.itu.int/rec/T-REC-Y.2060-201206-I.

9. Kosmos, E.A., Tselikas, N.D., & Boucouvalas, A.C. (2011). Integrating RFIDs and smart Objects into a unified internet of things architecture. *Advances in Internet of Things*, 1(1), 5–12.

10. Gartner. (2013). Gartner says the Internet of Things installed base will grow to 26 billion units by 2020. Retrieved January 20, 2017, from https://attivonetworks.com/gartner-says-the-internet-of-things-installed-basewill-grow-to-26-billion-units-by-2020/.

11. Vermesan, O., Friess, P., Guillemin, P. Gusmeroli, S., Sundmaeker, H., Bassi, A., Jubert, I.S., Mazura, M., Harrison, M., Eisenhauer, M., & Doody, P. (2011). Internet of things strategic research roadmap. *Internet of Things – Global Technological and Societal Trends*, 1, 9–52.

12. Malik, A., Gautam, S., Abidin, S., & Bhushan, B. (2019). Blockchain Technology-Future of IoT: Including Structure, Limitations and Various Possible Attacks. *2019 2nd International Conference on Intelligent Computing, Instrumentation and Control Technologies (ICICICT).* doi: 10.1109/icicict46008.2019.8993144

13. Lopez, T.S., Ranasinghe, D.C., Harrison, M., & McFarlane, D. (2012). Adding sense to the internet of things. *Personal and Ubiquitous Computing*, 16(3), 291–308.

14. Kawamoto, Y., Nishiyama, H., Kato, N., Shimizu, Y., Takahara, A., & Jiang, T. (2015). Effectively collecting data for the location-based authentication in the Internet of Things. *IEEE Systems Journal*, 11(3), 1403–1411.

15. Lee, G.M., Park, J., Kong, N., & Crespi, N. (2012). The Internet of Things – concept and problem statement. *Internet Research Task Force*, 19 pages.

16. Zhang, J., & Sun, Y. (2012). Managing Resources in the Internet of Things with Semantic Hyper-Network Model. *Proceedings of the International Workshop on Enabling Technologies: Infrastructure for Collaborative Enterprises.* Hammamet, Tunisia, 318–323.

17. National Intelligence Council. (2008). Disruptive civil technologies: Six technologies with potential impacts on the US interests out to 2025. Retrieved April 22, 2017, from https://fas.org/irp/nic/disruptive.pdf.

18. Varshney, T., Sharma, N., Kaushik, I., & Bhushan, B. (2019). Authentication & Encryption Based Security Services in Blockchain Technology. *2019 International Conference on Computing, Communication, and Intelligent Systems (ICCCIS).* doi: 10.1109/icccis48478.2019.8974500

19. Jindal, M., Gupta, J., & Bhushan, B. (2019). Machine Learning Methods for IoT and their Future Applications. *2019 International Conference on Computing, Communication, and Intelligent Systems (ICCCIS)*. doi: 10.1109/icccis48478.2019.8974551.

20. ABI Research. (2013). More than 30 billion devices will wirelessly connect to the internet of everything in 2020. Retrieved March 20, 2017, from https://www.abiresearch.com/press/more-than-30 billion-deviceswill-wirelessly-conne/.

21. Pew Research Internet Project. (2017). The fate of online trust in the next decade. Retrieved August 10, 2017, from https://www.pewresearch.org/internet/2017/08/10/the-fate-of-online-trust-in-the-next-decade/.

22. Hendricks, D. (2015). The trouble with the internet of things. Retrieved December 24, 2016, from http://data.london.gov.uk/blog/the-trouble-with-theinternet-of-things/.

23. Europen Research Cluster (IERC). (2014). Internet of things. Retrieved March 21, 2017, from http://www.internet-of-things-research.eu/about_iot.htm.

24. Miorandi, D., Sicari, S., De Pellegrini, F., & Chlamtac, I. (2012). Internet of things: Vision, applications and research challenges. *Ad Hoc Networks*, 10(7), 1497–1516.

25. Castellani, A.P., Bui, N., Casari, P., Rossi, M., Shelby, Z., & Zorzi, M. (2010). Architecture and Protocols for the Internet of Things: A Case Study. *Proceedings of 8th IEEE International Conference on Pervasive Computing and Communication Workshops (PERCOM)*. Mainheim, Germany, 678–683.

26. Yan, L., Zhang, Y., Yang, L.T., & Ning, H. (2008). *Internet of Things: From RFID to the Next Generation Pervasive Networked Systems* (1st ed.). Boca Raton, FL: Auerbach Publications.

27. Beier, S., Grandison, T., Kailing, K., & Rantzau, R. (2006). Discovery Services – Enabling RFID Traceability in EPCglobal Networks. *Proceedings of the 13th International Conference on Management of Data (COMAD)*, 4 pages.

28. Gronbaek, I. (2008). Architecture for the Internet of Things (IoT): API and Interconnect. *Proceedings of 2nd International Conference on Sensor Technologies and Applications (SENSORCOMM)*. Cap Casterel, France, 802–807.

29. Dai, G., & Wang, Y. (2012). Design on the architecture of the Internet of Things. In: Jin D., and Lin S. (eds.), *Advances in Computer Science and Information Engineering*, Advances in Intelligent and Soft Computing, vol. 168. Berlin, Heidelberg: Springer, 1–7. https://doi.org/10.1007/978-3-642-30126-1_1.

30. Tan L., & Wang, N. (2010). Future Internet: The Internet of Things. *Proceedings of 3rd International Conference on Advanced Computing Theory Engineering (ICACTE)*. Chengdu, China, 5, 376–380.

31. Ma, H.-D. (2011). Internet of things: Objectives and scientific challenges. *Journal of Computer Science and Technology*, 26, 919–924.

32. Sarkar, C., Nambi, A.U.S.N., Prasad, R.V., Rahim, A., Neisse, R., & Baldini, G. (2015). DIAT: A scalable distributed architecture for IoT. *Internet of Things Journal*, 2(3), 230–239.

33. Kovatsch, M., Mayer, S., & Ostermaier, B. (2012). Moving Application Logic from the Firmware to the Cloud: Towards the Thin Server Architecture for the Internet of Things. *Proceedings of 6th International Conference on Innovative Mobile and Internet Services in Ubiquitous Computing (IMIS)*. Palermo, Italy, 751–756.

34. Baraniuk, R.G. (2011). More is less: Signal processing and the data deluge. *Science*, 331(6018), 717–719.

35. Sinha, P., Rai, A.K., & Bhushan, B. (2019). Information Security Threats and Attacks with Conceivable Counteraction. *2019 2nd International Conference on Intelligent Computing, Instrumentation and Control Technologies (ICICICT)*. doi: 10.1109/icicict46008.2019.8993384

36. Chen, S., Xu, H., Liu, D., Hu, B., & Wang, H. (2014). A vision of IoT: Applications, challenges, opportunities with China perspective. *IEEE Internet of Things Journal*, 1(4), 349–359.

37. Sundmaeker, H., Guillemin, P., Friess, P., & Woelffle, S. (2010). Vision and challenges for realising the internet of things (CERP-IoT). *The Cluster of European Research Projects on the Internet of Things, European Commission*, 3(3), 34–36.

38. Lopez, T.S., Ranasinghe, D.C., Harrison, M., & McFarlane, D. (2012). Adding sense to the internet of things. *Personal and Ubiquitous Computing*, 16(3), 291–308.

39. Arora, A., Kaur, A., Bhushan, B., & Saini, H. (2019). Security Concerns and Future Trends of Internet of Things. *2019 2nd International Conference on Intelligent Computing, Instrumentation and Control Technologies (ICICICT)*. doi: 10.1109/icicict46008.2019.8993222

40. Soni, S., & Bhushan, B. (2019). A Comprehensive Survey on Blockchain: Working, Security Analysis, Privacy Threats and Potential Applications. *2019 2nd International Conference on Intelligent Computing, Instrumentation and Control Technologies (ICICICT)*. doi: 10.1109/icicict46008.2019.8993210

41. Varshney, T., Sharma, N., Kaushik, I., & Bhushan, B. (2019). Architectural Model of Security Threats & their Countermeasures in IoT. *2019 International Conference on Computing, Communication, and Intelligent Systems (ICCCIS)*. doi: 10.1109/icccis48478.2019.8974544

42. Goel, A.K., Rose, A., Gaur, J., & Bhushan, B. (2019). Attacks, Countermeasures and Security Paradigms in IoT. *2019 2nd International Conference on Intelligent Computing, Instrumentation and Control Technologies (ICICICT)*. doi: 10.1109/icicict46008.2019.8993338

43. Tsai, C.-W., Lai, C.-F., & Vasilakos, A.V. (2014). Future internet of things: Open issues and challenges. *Wireless Networks*, 20(8), 2201–2217.

44. Farooq, M.U., Waseem, M., Khairi, A., & Mazhar, S. (2015). A critical analysis of the security concerns of the Internet of Things (IoT). *International Journal of Computer Applications*, 111(7), 1–6.

45. Jing, Q., Vasilakos, A.V., Wan, J., Lu, J., & Qiu, D. (2014). Security of the Internet of Things: Perspectives and challenges. *Wireless Networks*, 20(8), 2481–2501.

46. Matta, P., Pant, B., & Arora, M. (2017). All You Want to Know about the Internet of Things (IoT). *Proceedings of 4th IEEE International Conference on Computing, Communication and Automation (ICCCA)*. Greater Noida, India, 1306–1311.

47. Fazio, M., Celesti, A., Puliafito, A., & Villari, M. (2014). An integrated system for advanced multi-risk management based on cloud for IoT. In: Gaglio, S., and Lo Re, G. (eds.), *Advances onto the Internet of Things. Advances in Intelligent Systems and Computing*, vol. 260. Cham: Springer. doi:10.1007/978-3-319-03992-3_18.

48. Hardy, Q. (2017). Tim O'Reilly explains the Internet of Things. Retrieved February 25, 2019, from https://bits.blogs.nytimes.com/2015/02/04/timoreilly-explains-the-internet-of-things/.

49. Lee, I., & Lee, K. (2015). The Internet of Things (IoT): Applications, investments, and challenges for enterprises. *Business Horizon*, 58(4), 431–440.

50. Arseni, S.-C., Halunga, S., Fratu, O., Vulpe, A., & Suciu, G. (2015). Analysis of the Security Solutions Implemented in the Current Internet of Things Platforms. *Proceedings of IEEE International Conference on Grid, Cloud and High-Performance Computing in Science (ROLCG)*. Cluj-Napoca, Romania, 1–4.

51. Helsinger, A., Thome, M., & Wright, T. (2014). Cougar: A Scalable, Distributed Multi-Agent Architecture. *Proceedings of IEEE International Conference on Systems, Man and Cybernetics*. The Hague, Netherlands, Vol. 2, 1910–1917.

52. Steinberg, J. (2014). These devices may be spying on you (even in your own home). Retrieved April 2, 2017, from http://www.forbes.com/sites/josephsteinberg/2014/01/27/these-devices-may-be-spying-on-you-even-inyour-own-home/#3be3cba66376.

53. Stankovic, J.A. (2014). Research directions for the internet of things. *IEEE Internet of Things Journal*, 1(1), 3–9.

54. Aggarwal, R. and Das, M.L. (2012). RFID Security in the Context of "Internet of Things". *Proceedings of First International Conference on Security of Internet of Things*. New York, United States of America, 51–56.

55. Vylegzhanina, V., Schmidt, D.C., & White, J. (2015). Gaps and Future Directions in Mobile Security Research. *Proceedings of 3rd International Workshop on Mobile Development Lifecycle*. New York, United States of America, 49–50.

56. Mohanty, S., Das, S.K., Barik, S., & Rout, M.M. (2016). A Survey on Big Data and Its Challenges Related to IoT. *Proceedings of International Interdisciplinary Conference on Engineering Science and Management*. Goa, India, 331–334.

57. Asplund, M., & Nadjm-Tehrani, S. (2016). Attitudes and perceptions of IoT security in critical societal services. *Special Section on the Plethora of Research in Internet of Things (IoT)*, 4, 2130–2138.

58. Mihovska, A., & Sarkar, M. (2018). Smart connectivity for the Internet of Things (IoT) applications. *New Advances in the Internet of Things*, 751, 105–118.

59. Bogdan, P., Pajic, M., Pande, P.P., & Raghunathan, V. (2016). Making the Internet-of-Things a Reality: From Smart Models, Sensing and Actuation to Energy-Efficient Architectures. *Proceedings of the International Conference on Hardware/Software Codeware and System Synthesis (CODES/ISSS)*. Pittsburgh, PA, United States of America, 1–10.

60. Jung, M., Park, D., & Cho, J. (2015). Efficient Remote Software Execution Architecture Based on Dynamic Address Translation for Internet-of-Things Software Execution Platform. *Proceedings of 18th IEEE International Conference on Network-Based Information Systems*. Taipei, Taiwan, 371–378.

61. Rehman, H.U., Asif, M., & Ahmad, M. (2017). Future Applications and Research Challenges of IOT. *2017 International Conference on Information and Communication Technologies (ICICT)*. Karachi, , 68–74, doi: 10.1109/ICICT.2017.8320166

2 Internet of Things Development in Polish Enterprises

M. Mądra-Sawicka and J. Paliszkiewicz
Warsaw University of Life Sciences

A. Koohang
Middle Georgia State University

J. Horn Nord
Oklahoma State University

M. Gębska
Warsaw University of Life Sciences

CONTENTS

2.1 INTRODUCTION

Mobile devices and computers have become particularly important in our lives. The devices are produced with the possibility to connect to the Internet, which creates an appropriate environment to develop the Internet of Things (IoT) (Liu and Yan 2015). The IoT enables communication across objects that are connected to the Internet to realize defined goals (Ghallab et al. 2019). The IoT groups the new technology that provides many applications by connecting things with people, things with things, and people with things through the Internet (Nord et al. 2019). The main IoT research focuses on the following aspects: the formulation of standards, the breakthrough of key technologies, and the development of commercial applications (Liu et al. 2020). The IoT is an emerging technology that brings added value to businesses, consumers, and governments.

The challenges and benefits of IoT cover not only business but also more global effects on the economy. Thus, the IoT challenges and benefits can be perceived from two perspectives of the organizational environment: internal and external. The challenges of IoT are oriented not only on the technical approach but also on numerous different IoT implementations.

This chapter reveals the importance of IoT challenges concerning the lack of international norms and standards. Thus, the study underlines the future potential of IoT that could be limited due to the patterns and standards gap. It could be a limiting factor in the creation of new business models.

This chapter aims to present the applications, challenges, and benefits of IoT development in Polish enterprises with the assessment of IoT in the internal and external operating environments. The following are the research questions for this study:

Research Question 1: To what extent the IoT is used internally and exter- nally by organizations.

Research Question 2: Do the challenges of IoT differentiate between its specific applications among organizations.

Research Question 3: Which of the IoT challenges are the most demanding for companies in the case of IoT implementation and utilization.

The first part of the chapter includes an analysis of the IoT with proposed defi- nitions, applications, and challenges. The second part presents the theoretical issues concerning the IoT framework. The third part describes its use, benefits, and development. The fourth part discusses the research methodology of the study. The fifth part illustrates the analysis and results. The sixth part sums up the empirical results. The seventh part presents conclusions with limitations and avenues for further research.

2.2   IOT – FRAMEWORK

The IoT is described as one of the disruptive technologies (Alkhatib et al. 2014) and has drawn the attention of academia, society, and industry (Ibarra-Esquer et al. 2017). Defining IoT from a different perspective of its use appears in the literature (Ben-Daya et al. 2019; Nord et al. 2019). The IoT was first mentioned in 1999 by Ashton, and it was defined as "a group of smart objects connected via radio fre- quency identification (RFID) technology" (Ashton 2009). The description of the IoT is still developing. Mattern and Floerkemeier (2010, p. 242) define IoT as the items connected to the virtual world where they are "controlled remotely and can act as physical access points to the Internet services." Gluhak et al. (2011) define IoT as an environment in which objects are equipped with specific identifiers capable of transmitting data over the Internet network without mutual interaction between each other. Mayordomo et al. (2011) described the IoT as a new dynamic network of systems where every daily object can communicate with each other. In their work, Gubbi et al. (2013) described the IoT as the interconnection of devices that provides the possibility to exchange information across platforms through a unified frame- work. With a different perspective, Fortino and Trunfio (2014) pictured the IoT as a loosely coupled, decentralized system of cooperating intelligent objects.

Xu et al. (2014) described the IoT as a dynamic global network infrastructure with self-configuring capabilities. Ben-Daya et al. (2019, p. 4721) presented a defi- nition related to the IoT and supply chain management. According to them,

the IoT is a network of physical objects that are digitally connected to sense, moni- tor, and interact within a company and between the company and its supply chain enabling agility, visibility, tracking and information sharing to facilitate timely planning, control, and coordination of the supply chain processes.

Based on these definitions, the key features of the IoT can be described. First, there is a requirement for the digital connectivity of the objects. Second, the nature of this communication is proactive, enabling data storage, analysis, and sharing. Third, the IoT will facilitate control, planning, and coordination. The IoT can be viewed from two perspectives as external use by enterprises or internally in business operations.

2.3   THE EXTERNAL APPLICATION OF THE IOT IN A COMPANY ENVIRONMENT

The IoT has great potential in the future development of Internet services. IT companies are keen to explore the possibilities of IoT (Stoces et al. 2016). The IoT is used almost in every area of modern society (Vermesan and Friess 2013). Its usage areas include information technology, retail, transportation and logistics, financial services, supply chain, personal and social use, energy, manufacturing, smart grid, healthcare, smart city, wearables, and smart agriculture. Examples of IoT application areas are listed in Table 2.1 and are described below.

2.3.1   INFORMATION TECHNOLOGY

Sezer et al. (2018) analyzed context-aware systems for IoT and provided context information concerning different methods. Matos et al. (2020) examined the requirements for sharing context information for the IoT, then reviewed the relevant literature for context sharing and classified them based on their needs and characteristics. Diène et al. (2020) identified the most relevant concepts of data management in IoT.

2.3.2   RETAIL

The retail industry, at the forefront of the IoT, modified customers' behavior in how they view shopping (Nguyen et al. 2017). IoT application in retail can help control the quality of food products, manage the temperature, and enable the reduction of energy consumption. The IoT can plan waste management and monitor items that have exceeded their shelf life. In their study, Kamble et al. (2019) attempted to describe the barriers that affect the adoption of IoT in the retail supply chain.

According to Caro and Sadr (2019), IoT can play a fundamental role in integrating retail channels, thus enabling companies to balance supply and demand. The ongoing IoT development in these sphere concerns a new digital ecosystem which will use connected devices to offer new products and service (Gregory 2015; Kamble et al. 2019).

TABLE 2.1

Internet of Things Applications

IoT Application	Authors
Information technology	Sezer et al. (2018), Matos et al. (2020), and Diène et al. (2020)
Retail	Gregory (2015), Balaji and Roy (2017), Kamble et al. (2019), and Caro and Sadr (2019)
Transportation and logistics	Karakostas (2013), Guerrero-Ibanez et al. (2015), Liu et al. (2019), and Poenicke et al. (2019)
Financial services	Giudice et al. (2016), Shepherd et al. (2017), Cuomo et al. (2018), and Wang et al. (2020)
Supply chain	Yan et al. (2018), Li and Li (2017), Ben-Daya et al. (2019), and Mostafa et al. (2019)
Personal and social	Venkatesh (2017), Dash et al. (2019), and Karimova and Shirkhanbeik (2015)
Energy	Machado et al. (2013), Baker et al. (2017), Afghan and Géza (2019), Bello et al. (2019), and Yue et al. (2019)
Manufacturing	Bi et al. (2016), Mourtzis et al. (2016), Yang et al. (2016), Heinis et al. (2018), and Li et al. (2018)
Smart grid	Bekara (2014), Hussain et al. (2018), and Soni and Talwekar (2019)
Healthcare	Catarinucci et al. (2015), Rghioui and Oumnad (2018), Singh (2018), and Ansari et al. (2019)
Smart city	Liu et al. (2012), Zanella et al. (2014), Khajenasiri et al. (2017), Mijac et al. (2017), Alavi et al. (2018), Chowdhry et al. (2019), and Wener (2019)
Wearables	Jiang (2020), Sivathanu (2018), Adiputra et al. (2018), Bayo-Monton et al. (2018), Eskofier et al. (2017), and Tahir (2018)
Smart agriculture	Qiu et al. (2013), Ferrandez-Pastor et al. (2016), Stoces et al. (2016), Navulur and Prasad (2017), and Mushtaq (2018)

2.3.3 TRANSPORTATION AND LOGISTICS

Developing an intelligent transportation system requires emerging technologies such as cloud computing, connected vehicles, and IoT (Guerrero-Ibanez et al. 2015). Liu et al. (2019) showed that IoT application contributes to improving vehicles' utilization rate, reducing logistics cost, reducing fuel consumption, and achieving real-time logistics services with high efficiency. Poenicke et al. (2019) depict a method of IoT that aims at the early planning stage to integrate technologies in logistics processes. Karakostas (2013) proposed a Domain Name System (DNS) architecture for the IoT and described a case study in transport logistics.

2.3.4 FINANCIAL SERVICES

IoT application and service can be applied in banking transitions, nonbanking financial products, and services such as insurance or frictionless payments (Shepherd et al. 2017). Cuomo et al. (2018) presented an application of the one-factor HullWhite model in an IoT financial scenario. Wang et al. (2020) studied the link between the IoT and blockchain. Giudice et al. (2016) presented the relation of products offered by the banks of things and their possible increase of return on equity. It highlights the benefits of banks of things and IoT usage across financial services.

2.3.5 SUPPLY CHAIN

The wide supply chain network that is built on multiple businesses and relationships demands changes in technology in the area of dynamic environment change, as well as due to natural factors that impact their operations (Ben-Daya et al. 2019). In their article, Mostafa et al. (2019) described impacts of the IoT on supply chains. Factors affecting IoT implementation in supply chains were described by Yan et al. (2018). Li and Li (2017) presented the link between the IoT and supply chain innovation.

2.3.6 PERSONAL AND SOCIAL

Dash et al. (2019) explored and assimilated the viewpoints regarding the new standards and challenges of the IoT-enabled human resource management (HRM) during Industry 4.0. The study proposes to implement a reusable IoT-based skeleton for automating HRM systems and processes. It refers to the comprehensive development of skills regarding the smooth integration of existing HRM systems through the implementation of sensors, actuators, and IoT devices. Karimova and Shirkhanbeik (2015) described an alternative vision of the IoT, mainly the transformation of the IoT into a community or society of things. Venkatesh (2017) believed that the application of the IoT in an organization could be managed with smart and efficient workspaces such as real-time monitoring or intelligent meeting rooms.

2.3.7 ENERGY

IoT services can be used in many energy technology solutions, that is, in creating an energy-aware composition plan to fulfill user requirements (Baker et al. 2017). The literature has documented many studies that focus on energy issues and IoT (Afghan and Géza 2019; Bello et al. 2019; Dash et al. 2019; Yue et al. 2019).

2.3.8 MANUFACTURING

The IoT helps to transform the traditional manufacturing systems into modern digitalized ones, and thus support the reshaping of the industries. According to the literature, industrial IoT provides more possibilities for enterprises to adopt new data-driven strategies, and thus increase their competitive advantage in the market (Mourtzis et al. 2016). Three core manufacturing technologies play vital roles in IoT: radio-frequency identification, wireless sensor networks and cloud computing, and big data (Yang et al. 2016). Other authors like Bi et al. (2016) described a visualization platform for the IoT in manufacturing applications. Heinis et al. (2018) presented an empirical study on the factors motivating and inhibiting innovation in IoT applications for industrial manufacturing companies. Li et al. (2018) proposed a framework for active sensing and processing of multiple events on the Internet of Manufacturing Things (Li et al. 2018).

2.3.9 SMART GRID

The smart grid technology is linked to the IoT for energy and uses smart meters, smart appliances, and even renewable energy resources. The smart grid benefits from the IoT because smart objects are deployed alongside the energy path – from the generation plant to the end customer (Baker et al. 2017). Soni and Talwekar (2019) described an overview of the IoT in smart grid. The architecture of a smart grid was proposed by Hussain et al. (2018).

2.3.10 HEALTHCARE

IoT technology finds use in healthcare services. It is known as IoT-aware that creates the system of automatic monitoring and tracking of patients, personnel, or biomedical devices by creating a smart hospital system (Catarinucci et al. 2015). IoT-based healthcare applications were researched by Ansari et al. (2019). Challenges and opportunities of the IoT in healthcare were described by Rghioui and Oumnad (2018) and Singh (2018).

2.3.11 SMART CITY

The IoT in a smart city concept, according to Arasteh et al. (2016), is being used in smart homes, smart parking lots, weather and water systems, vehicular traffic, environmental pollution, or surveillance systems. The concept of the smart city gained popularity and attracted significant research attention (Alavi et al. 2018; Mijac et al. 2017; Zanella et al. 2014). Researchers have described architectures, models, and platforms for smart city applications (e.g., Chowdhry et al. 2019; Wener 2019). Smart city incorporates smart homes, which consist of IoT-enabled home appliances, air-conditioning/heating system, television, audio streaming devices, and security systems. These IoT services provide the best comfort and security, as well as reduced energy consumption and cost (Khajenasiri et al. 2017). Another category that has been studied within a smart city is smart vehicles equipped with intelligent devices (Liu et al. 2012).

2.3.12 WEARABLES

Jiang (2020) presented a wearable sports rehabilitation system for monitoring various physiological parameters during rehabilitation training. The adoption of the IoT-based wearables for healthcare has received attention in the literature (e.g., Adiputra et al. 2018; Bayo-Monton et al. 2018; Eskofier et al. 2017;

Sivathanu 2018). Tahir (2018) studied the importance of the security of consumer wearable devices related to the IoT.

2.3.13 SMART AGRICULTURE

Agriculture plays a significant role in most countries, and there is an enormous need to combine agriculture with technology so that production can be improved efficiently. IoT in agriculture has received the attention of many researchers. For example, Stoces et al. (2016) described this topic in general. Navulur et al. (2017) focused on agriculture management through the IoT. Ferrandez-Pastor et al. (2016) described IoT application in precision agriculture. Mushtaq (2018) also investigated intelligent agriculture systems based on IoT and image processing. Qiu et al. (2013) pronounced an intelligent platform for monitoring IoT-based facilities of the agriculture ecosystem.

2.4 THE IOT INTERNAL APPLICATION IN ENTERPRISES' BUSINESS OPERATIONS

The application of the IoT in enterprises was investigated in nine areas evaluated from the perspective of the internal business processes: asset management, customer experience, environment, finance, manufacturing, product development, safety, supply chain, warehousing, and logistics (see Table 2.2).

TABLE 2.2
Internet of Things Applications

IoT Application	Authors
Asset management	Wang et al. (2015), Backman and Helaakoski (2016), Brous et al. (2017), and Kinnunen et al. (2018)
Customer experience	Ceipidor et al. (2011), Jamison and Snow (2014), Lee and Lee (2015), and Nguyen and Simkin (2017)
Environment	Bandyopadhyay and Sen (2011) and Haller et al. (2008)
Finance	Dineshreddy and Gangadharan (2016) and Shepherd et al. (2017)
Manufacturing	Tao et al. (2014), Bi et al. (2014), Zhao et al. (2015), Kang et al. (2016), and Mourtzis et al. (2016)
Product development	Monostori (2014), Mourtzis et al. (2016), and Wang et al. (2019)
Safety	Riahi et al. (2013), Mourtzis et al. (2016), Wolfs (2017), and Hou et al. (2019)
Supply chain	Angeles (2005), Barratt and Oke (2007), Choy et al. (2017), Ben-Daya et al. (2019), and Simec (2019)
Warehousing and logistics	Ding (2013), Reaidy et al. (2015), Lee et al. (2018), and Buntak et al. (2019)

2.4.1 Asset Management

According to Brous et al. (2017), the application of IoT data for making asset management decisions stayed at a low level. Furthermore, IoT technologies have a high potential to be used in managing various asset groups (Kinnunen et al. 2018). The development of business and technical perspectives for evaluating IoT platforms in asset management is attracting growing attention (Backman and Helaakoski 2016). Asset management in enterprises can include different technologies used in intelligent control and monitoring of the inventory of warehouse management to reduce inventory resources and improve the level of management (Wang et al. 2015).

2.4.2 Customer Experience

The IoT has changed customer experience with shopping in the retail industry. Customer interaction with IoT retail technology results in value co-creation that reveals ease of use of retail technology (Nguyen et al. 2017). Customer support also includes video management software designed to provide customers a better shopping experience (Lee and Lee 2015).

2.4.3 Environment

The appropriate designing environment in enterprises demands IoT architecture to incorporate and use factors such as networking, communication, business models, and processes, as well as security issues. Furthermore, the IoT is used to improve the quality of buildings and reduce wastes (Zhao et al. 2015).

2.4.4 Finance

IoT devices enable unique payment methods, improve customer profiling, and establish the price of offered financial services and products. IoT application uses financial data in a more effortless manner, for example, by adding stock tickers and using online trading platforms or instruments to collect data about asset prices (Shepherd et al. 2017). The IoT is widely used by various applications, including retail banking, insurance, and investments (Dineshreddy and Gangadharan 2016).

2.4.5 Manufacturing

IoT usage in manufacturing creates more efficient and optimal ways to allocate various manufacturing resources and capabilities (Tao et al. 2014). Changes and development in manufacturing processes generate demand for monitoring devices to enable the distribution of sensors connected to high-speed wireless networks (Zhao et al. 2015).

2.4.6 Product Development

IoT adoption allows new ways for businesses to connect and co-create value. It gathers and analyzes data through the entire product lifecycle. The IoT can change industries into "cyber production systems" that are more flexible and adaptive for production conditions, as well as product changes and development (Mourtzis et al. 2016).

2.4.7 Safety

Safety is the tension that ties the person with the process. A company environment permeated with intelligent objects is expected to cope with many security challenges. IoT devices ensure safety in case failure occurs in any system component. These aspects of safety can reduce the role of IoT devices in case of damage (Riahi et al. 2013). Furthermore, the IoT systems improve the safety and security of cyber-physical systems (Wolfs 2017).

2.4.8 Supply Chain

IoT application can support real-time information for the entire supply chain that enhances the manufacturing value chain in production and distribution that can respond as quickly as needed (Ding 2013). Furthermore, supply chains are operating under an ever-changing environment that allows for effective internal operations and more efficient collaboration with suppliers and customers (Ben-Daya et al. 2019).

2.4.9 Warehousing and Logistics

Warehouse operations need to be adjusted and developed due to the increasing variety of customer orders (Lee et al. 2018), and the intelligent logistics warehouse management system based on IoT systems can help (Reaidy et al. 2015). The IoT system uses hardware, electronic label, or barcode that reflects the variety of storage and the flow of materials. IoT systems improve the efficiency of warehouse management and reduce the error rate for enterprises (Ding 2013).

2.5 IOT CHALLENGES, BENEFITS, AND TRENDS IN DEVELOPMENT

The IoT offers a secure platform to access the facilities without any computational and programming complexities. IoT applications provide more advantages, for example, saving valuable user time, providing learning opportunities, and improving existing communication infrastructures.

The development of IoT knowledge is very important among scientists and practitioners. The IoT is a fundamental concept of a new technology that promises to play significant roles in virtually all aspects of life. The benefits of the IoT, that is, convenience, earning new revenues, making the company more efficient, and saving money, have been identified in the literature (Dachyar et al. 2019; Nord et al. 2019; Perera et al. 2014).

IoT-based systems play essential roles in all aspects of human lives. Multifarious technologies are included in data transfer among settled network devices. The IoT has given rise to several challenges such as access control, authentication, cooperation, integration among technologies, mobile security, networking challenges, policy enforcement, lack of international norms and standards, privacy, security, resources, return on investment, best practices, and trust (Bao et al. 2013; HaddadPajouh et al. 2020; Hao et al. 2015; Huo and Wang 2016; Kumar et al. 2019; Ndibanje et al. 2014; Pereira and Aguiar 2014; Roman et al. 2011; Simec 2019; Tikk-Ringas 2016; Wolfs 2017; Yan et al. 2014).

Authentication and access control are important issues of the IoT that strengthen security. They can prevent unauthorized users from gaining access to resources and can prevent legitimate users from accessing resources in an unauthorized manner. When developing an IoT infrastructure, it is crucial to take into consideration efficiency, security, scalability, market-oriented computing, power resource, and storage features for the best quality of services for users (Ndibanje et al. 2014).

The development of IoT networks rises to complex issues of integration and cooperation among different technologies. The IoT systems consist of a multitude of devices and sensors that communicate with each other and transfer huge amounts of data over the Internet (Hao et al. 2015). The challenges/issues with the IoT expansions are maintenance (Wolfs 2017) and support cooperation, and integration of many devices with different memory, processing, storage bandwidth, or power supply (Pereira and Aguiar 2014). Mobile security and research in this area concern the technology that will secure transferred information (Kumar et al. 2019). Addressing and analyzing IoT security involves potential solutions that would assist enterprises in finding appropriate solutions to tackle specific threats and provide the best possible IoT-based services (HaddadPajouh et al. 2020).

The IoT faces everlasting networking challenges that include the quality of service delivered in a specific area, security, cost, reliability, energy consumption, availability, and service time (Huo and Wang 2016). The policy is another issue for IoT developers. There are specific regulations to maintain the standard to prevent people from violating them. It will be important to create international norms and standards for the IoT. Governments need to collaborate with businesses to harmonize compliance requirements in data and liability laws (Simec 2019; Tikk-Ringas 2016). Organizations and governments must take appropriate measures to improve their capabilities against cyberattacks. Privacy is an important concern that allows users to feel secure while using IoT technologies. It is crucial to maintain the authorization and authentication over a secure network to establish communication among trusted parties (Roman et al. 2011).

Furthermore, each object should have the possibility to verify and check the privacy policies of other objects in the IoT system before transmitting the data. An important issue of IoT that requires attention is security (Kumar Tiwari and

Zymbler 2019). The IoT is easy to hack and difficult to survive various cyberattacks (Falco et al. 2018). Security issues are fundamental parameters to develop trust in IoT (Nord et al. 2019).

The availability of resources for the IoT is a challenge that requires training, education, and the development of experts. Return on investment is another challenge. One of the main problems concerns the impact on the environment in case of energy consumption by IoT devices. Operational safety and security practices in the IoT vary significantly across industries. It is essential to document and analyze the best practices in business. It supports the identification of requirements for potential innovation (Simec 2019). The challenge creates trust in society to IoT Technologies (Bao et al 2013; Yan et al. 2014).

The development of the IoT is related to virtually all industries. The IoT is promising to create new jobs and offers opportunities to social development, including the promotion of employment and economic growth (Simec 2019).

2.6 METHODS

2.6.1 INSTRUMENT

The instrument survey used for this study consisted of the following four parts:

(1) The IoT external application usage consisted of 12 areas, that is, energy, financial services, healthcare, information technology, manufacturing, personal and social issues, retail, smart agriculture, smart city, smart grid (smart environment), supply chain, transportation and logistics, and wearables (a five-point Likert-type scale: 1 – extremely high use, 2 – high use, 3 – neutral, 4 – low use, and 5 – extremely low use).

(2) The IoT internal application usage consisted of nine areas, that is, asset management, customer experience, environment, finance, manufacturing, product development, safety, supply chain, warehousing, and logistics (a five-point Likert-type scale: 1 – extremely high priority, 2 – high priority, 3 – neutral, 4 – low priority, 5 –extremely low priority).

(3) The IoT benefits consisted of four main benefits for the organization through IoT utilization, that is, convenience, earning new revenues, making the company more efficient, and saving money (a five-point Likert-type scale: 1 – extremely high priority, 2 – high priority, 3 – neutral, 4 – low priority, 5 –extremely low priority).

(4) The IoT challenges consisted of 14 areas, that is, access control, authentication, cooperation among departments, integration among technologies, lack of international norms and standards, mobile security, networking challenges, policy enforcement, privacy, resources (skilled personnel), return on investment, security, too few best practices, and trust (a five-point Likert-type scale: 1 – extremely high priority, 2 – high priority, 3 – neutral, 4 – low priority, 5 –extremely low priority).

2.6.2 SAMPLE AND PROCEDURE

The survey was translated into the Polish language and administered online from the United States to approximately 200 enterprises. A total of 102 enterprises in Poland completed the survey. All responses were translated into English. These enterprises had ongoing cooperation with the Warsaw University of Life Sciences (WULS). The study was conducted in 2019. Participation in completing the survey was completely voluntary and the companies were assured confidentiality and anonymity.

2.6.3 DATA ANALYSIS

Descriptive statistics were used to identify IoT internal and external application assessment, as well as challenges and benefits. Pearson correlation and multiple linear regression analysis were used to infer causal relationships between the dependent variables (IoT internal and IoT external application) and independent variables defined as listed challenges (access control, authentication, cooperation among departments, integration among technologies, lack of international norms and standards, mobile security, networking challenges, policy enforcement, privacy, resources, return on investment, security, too few best practices and trust). Regression analysis (backward stepwise method) was used to estimate the model fit.

2.7 RESULTS

2.7.1 SAMPLE DEMOGRAPHICS

The enterprises represent five areas of operation (see Figure 2.1). The largest group of companies was classified as service companies (45%), the second was retail companies (29%), and the third was manufacturing (14%). Ninety-nine percent of the companies were located in the Mazovia Province in Poland. Government organizations were the smallest group of investigated enterprises (4%). Among 102 companies, 88% were using IoT, the rest of the companies took part in the study by assessing future IoT challenges (Figure 2.2).

Figure 2.3 presents the total annual revenue of companies. The largest group of the surveyed companies reached the revenue level higher than $1M (45%). The second group generated revenues from $251,000 to $500,000 and represented 30% of the companies. The smallest group of surveyed enterprises were entities with the lowest revenues, whose share in the studied group was 5% (less than $250,000).

2.7.2 IoT EXTERNAL APPLICATION USAGE

In the surveyed enterprises, six areas of IoT usage reached higher than neutral application (see Figure 2.4). The highest extensive usage was in information technology that reached 2.1 ranks, followed by retail (2.2), transportation and logistics (2.3), financial services (2.4), supply chain (2.5), and personal and social (2.7).

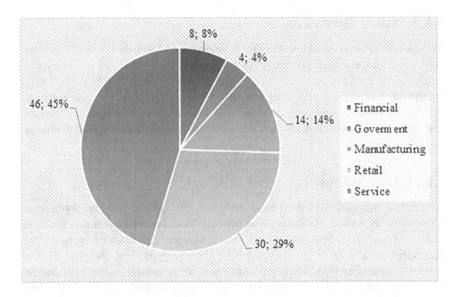

FIGURE 2.1   Organizations operating area.

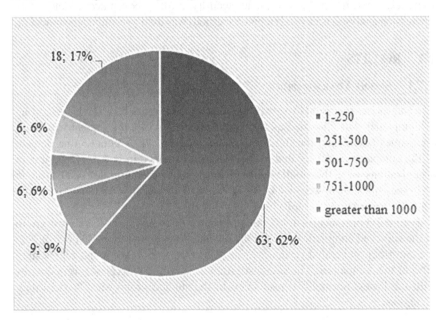

FIGURE 2.2   Number of employees within surveyed companies.

The application of IoT in the energy sector was assessed as neutral (3.0). The lowest usage was identified for information smart agriculture in Poland 4.5 – showed as extremely low use. This rank could be explained by limited agriculture enterprise representatives in the survey sample.

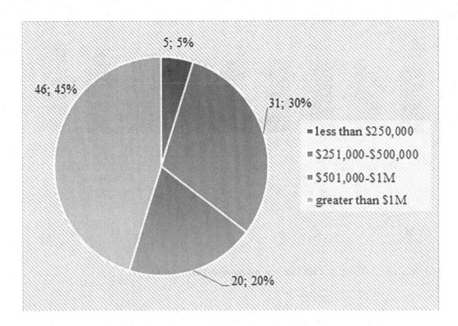

FIGURE 2.3   The total annual revenue of the organization.

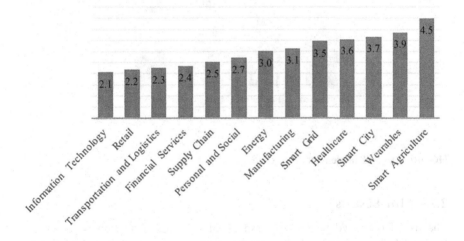

FIGURE 2.4   IoT application usage assessment rank.

2.7.3 IoT Internal Application and Services Priority Usage

Among investigates companies, the IoT application (see Figure 2.5) reached the highest priority of utilization in finance and safety (both 2.0), followed by warehousing and logistics (2.2). Three applications of IoT were assessed on the same level of priority (2.3): customer experience, product development, and supply chain. The lowest priority was found in the case of manufacturing applications.

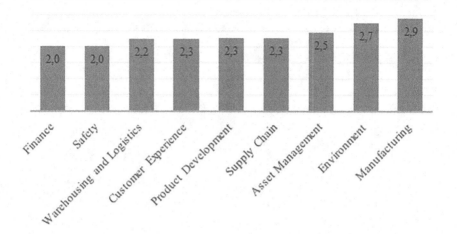

FIGURE 2.5   IoT application internal usage assessment rank.

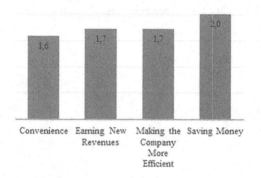

FIGURE 2.6   IoT implementation benefits.

2.7.4   IoT Benefits

The list of predicted benefits assessment of IoT implementation is shown in Figure 2.6. The extent of the benefit provided to the organization through the utilization of the IoT noticed the most extremely beneficial assessment for convenience (1.6). Earning new revenues and making the company more efficient were 1.7. The use of IoT to improve the convenience of working conditions is the main benefit that can be both important for the organization itself and its environment.

Table 2.3 presents descriptive statistics listed in the external study application of IoT in the company. The indicated answers were slightly variable in the scope of opinions expressed by the respondents. A slightly higher standard deviation

TABLE 2.3
Descriptive Statistics for IoT Benefits

Challenges	Standard Deviation (SD)	Coefficient of Variation (%)
Convenience	0.70	40.6
Earning new revenues	0.78	45.4
Making the company more efficient	0.67	42.5
Saving money	0.97	48.2

(SD) level occurred in the benefit defined as earning new revenues (0.67). However, the highest coefficient of variance was recorded for saving money (48.2%). These may indicate a diverse approach to assessing the cost-effectiveness of the IoT benefits.

2.8 IOT UTILIZATION AND IMPLEMENTATION CHALLENGES

The list of challenges faced by companies when implementing and utilizing the IoT is presented in Figure 2.7. All unique challenges in IoT noticed the assessment on an average level lower than 3.0. It explains that all challenges were assessed as "extremely high" or "high." The highest challenged rank was noticed for security, privacy, and resource issues in companies (1.8). The authentication and integration among technologies reached a very similar rank of 1.9, which also expresses a high level of challenges in IoT implementation. The same average rank (2.0) was also achieved in four challenges: access control, mobile security, trust, and return on investment. The closest to natural rank (3.0) in challenges assessment was noticed for "too few best practices" (2.9).

Table 2.4 presents the descriptive statistics listed in the internal study application of IoT in the company. Although most of the listed challenges had a similar average rating rank, SD was the highest in the case of the lack of international norms and standards (1.06), for which the coefficient of variation was 40.2%. These demonstrate the diverse assessment of this challenge in the application and implementation of IoT in enterprises. Moreover, high SD was noticed for networking challenges (0.93), in which the variation amounted to 41.3%. This issue could be understood as visibility and troubleshooting capabilities and usage of internal and external cloud resources, which may be related to employees being afraid of sharing knowledge. The lowest SD was noticed for authentication (0.68), which also had a low coefficient of variation (35.2%). These indicate a convergent assessment of these IoT challenges in the opinion of respondents. The highest variation was noticed in the case of IoT challenges such as privacy (47.0%), security (45.4%), and cooperation among departments (45.1%). The lowest was recorded for "too few best practices" challenge, which was similar among the investigated sample of companies.

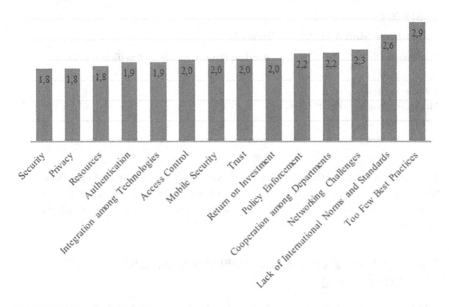

FIGURE 2.7 IoT challenges.

TABLE 2.4
Descriptive Statistics for IoT Challenges

Challenges	Standard Deviation (SD)	Coefficient of Variation (%)
Access control	0.74	37.1
Authentication	0.68	35.2
Cooperation among departments	0.98	45.1
Integration among technologies	0.84	43.5
Lack of international norms and standards	1.06	40.2
Mobile security	0.80	39.4
Networking challenges	0.93	41.3
Policy enforcement	0.85	39.5
Privacy	0.84	47.0
Resources	0.75	40.4
Return on investment	0.83	40.8
Security	0.81	45.4
Too few best practices	0.85	29.1
Trust	0.86	42.5

2.9 IOT EXTERNAL AND INTERNAL APPLICATION USAGE – ORGANIZATION APPROACH

Table 2.5 shows the correlations between the main areas of IoT external application for the listed challenges. Access control, privacy, trust, and return on investment did not have a significant correlation with any of the listed external applications of IoT. These could be assessed as challenges that are not important at this stage of IoT development in Poland. The most important challenges of IoT, according to external IoT application, was recognized as the lack of international norms and standards. Furthermore, a low but significant relationship was also noticed for IoT applications in energy, manufacturing, personal and social issues, retail, and smart agriculture. There was a negative and significant relationship between integration among technologies and information technology. The listed challenges not directly identified with external IoT application was healthcare, which could be a result of no representation in the sample. Moreover, the implementation of the IoT in the smart city and smart grid does not see further IoT applications. A similar situation appears in the case of a smart grid.

Table 2.6 presents the correlations between the main areas of IoT internal application for the listed challenges. The highest level of challenges was noticed for securities concerning privacy (0.606). The second were networking challenges (0.442) and mobile security (0.430) related to finance IoT applications and services. The most important challenges in investigated enterprises sample were recorded for security issues. These challenges noticed significant correlation with IoT application in asset management (0.284), customer experience (0.209), environment (0.404), finance (0.353), and safety issues (0.606). For this IoT application, relationships had a low level of correlation (see Table 2.6).

2.10 IOT INTERNAL APPLICATION USAGE – ORGANIZATION CHALLENGES

Table 2.7 presents the results of standardized regression models for external IoT utilization and application versus its challenges issues. The multiple regression model was constructed for combining the most important variables with their overall explanatory power. The model formula can be presented as follows:

$$Y = \beta 0 + \beta 1 X1 + \beta 2 X2 + \cdots + \beta k\ Xk + E \qquad (2.1)$$

The Shapiro–Wilk test checked the normality of the residuals in the eliminated model.

The value ("R-squared" row) represents the proportion of variance in the dependent variable that was explained by our list of independent variables.

The results indicate that few IoT applications did not record significant relationships with listed IoT challenges such as financial services, healthcare, manufacturing, personal and social, retail, smart agriculture, smart grid, and wearables. They were not included in the results in Table 2.7.

TABLE 2.5

Correlation Matrix for Challenges of IoT and External Application

	Authentication	Cooperation among Departments	Integration among Technologies	Lack of International Norms and Standards	Mobile Security	Networking Challenges	Policy Enforcement	Resources	Security	Too Few Best Practices
Energy	−0.025	0.140	0.003	0.312*	0.071	0.103	0.263*	0.136	−0.002	0.159
Supply chain	−0.099	0.094	−0.017	0.184	0.065	0.056	0.224*	0.029	−0.001	0.118
Transportation and logistics	0.248*	0.348*	0.311*	0.120	0.271*	0.217*	0.038	0.206*	0.258*	0.222*
Wearables	−0.036	0.275*	−0.074	0.128	−0.068	0.066	0.178	0.236*	0.049	0.041
Financial services	0.204*	0.224*	−0.022	0.144	0.144	0.189	0.101	0.254*	0.066	0.082
Information technology	−0.073	0.007	−0.264*	0.097	−0.015	−0.085	0.081	−0.010	−0.009	0.028
Manufacturing	0.087	0.097	−0.001	0.327*	0.171	0.166	0.298*	0.193	0.198*	0.302*
Personal and social	0.086	0.139	−0.003	0.238*	0.101	0.056	0.160	0.280*	0.124	0.234*
Retail	0.021	0.115	0.065	0.321*	−0.141	0.021	0.123	0.182	0.047	0.082
Smart agriculture	0.065	0.064	0.103	0.335*	−0.068	0.034	0.177	0.183	0.026	0.052

*p-value < 0.05.

TABLE 2.6

Correlation Matrix for Challenges of IoT and Internal Application

	Access Control	Authentication	Cooperation among Departments	Integration among Technologies	Lack of International Norms and Standards	Mobile Security	Networking Challenges	Policy Enforcement	Privacy	Resources	Return on Investment	Security	Too Few Best Practices	Trust
Asset management	0.203*	0.123	0.288*	0.234*	0.065	0.081	0.026	0.079	0.058	0.264*	0.170	0.284*	0.113	0.156
Customer experience	−0.015	0.105	0.125	0.149	0.134	0.128	0.188	0.151	0.019	0.045	0.051	0.209*	0.147	0.140
Environment	0.046	0.088	0.251*	0.205*	0.123	0.079	0.074	0.349*	0.258*	0.223*	0.110	0.404*	0.070	0.037
Finance	0.159	0.134	0.150	0.171	0.058	0.430*	0.442*	0.173	0.184	0.103	0.254*	0.353*	0.061	0.168
Manufacturing	0.130	−0.082	0.285*	−0.018	0.148	−0.068	0.042	0.126	0.057	0.245*	0.161	0.137	0.124	−0.024
Product development	0.095	0.084	0.355*	0.188	0.097	0.019	0.220*	0.166	0.089	0.150	0.169	0.096	0.114	−0.033
Safety	0.354*	0.352*	0.292*	0.299*	0.043	0.216*	0.221*	−0.027	0.284*	0.184	0.069	0.606*	0.161	0.000
Supply chain	0.050	0.129	0.132	0.192	0.299*	0.064	0.109	0.168	−0.013	0.244*	0.056	0.127	0.157	0.111
Warehousing and logistics	0.123	0.135	0.044	0.239*	0.346*	0.030	0.080	0.044	0.009	0.205*	−0.008	0.180	0.236*	0.226*

*p-value < 0.05.

TABLE 2.7

External IoT Application and Services Versus Its Challenges of Utilization

Dependent Variable	Variables – Challenges	Estimate b^*	Std. Error	$t(97)$	p-Value	
Energy	Intercept			8.005	0.000	*
	Lack of international norms and standards	0.312	0.096	3.236	0.001	*
	$R=0.312\ R^2=0.097$					
	$F(1,97)=10.474\ p<0.001$ Standard error estimation: 1.020					
Information technology	Intercept			5.687	0.000	*
	Cooperation among departments	0.347	0.095	3.650	0.001	*
	$R=0.347\ R^2=0.121$					
	$F(1,97)=13.329\ p<0.001$ Standard error estimation: 0.931					
Smart city	Intercept			8.847	0.000	*
	Lack of international norms and standards	0.327	0.096	3.406	0.001	*
	$R=0.3268,\ R^2=0.107$					
	$F(1,97)=11.603\ p<0.001$ Standard error estimation: 1.121					
Supply chain	Intercept			4.421	0.000	*
	Lack of international norms and standards	0.320	0.096	3.333	0.001	*
	$R=0.320\ R^2=0.103$					
	$F(1,97)=11.114\ p<0.001$ Standard error estimation: 1.211					
Transportation and logistics	Intercept			3.899	0.000	*
	Lack of international norms and standards	0.335	0.095	3.507	0.000	*
	$R=0.335\ R^2=0.112$					
	$F(1,97)=12.300\ p<0.001$ Standard error estimation: 1.142					

$^*p$-value < 0.05.

The estimated models were significant for external IoT utilization and application in energy, smart city, supply chain, transportation, and logistics. The estimated multiple regression model identifying the challenges is significant for IoT application in energy ($F(1,97)=10.474\ p<.001$). However, the model predictor accounted for 9.7% of the dependent variable variance. A significant effect on the determination of given challenges had IoT challenges defined as "lack of international norms and standards" ($b=0.312$, $t(97)$, $p<.001$).

The next IoT application for which the model was estimated was information technology for which the cooperation among departments was noticed as a significant challenge. The regression model was significant for $F(1,97)=13.329$ $p<.001$, and the model predictor explained 12.1% of the variance of the dependent variable (energy IoT application) only in. The defined challenge as cooperation among department reached the level of beta 0.347 for $t(97)$ and $p < .001$.

The estimated multiple regression model was also significant for the smart city IoT application ($F(1,97) = 11.603$ $p < .001$); only one predictor was included as significant in the model, and it explained 10.7% of the dependent variable variance. A significant effect on the smart city application was related to "lack of international norms and standards" (beta 0.327, $t(97)$ and $p < 0.01$). This predictor was also significant in the case of IoT application in energy.

The multiple regression model was estimated for the supply chain IoT application ($F(1,97) = 11.114$ $p < .001$); only one predictor was included as significant in these models, and it explained 10.3% variance of the dependent variable. A significant effect on the supply chain application was related to "lack of international norms and standards" (beta 0.322, $t(97)$ and $p < .001$). This predictor was also significant in the case of IoT application in energy and smart city.

The multiple regression model for transportation and logistics IoT application ($F(1,97) = 12.300$ $p < .001$) recorded only one predictor. The model explained 11.2% variance of the dependent variable. A significant effect on the transportation and logistics application was related to "lack of international norms and standards" (beta 0.32, $t(97)$ and $p < .001$). The external IoT application and utilization were mostly connected with the highest score assessment to "lack of international norms and standards" as the main challenge.

Table 2.8 presents the results of standardized regression models for internal IoT utilization and application internally in enterprises versus its challenges issues. The results indicate that few IoT internal applications did not record significant relationships with listed IoT challenges such as asset management, customer experience, manufacturing, and supply chain. They were not included in the results in Table 2.8.

The estimated multiple regression model was identified for environment IoT application $F(2,96) = 16.684$ $p < .000$, the model predictors accounted for a total of 25.8% of the dependent variable. Significant effect on the determination of a given challenges had IoT challenge defined as "policy enforcement" ($b = 0.309$, $t(97)$, $p < .001$) and "security" ($b = 0.371$, $t(97)$, $p < .001$). The next estimated multiple regression model was identified for finance IoT application $F(1,97) = 23.574$ $p < .000$, the model predictors included accounted for a total of 19.5% of the dependent variable. A significant effect on the determination of given challenges had IoT challenges defined as "networking challenges" ($b = 0.312$, $t(97)$, $p < .000$).

The multiple regression model was identified for "product development" IoT application $F(1,97) = 13.981$ $p < .001$, the model predictors included accounted for a total of 12.5% of the dependent variable. A significant effect on the determination of given challenges had an IoT challenge defined as "cooperation among departments" ($b = 0.354$, $t(97)$, $p < .000$).

The next regression model concerns the "safety" of the IoT application and utilization. These IoT challenges were related to the "security" challenge ($b = 0.606$, $t(97)$, $p < .000$). The model predictors included account for a total of 36.7% of the dependent variable.

TABLE 2.8

Internal IoT Application and Services versus Its Challenges of Utilization

Dependent Variable	Variables – Challenges	Estimate b^*	Std. Error	$t(97)$	p-Value	
Environment	Intercept			4.776	0.000	*
	Policy enforcement	0.309	0.088	3.497	0.000	*
	Security	0.371	0.088	4.199	0.000	*
	$R=0.508\ R^2=0.258$					
	$F(2,96)=16.684\ p<0.000$ Standard error estimation: 0.787					
Finance	Intercept			6.260	0.000	*
	Networking challenges	0.442	0.091	4.855	0.000	*
	$R=0.442\ R^2=0.195$					
	$F(1,97)=23.574\ p<0.000$ Standard error estimation: 0.699					
Product development	Intercept			6.362	0.000	*
	Cooperation among departments	0.354	0.094	3.739	0.000	*
	$R=0.355\ R^2=0.125$					
	$F(1,97)=13.981\ p<0.001$ Standard error estimation: 0.947					
Safety	Intercept			4.644	0.000	*
	Security	0.606	0.081	7.503	0.000	*
	$R=0.606\ R^2=0.367$					
	$F(1,97)=56.302\ p<0.000$ Standard error estimation: 0.713					
Warehousing and logistics	Intercept			4.284	0.000	*
	Lack of international norms and standards	0.345	0.095	3.626	0.000	*
	$R=0.345\ R^2=0.119$					
	$F(1,97)=13.154\ p<0.001$ Standard error estimation: 1.0582					

*p-value < 0.05.

The last IoT application regression model was identified for "warehousing and logistics" $F(1,97)=13.154\ p<.001$, the model predictors included accounted for a total of 11.9% of the dependent variable. A significant effect on the determination of given challenges had IoT challenges defined as "lack of international norms and standards" ($b=0.\ 345$, $t(97)$, $p<.001$).

The estimated models were mainly characterized by one predictor of challenges concerning selected internal areas of IoT application assessment. These indicate a significant diversity of assessments of given challenges in the examined group of enterprises.

2.11　DISCUSSION

The purpose of this chapter was to establish the challenges of IoT business applications. The first research question concerns the scope of the IoT external and internal usage by the organizations. The external IoT system usage was noticed as a priority in the case of information technology, retail, transportation and logistics, and financial services. The IoT applications in business operations were assessed as a priority in the case of finance, safety and warehousing, and logistics. Both perspectives indicate perspectives on the use of IoT in business. However, the challenged assessment was quite diverse in the study.

The second research question concerns the differentiation of challenges in specific IoT applications. The analysis for external IoT services noticed insignificant relations in the case of financial services, healthcare, manufacturing, personal and social, retail, smart agriculture, smart grid, and wearables. The multiple regression analysis for external IoT services identified business applications that noticed insignificant relationships in the case of asset management, customer experience, manufacturing, and supply chain.

The regression models of the IoT external services were estimated for energy, smart city supply chain, transportation, and logistics. For this IoT application, the lack of international norms and standards was a significant predictor of application assessment. These gaps and requirements for IoT standards were also underlined by Simec (2019) and Liu et al. (2020). The challenge of international norms and standards created for the IoT were noticed in the research of Tikk-Ringas (2016), Ndibanje et al. (2014), Hao et al. (2015), Wolfs (2017), Pereira and Aguiar (2014), HaddadPajouh et al. (2020), Huo and Wang (2016), Kumar et al. (2019), Yan et al. (2014), and Bao et al. (2013).

In the case of the model constructed for information technology applications, cooperation among departments was looked at as the predictor. Implementing new IoT solutions should be beneficial for all cooperating departments, which was confirmed by Pereira and Aguiar (2014).

The regression models for the IoT internal applications were estimated for the environment, finance, product development, safety, warehousing, and logistics. For every IoT internal business application, different challenges were assigned by responders. In the case of an enterprise environment, policy enforcement and security became two of the most important challenges. The IoT finance application meets with networking challenges; for product development, it was cooperation among departments; for safety, the predictor was security; and for warehousing and logistics, a lack of international norms and standards.

The most demanding IoT challenges for companies were security, privacy, and resource issues in companies (research question 3). This was followed by authentication and integration among technologies. However, the highest SD was noticed for the lack of international norms and standards, which could be perceived by responders differently according to their primary field of operation.

2.12 CONCLUSION

Recent advancements in the IoT have drawn the attention of researchers and developers worldwide. Although the IoT innovations have been deployed in numerous industries and use cases, the results show that it is still at an early stage of adoption and implementation, and many challenges remain regarding the applications. The results of this study underline the important IoT challenges concerning the lack of international norms and standards. The IoT future potential could be limited due to the gap in norms and standards and the creation of new business models. This approach sheds light on future research needs across different disciplines in theory and methodologies to support IoT implementation in organizations.

The empirical examination in this chapter is limited to the responses and characteristics of the sample. Further study could include the joint model that would join the external and internal IoT applications regarding its application and utilization challenges. Furthermore, in future studies, the authors plan to expand this research to enterprises in other countries.

REFERENCES

Adiputra, R. R., Hadiyoso, S., and Y. S. Hariyani. 2018. Internet of Things – Low cost and wearable SpO2 device for health monitoring. *International Journal of Electrical and Computer Engineering* 8(2): 939–945.

Afghan, S. A., and H. Géza. 2019. Modelling and analysis of energy harvesting in Internet of Things (IoT) – Characterization of a thermal energy harvesting circuit for IoT based applications with LTC3108. *Energies* 12(20): 1–13.

Alavi, A. H., Jiao, P., Buttlar, W. G., and N. Lajnef. 2018. Internet of Things-enabled smart cities – state-of-the-art and future trends. *Measurement* 129: 589–606.

Alkhatib, H., Faraboschi, P., Frachtenberg, E., Kasahara, H., Lange, D., Laplante, P., Merchant, A., Milojicic, D., and K. Schwan. 2014. *IEEE CS 2022 Report*. Washington, DC: IEEE Computer Society, 25–27.

Angeles, R. 2005. RFID technologies: Supply-chain applications and implementation issues. *Information Systems Management* 22(1): 51–65.

Ansari, S., Aslam, T., Poncela, J., Otero, P., and A. Ansari. 2019. Internet of Things-based healthcare applications. In *IoT Architectures, Models and Platforms for Smart City Applications*, ed. B. S. Chowdhry, F. K. Shaikh and N. A. Mahoto, 1–28. Hershey, PA: IGI Global.

Ashton, K. 2009. That 'Internet of Things' Thing. Retrieved from: http://www.rfidjournal.com/articles/view?4986 [5 January 2020].

Backman, J., and H. Helaakoski. 2016. Evaluation of internet-of-things platforms for asset management. In *Proceedings of the 10th World Congress on Engineering Asset Management (WCEAM 2015)*, eds. K. T. Koskinen, H. Kortelainen, J. Aaltonen, T. Uusitalo, K. Komonen, J. Mathew, and J. Laitinen, 97–104. Cham: Springer.

Baker, T., Asim, M., Tawfik, H., Aldawsari, B., and R. Buyya. 2017. An energy-aware service composition algorithm for multiple cloud-based IoT applications. *Journal of Network and Computer Applications* 89: 96–108.

Balaji, M. S., and S.K. Roy. 2017. Value co-creation with Internet of Things technology in the retail industry. *Journal of Marketing Management* 33(1–2): 7–31.

Bandyopadhyay, D., and J. Sen. 2011. Internet of Things: Applications and challenges in technology and standardization. *Wireless Personal Communications* 58(1): 49–69.

Bao, F., Chen, I. R., and J. Guo. 2013. Scalable, adaptive and survivable trust management for community of interest based internet of things systems. In Proc. IEEE 11th *International Symposium on Autonomous Decentralized Systems (ISADS)*, ed. G. Hernandez Lopez, 1–7. Mexico City: Institute of Electrical and Electronics Engineers.

Barratt, M., and A. Oke. 2007. Antecedents of supply chain visibility in retail supply chains: A resource-based theory perspective. *Journal of Operations Management* 25(6): 1217–1233.

Bayo-Monton, J. L., Martinez-Millana, A., Han, W., Fernandez-Llatas, C., Sun, Y., and V. Traver. 2018. Wearable sensors integrated with Internet of Things for advancing eHealth care. *Sensors* 18(6): 1851.

Bekara, C. 2014. Security issues and challenges for the IoT-based smart grid. *Procedia Computer Science* 34: 532–537.

Bello, H., Xiaoping, Z., Nordin, R., and J. Xin. 2019. Advances and opportunities in passive wake-up radios with wireless energy harvesting for the Internet of Things applications. *Sensors* 19(14): 1–33.

Ben-Daya, M., Hassini, E., and Z. Bahroun. 2019. Internet of Things and supply chain management: A literature review. *International Journal of Production Research* 57(15–16): 4719–4742.

Bi, Z., Wang, G., and X. Li Da. 2016. A visualization platform for Internet of Things in manufacturing applications. *Internet Research* 26(2): 377–401.

Bi, Z., Da Xu, L., and C. Wang. 2014. Internet of Things for enterprise systems of modern manufacturing. *IEEE Transactions on Industrial Informatics* 10(2): 1537–1546.

Brous, P., Janssen, M., Schraven, D., Spiegeler, J., and B. C. Duzgun. 2017. Factors influencing adoption of IoT for data-driven decision making in asset management organizations. In *IoTBDS*, eds. M. Ramachandran, V. M. Muñoz, V. Kantere; G. Wills; R. Waltes and V. Chang, 70–79. Porto: Science and Technology Publications, Lda

Buntak, K., Kovačić, M., and M. Mutavdžija. 2.019. Internet of Things and smart warehouses as the future of logistics. *Tehnički Glasnik* 13(3): 248–253.

Caro, F., and R. Sadr. 2019. The Internet of Things (IoT) in retail: Bridging supply and demand. *Business Horizons* 62: 47–54.

Catarinucci, L., De Donno, D., Mainetti, L., Palano, L., Patrono, L., Stefanizzi, M. L., and L. Tarricone. 2015. An IoT-aware architecture for smart healthcare systems. *IEEE Internet of Things Journal* 2(6): 515–526.

Ceipidor, U. B., Medaglia, C. M., Volpi, V., Moroni, A., Sposato, S., and M. Tamburrano. 2011. Design and development of a social shopping experience in the IoT domain: The ShopLovers solution. In *SoftCOM 2011, 19th International Conference on Software, Telecommunications and Computer Networks*, eds. N. Rožic and D. Beguši, 1–5. Split: Institute of Electrical and Electronics Engineers.

Chowdhry, B. S., Shaikh, F. K., and N. A. Mahoto. 2019. *IoT Architectures, Models and Platforms for Smart City Applications*. Hershey, PA: IGI Global.

Choy, K. L., Ho, G. T., and C. K. H. Lee. 2017. A RFID-based storage assignment system for enhancing the efficiency of order picking. *Journal of Intelligent Manufacturing* 28(1): 111–129.

Cuomo, S., Somma, V. D., and F. Sica. 2018. An application of the one-factor HullWhite model in an IoT financial scenario. *Sustainable Cities and Society* 38: 18–20.

Dachyar, M., Zagloel, T. Y. M., and L. Ranjaliba Saragih. 2019. Knowledge growth and development: Internet of Things (IoT) research, 2006–2018. *Heliyon* 5: e02264.

Dash, D., Farooq, R., Panda, J. S., and K. V. Sandhyavani. 2019. Internet of Things (IoT): The new paradigm of HRM and skill development in the fourth industrial revolution (industry 4.0). *IUP Journal of Information Technology* 15(4): 7–30.

Diène, B., Rodrigues, J. J. P. C., Diallo, O., Ndoye, E. H. M., and V. V. Korotaev. 2020. Data management techniques for Internet of Things. *Mechanical Systems and Signal Processing* 138: 106564.

Dineshreddy, V., and G. R. Gangadharan. 2016. Towards an "Internet of Things" frame-work for financial services sector. In *2016 3rd International Conference on Recent Advances in Information Technology (RAIT)*, eds. C. Kumar and H. Banka, 177–181. Dhanbad: Institute of Electrical and Electronics Engineers.

Ding, W. 2013. Study of smart warehouse management system based on the IOT. In *Intelligence Computation and Evolutionary Computation*, ed. Z. Du, 203–207. Berlin, Heidelberg: Springer.

Eskofier, B. M., Lee, S. I., Baron, M., Simon, A., Martindale, C. F., Gaßner, H., and J. Klucken. 2017. An overview of smart shoes in the Internet of health things: Gait and mobility assessment in health promotion and disease monitoring. *Applied Sciences* 7(10): 1–17.

Falco, G., Viswanathan, A., Caldera, C., and H. Shrobe. 2018. A master attack method-ology for an AI-based automated attack planner for smart cities. IEEE Access 6, 48360–48373. IEEE.

Ferrandez-Pastor, F., García-Chamizo, J. M., Nieto-Hidalgo, M., Mora-Pascual, J., and J. Mora-Martínez. 2016. Developing ubiquitous sensor network platform using Internet of Things: Application in precision agriculture. *Sensors* 16(7): 1141.

Fortino, G., and P. Trunfio. 2014. Preface of *Internet of Things based on smart objects: Technology*, middleware and applications. In *Internet of Things based on Smart Objects: Technology, Middleware and Applications*, ed. G. Fortino and P. Trunfio. Cham: Springer.

Giudice, M. D., Campanella, F., and L. Dezi. 2016. The bank of things. *Business Process Management Journal* 22(2): 324–340.

Ghallab, H., Fahmy, H., and M. Nasr. 2019. Detection outliers on Internet of Things using big data technology. *Egyptian Informatics Journal* https://doi.org/10.1016/j.eij.2019.12.001

Gluhak, A., Krc, S., Nati, M., Pfristerer, D., Mitton, N., and T. Razafindralambo. 2011. A sur-vey on facilities for experimental Internet of Things research. *IEEE Communications Magazine* 49(11): 58–67.

Gregory, J. 2015. The Internet of Things: Revolutionizing the retail industry. *Accenture Strategy*, available on on-line at iotone.com.

Gubbi, J., Buyya, R., Marusic, S., and M. Palaniswami. 2013. Internet of Things (IoT): A vision, architectural elements, and future directions. *Future Generation Computer Systems* 29(7): 1645–1660.

Guerrero-Ibanez, J. A., Zeadally, S., and J. Contreras-Castillo. 2015. Integration chal-lenges of intelligent transportation systems with connected vehicle, cloud comput-ing, and Internet of Things technologies. *IEEE Wireless Communications* 22(6): 122–128.

HaddadPajouh, H., Dehghantanha, A., Parizi, R. M., Aledhari, M., and H. Karimipour. 2020. A survey on Internet of Things security: Requirements, challenges, and solu-tions, *Internet of Things*, https://doi.org/10.1016/j.iot.2019.100129

Haller, S., Karnouskos, S., and C. Schroth. 2008. The Internet of Things in an enterprise context. In *Future Internet Symposium*, 14–28. Berlin, Heidelberg: Springer.

Hao, Q., Zhang, F., Liu, Z., and L. Qin. 2015. Design of chemical industrial park integrated information management platform based on cloud computing and IOT (the internet of things) technologies. *International Journal of Smart Home* 9(4): 35–46.

Heinis, T. B., Hilario, J., and M. Meboldt. 2018. Empirical study on innovation motivators and inhibitors of Internet of Things applications for industrial manufacturing enter-prises. *Journal of Innovation and Entrepreneurship* 7(1): 1–22.

Hou, J., Qu, L., and W. Shi. 2019. A survey on Internet of Things security from data per-spectives. *Computer Networks* 148: 295–306.

Huo, L., and Z. Wang. 2016. Service composition instantiation based on cross-modified artificial Bee Colony algorithm. *China Communication* 13(10): 233–244.

Hussain, M. M., Mohammad, S. A., and M. M. Sufyan Beg. 2018. Computational viability of fog methodologies in IoT enabled smart city architectures – A smart grid case study. *EAI Endorsed Transactions on Smart Cities* 3(7): 1–12.

Ibarra-Esquer, J., González-Navarro, F., Flores-Rios, B., Burtseva, L., and M. Astorga-Vargas. 2017. Tracking the evolution of the Internet of Things concept across different application domains. *Sensors* 17(6): 1379.

Jamison, J., and C. Snow. 2014. An architecture for customer experience management based on the Internet of Things. *IBM Journal of Research and Development* 58(5/6): 15-1–15-11.

Jiang, Y. 2020. Combination of wearable sensors and Internet of Things and its application in sports rehabilitation. *Computer Communications* 150: 167–176.

Kamble, S. S., Gunasekaran, A., Parekh, H., and S. Joshi. 2019. Modeling the Internet of Things adoption barriers in food retail supply chains. *Journal of Retailing and Consumer Services* 48: 154–168.

Kang, H. S., Lee, J. Y., Choi, S., Kim, H., Park, J. H., Son, J. Y., and S. Do Noh. 2016. Smart manufacturing: Past research, present findings, and future directions. *International Journal of Precision Engineering and Manufacturing-Green Technology* 3(1): 111–128.

Karakostas, B. 2013. A DNS architecture for the Internet of Things: A case study in transport logistics. *Procedia Computer Science* 19: 594–601.

Karimova, G. Z., and A. Shirkhanbeik. 2015. Society of things: An alternative vision of Internet of Things. *Cogent Social Sciences* 1(1): 1–21.

Kinnunen, S. K., Ylä-Kujala, A., Marttonen-Arola, S., Kärri, T., and D. Baglee. 2018. Internet of Things in asset management: Insights from industrial professionals and academia. *International Journal of Service Science, Management, Engineering, and Technology (IJSSMET)* 9(2): 104–119.

Khajenasiri, I., Estebsari, A., Verhelst, M., and G. Gielen. 2017. A review on Internet of Things for intelligent energy control in buildings for smart city applications. *Energy Procedia* 111: 770–779.

Kumar, S., Tiwari, P., and M. Zymbler. 2019. Internet of Things is a revolutionary approach for future technology enhancement: A review. *Journal of Big Data* 6(1): 1–21.

Lee, I., and K. Lee. 2015. The Internet of Things (IoT): Applications, investments, and challenges for enterprises. *Business Horizons* 58(4): 431–440.

Lee, C. K. M., Lv, Y., Ng, K. K. H., Ho, W., and K. L. Choy. (2018). Design and application of Internet of Things-based warehouse management system for smart logistics. *International Journal of Production Research* 56(8): 2753–2768.

Li, B., and Y. Li. 2017. Internet of Things drives supply chain innovation: A research framework. *International Journal of Organizational Innovation* 9(3): 71–92.

Li, S., Chen, W., Hu, J., and J. Hu. 2018. ASPIE: A framework for active sensing and processing of complex events in the Internet of manufacturing things. *Sustainability* 10(3): 692.

Liu, J. and Z. Yan. 2015. Fusion – An aide to data mining in Internet of Things. *Information Fusion* 23(8): 1–2.

Liu, S., Zhang, Y., Liu, Y., Wang, L., and X. W. Wang. 2019. An 'Internet of Things' enabled dynamic optimization method for smart vehicles and logistics tasks. *Journal of Cleaner Production* 215: 806–820.

Liu, T., Yuan, R., and H. Chang. 2012. Research on the Internet of Things in the automotive industry. In *ICMeCG 2012 International Conference on Management of e-Commerce and e-Government*, 230–233. Beijing, China, 20–21 October 2012.

Liu, X., Jia, M., and H. Ding. 2020. Uplink resource allocation for multicarrier grouping cognitive Internet of things based on K-means Learning. *Ad Hoc Networks* 96: 102002.

Machado, K., Rosario, D., Cerqueira, E., Loureiro, A. A. F., Neto, A., and J. N. Souza. 2013. A routing protocol based on energy and link quality for Internet of Things applications. *Sensors* 13(2): 1942–1964.

Mattern, F., and C. Floerkemeier. 2010. From the internet of computers to the Internet of Things. In *From Active Data Management to Event-Based Systems and More*, ed. K. Sachs, I. Petrov and P. Guerrero, 6462: 242–259. Berlin/Heidelberg: Springer.

De Matos, E., Tiburski, R.T., Moratelli C. R., Filho, S. J., Amaral, L. A., Ramachandran, G., Krishnamachari, B., and F. Hessel. 2020. Context information sharing for the Internet of Things: A survey. *Computer Networks* 166: 106988.

Mayordomo, I., Spies, P., Meier, F., Otto, S., Lempert, S., Bernhard, J., and A. Pflaum. 2011. Emerging technologies and challenges for the Internet of Things. In *Proceedings of the 2011 IEEE 54Th International Midwest Symposium on Circuits and Systems (MWSCAS)*, 1–4. Seoul, Korea, 7–10 August 2011.

Mijac, M., Androcec, D., and R. Picek. 2017. Smart city services driven by IoT: A systematic review. *Journal of Economic and Social Development* 4(2): 40–50.

Monostori, L. 2014. Cyber-physical production systems: Roots, expectations and R&D challenges. *Procedia Cirp* 17: 9–13.

Mostafa, N., Hamdy, W., and H. Alawady. 2019. Impacts of Internet of Things on supply chains: A framework for warehousing. *Social Sciences* 8(3): 1–10.

Mourtzis, D., Vlachou, E., and N. Milas. 2016. Industrial Big Data as a result of IoT adoption in manufacturing. *Procedia Cirp* 55: 290–295.

Mushtaq, S. 2018. Smart agriculture system based on IoT and image processing. *International Journal of Advanced Research in Computer Science* 9(1): 351–353.

Navulur, S., Sastry, A, S. C. S., and M. N. Giri Prasad. 2017. Agricultural management through wireless sensors and Internet of Things. *International Journal of Electrical and Computer Engineering* 7(6): 3492–3499.

Ndibanje, B., Lee, H., and S. Lee. 2014. Security analysis and improvements of authentication and access control in the Internet of Things. *Sensors* 14(8): 14786–14805.

Nguyen, B., Simkin, L., Balaji, M. S., and S. K. Roy. 2017. The Internet of Things and marketing: The state of play, future trends and the implications for marketing. *Journal of Marketing Management* 33(1–2): 1–6.

Nord, J. H., Koohang, A., and J. Paliszkiewicz. 2019. The Internet of Things: Review and theoretical framework. *Expert Systems with Applications* 133: 97–108.

Qiu, T., Xiao, H., and P. Zhou. 2013. Framework and case studies of intelligent monitoring platform in facility agriculture eco-system. In *Proc. 2013 Second International Conference on Agro-Geoinformatics (Agro-Geoinformatics)*, Fairfax, VA, USA, 12–16 August 2013. IEEE.

Pereira, C., and A. Aguiar. 2014. Towards efficient mobile M2M communications: Survey and open challenges. *Sensors* 14(10): 19582–19608.

Perera, C., Zaslavsky, A., Christen, P., and D. Georgakopoulos. 2014. Context aware computing for the Internet of Things: A survey. *IEEE Communications Surveys & Tutorials* 16: 414–454.

Poenicke, O., Groneberg, M., and K. Richter. 2019. Method for the planning of IoT use cases in Smart Logistics Zones. *IFAC PapersOnLine* 52(13): 2449–2454.

Reaidy, P. J., Gunasekaran, A., and A. Spalanzani. 2015. Bottom-up approach based on Internet of Things for order fulfillment in a collaborative warehousing environment. *International Journal of Production Economics* 159: 29–40.

Rghioui, A., and A. Oumnad. 2018. Challenges and opportunities of Internet of Things in healthcare. *International Journal of Electrical and Computer Engineering* 8(5): 2753–2761.

Riahi, A., Challal, Y., Natalizio, E., Chtourou, Z., and A. Bouabdallah. 2013. A systemic approach for IoT security. In *2013 IEEE International Conference on Distributed Computing in Sensor Systems*, eds. P. Chen, E. Feig, and L. Moser, 351–355. Santa Clara, CA: Institute of Electrical and Electronics Engineers.

Roman, R., Najera, P., and J. Lopez. 2011. Securing the Internet of Things. *Computer* 44(9): 51–58.

Sezer, O. B., Dogdu, E., and A. M. Ozbayoglu. 2018. Context-aware computing, learning, and big data in Internet of Things: A Survey. *IEEE Internet of Things Journal* 5(1): 1–27.

Shepherd, C., Petitcolas, F. A., Akram, R. N., and K. Markantonakis. 2017. An exploratory analysis of the security risks of the Internet of Things in finance. In *International Conference on Trust and Privacy in Digital Business*, eds. J. Lopez, S. Fischer-Hübner, and C. Lambrinoudakis, 164–179. Cham: Springer.

Simec, A. 2019. *Cyber-Attacks and Internet of Things as a Threat to Critical Infrastructure*. Varazdin: Varazdin Development and Entrepreneurship Agency (VADEA).

Singh, P. 2018. Internet of Things based health monitoring system: Opportunities and challenges. *International Journal of Advanced Research in Computer Science* 9(1): 224–228.

Sivathanu, B. 2018. Adoption of Internet of Things (IOT) based wearables for healthcare of older adults – A behavioural reasoning theory (BRT) approach. *Journal of Enabling Technologies* 12(4): 169–185.

Soni, U. S., and R. H. Talwekar. 2019. Internet of Things in smart grid: An overview. *I-Manager's Journal on Communication Engineering and Systems* 8(2): 28–36.

Stoces, M., Vanek, J., Masner, J., and J. Pavlík. 2016. Internet of Things (IoT) in agriculture – Selected aspects. *AGRIS on-Line Papers in Economics and Informatics* 8(1): 83–88.

Tahir, H., Tahir, R., and K. McDonald-Maier. 2018. On the security of consumer wearable devices in the Internet of Things. *PLoS One* 13(4): e0195487.

Tao, F., Cheng, Y., Da Xu, L., Zhang, L., and B. H. Li. 2014. CCIoT-CMfg: Cloud computing and Internet of Things-based cloud manufacturing service system. *IEEE Transactions on Industrial Informatics* 10(2): 1435–1442.

Tikk-Ringas, E. 2016. International cyber norms dialogue as an exercise of normative power. *Georgetown Journal of International Affairs* 17(3): 47–59.

Venkatesh, A. N. 2017. Connecting the dots: Internet of Things and human resource management. *American International Journal of Research in Humanities, Arts and Social Sciences*, 17(109): 2328–3734.

Vermesan, O., and P. Friess. 2013. Internet of Things: Converging *Technologies* for *Smart Environments* and *Integrated Ecosystems*. Aalborg: River Publishers.

Wang, M., Tan, J., and Y. Li. 2015. Design and implementation of enterprise asset management system based on IOT technology. In *2015 IEEE International Conference on Communication Software and Networks (ICCSN)*, ed. Yang Xiao, 384–388. Chengdu: Institute of Electrical and Electronics Engineers.

Wang, Q., Zhu, X., Ni, Y., Gu, L., and H. Zhu. 2020. Blockchain for the IoT and industrial IoT: A review. *Internet of Things*, https://doi.org/10.1016/j.iot.2019.100081

Wang, Y., Lin, Y., Zhong, R. Y., and X. Xu. 2019. IoT-enabled cloud-based additive manufacturing platform to support rapid product development. *International Journal of Production Research* 57(12): 3975–3991.

Wener, K. 2019. Can smart cities really deliver urban sustainability? Governance networks, sensor-based big data applications, and the citizen-driven Internet of Things. *Geopolitics, History and International Relations* 11(1): 104–109.

Wolfs, E. L. M. 2017. Developing rewarding, safe and sustainable IoT product experiences. *Archives of Design Research* 30(1): 73.

Xu, L. D., He, W., and S. Li. 2014. Internet of Things in industries: A survey. *IEEE Transactions on Industrial Informatics* 10(4): 2233–2243.

Yan, B., Jin, Z., Liu, L., and S. Liu. 2018. Factors influencing the adoption of the Internet of Things in supply chains. *Journal of Evolutionary Economics* 28(3): 523–545.

Yan, Z., Zhang, P., and A. V. Vasilakos. 2014. A survey on trust management for Internet of Things. *Journal of Network and Computer Application* 42: 120–134.

Yang, C., Shen, W., and X. Wang. 2016. Applications of Internet of Things in manufacturing. In *2016 IEEE 20th International Conference on Computer Supported Cooperative Work in Design (CSCWD)*, ed. S. Reiff-Marganiec, 670–675. San Francisco, CA: Institute of Electrical and Electronics Engineers.

Yue, J., Hu, Z., He, R., Zhang, X., Dulout, J., Li, C., and J. M. Guerrero. 2019. Cloud-fog architecture based energy management and decision-making for next-generation distribution network with prosumers and Internet of Things devices. *Applied Sciences* 9(3): 372.

Zanella, A., Bui N, Castellani, A, Vangelista, L., and M. Zorgi. (2014). Internet of Things for smart cities. *IEEE Internet of Things Journal* 1(1): 22–32.

Zhao, K., Xie, Y., Tsui, K. L., Wei, Q., Huang, W., Jiang, W., and J. Shi. 2015. System informatics: From methodology to applications. *IEEE Intelligent Systems* 30(6): 12–29.

3 Information Security of Weather Monitoring System with Elements of Internet Things

M. Hajder
University of Information Technology and Management

M. Nycz
Rzeszow University of Technology

P. Hajder
AGH University of Science and Technology

M. Liput
University of Information Technology and Management

CONTENTS

3.1   INTRODUCTION

Protecting nature that surrounds us is an extremely complex and multilayered task. Its degradation is not always the result of planned human actions. To react properly, it is necessary to observe nature, which is one of the most widely used cognitive methods, consisting of a relatively long, targeted, and planned apperception of objects and real-world phenomena (Such and Szcześniak 2007). Monitoring the use of resources allows predicting the negative effects of human interference and forecasting the occurrence of dangerous natural disasters.

One of the effects of systematic improvement in people's quality of life is the worrying deterioration of the surrounding environment. In many places around the world, air, water, or soil pollution endangers the health of the people who live and work there. Therefore, in the last decade, considerable attention has been devoted to counteracting the negative effects of urbanization and industrialization of further areas. One of the methods used for toward this goal is monitoring the state of the environment, allowing to track harmful phenomena occurring in the environment and take appropriate actions in a timely manner (Patnaik 2010; Wiersma 2004).

Generally, environment monitoring defines a system for observing and controlling the state of the environment, supporting the rational use of natural resources, nature protection, and ensuring the stable functioning of various economic systems (Canter 2018). According to another commonly used definition, monitoring is a system of permanent observation of the environment components and biosphere by measuring their selected characteristics (Zhu, Song and Dong 2011).

Threat analysis is a fundamental scientific discipline which is called general security theory (GST) (Davis 2009; Gunn 2008). In GST, it is important to identify ways of estimating and measuring the scale of threats and level of protection of objects and areas. Based on this, quantitative and qualitative parameters are determined. They help in making decisions and creating the content of normative documents. Additionally, they provide conclusions for commissions investigating the effects of failures and disasters. In the GST, the most generalized assessment criteria are the level of risk to human health and activity, as well as to quality and life (Lei et al. 2017).

Environmental monitoring systems (EMSs) can work in extreme conditions, and can also be the target of cyberattacks. Therefore, when designing such systems, special attention should be paid to ensuring their continuity of operation and resistance to attacks aimed at falsifying forecasts of changes in weather or the environment. This is particularly important because to improve the accuracy of generated forecasts by monitoring systems, measurement elements owned by private individuals and appearing as components of the Internet of Things (IoT) are increasingly being used (Lee, Kim and Ha 2015; Manfreda, McCabe and Miller 2018).

3.1.1   CONTRIBUTION

This chapter presents the design, construction, and operation of a regional environment monitoring system. The most important contributions include:

1. Development, implementation, and deployment of an EMS intended for regional use. IoT was used to improve the accuracy and resilience of forecasts. First, known solutions register and forecast environmental phenomena occurring in larger areas, creating mainly medium and long-term forecasts. As a result, rapid regional phenomena are not included in the forecasts. Second, the use of weather stations compacts the measuring network without a noticeable increase in construction and operating costs. For this purpose, forecasting methods have been developed to eliminate incorrect indications of IoT weather stations and guarantee the security of sending measurement data.

2. Existing EMSs have a hierarchical organization. However, the organization of the hierarchy was not the result of formal methods but only the professional experience of designers. In the proposed method, a set of components is formally selected and combined to ensure the best parameters of the system.

3. Algorithms that formalize both the measurement system design procedure and the measurement process itself are another important contribution. They can be adapted to the selected objective function, which can minimize the cost of unit measurement.

4. The last, intensively developed, contribution is to focus the system on the extensive use of blockchain to solve broadly understood security tasks: from authentication, through communication, to secure storage of measurement data. Until now, these tasks were solved based on cryptography methods.

3.1.2   CHAPTER STRUCTURE

This chapter has is structured as follows: in Section 3.2, we present the main concept of EMS, where the required features and functionalities are enumerated and associated with potential security threats. In Sections 3.3 and 3.4, the hierarchical structure of sensor networks is described, which are usually the basis of most monitoring systems. Section 3.5 presents a mathematical description of system components used in environment monitoring. In Section 3.6, optimization of key system features is performed, that is, computational load and measurement costs. In Sections 3.7 and 3.8, we provide broad description of the software and hardware architecture components and its synthesis method. This chapter ends with a proposal to use blockchain as a method to increase system security in already deployed solutions, along with a description of other future works.

3.2   THE CONCEPT OF ENVIRONMENTAL MONITORING

In normative acts, international standards, monographs, articles, and other resources, a lot of alternative definition of monitoring have been proposed (Acevedo 2013; Patnaik 2010; Wiersma 2004). According to some authors, monitoring is an information system designed to assess and forecast changes in the environment, which is built to distinguish the anthropogenic component from natural processes. Due to the scientific interests of the authors, the above definition will be considered as the main one. The functional architecture of the EMS in the form of an information system is shown in Figure 3.1.

In addition to permanent tracking and assessment of the state of the environment, the system in Figure 3.1 forecasts possible changes and estimates emerging threats. The frequency and precision of measurements are constantly adapted to the current level of threats – the relationship between them is directly proportional. The increase in accuracy is usually initiated by the monitoring and forecasting subsystem components. It can also be forced by the control subsystem whose additional task is to run emergency safety procedures.

In Figure 3.1, there are several potential security information risks. First, the components of the system responsible for the measurements have relatively low computing power to ensure data security using classical protection methods. Second, in situations where continuity of communication is threatened, it may be necessary to use public channels of communication to deal with all the risks involved. Even in this case, applying classical security techniques to protection may be difficult or even impossible.

The primary function of monitoring systems is collecting information about physical phenomena occurring in the environment. The structure of the activation and data collection procedures is presented graphically in Figure 3.2. *A measuring sensor* is a device used to obtain information about an object or a physical process,

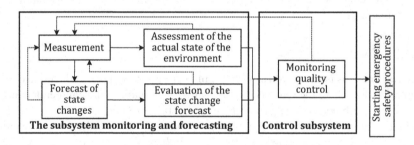

FIGURE 3.1   Architecture of environmental monitoring systems.

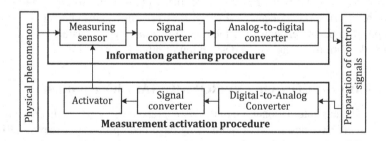

FIGURE 3.2   Procedures for activating and collecting measurement data.

usually related with temperature, pressure, or humidity of the object or the environment. *A converter* is a device used to transform a type of energy representing information. *Sensor* is a type of converter that changes physical information into an electrical signal. *Analog-to-digital* and *digital-to-analog* converters are special *converters* that transform a continuous electrical signal (current or voltage) into a discrete one and vice versa (Fraden 2013; Hać 2013; Ripka and Tipek 2007).

Analyzing the procedures that ensure the functioning of the monitoring system, we detail the information security risks previously defined. False measurement of environmental parameters can be due to an attack on information gathering procedures. As a result, the threat estimation system can be deprived of information about the state of the environment, and can have incorrect data. In both cases, interpretation of the environmental state can be misleading. Attack of this type is more possible because some of the measuring nodes can be part of the IoT. The owner is responsible for such security threats. Owner's knowledge in the field of cybersecurity may not always meet the requirements. In turn, attacking measurement activation procedures can lead to their complete termination – the monitoring system will not perform any environmental measurements.

The basic technology of modern monitoring is wireless communication network with particular architecture and functions, usually called a sensor network. The basic tasks include:

- Ensuring communication between system components;
- Preprocessing of measurement data;
- Making decisions in critical situations.

The nodes of the sensor network are densely located over the monitored area. This allocation is imposed by the architecture of the devices used based on elements with low computing performance, low energy consumption, and communication capabilities of 100 meters. The universal character of nodes allows their use to solve a wide range of tasks, in particular related with environment monitoring and drivers in inaccessible and dangerous areas. Special features of sensor networks also include advanced miniaturization, power supply from autonomous sources, using retranslation to transfer information between low power devices over long distances, and simplicity of installation and ease of expansion, allowing modification of the network without interrupting its operation and the ability to self-repair and self-organize (Hać 2013; Mao and Fidan 2009; Santi 2005; Wang 2010).

Sensor networks, especially in communications, can be the target of successful cyberattacks. Data transfer channels between nodes can be easily jammed or distorted. It can be assumed that when the 5G network is introduced, the role of the above organizations will reduce, which will likely improve the security of the monitoring network.

IT and ICT components play a special role in the design, construction, and operation of EMSs. The effectiveness of monitoring and forecasting depends on the effectiveness of the information and the analytical tools they offer. Usually, the information provided is used to improve crisis management. For such management to be effective, information on the state of the environment must be reliable and appear in the shortest time after the threat is detected (Mao and Fidan 2009; Slyke 2001; Wiersma 2004).

Two alternative concepts clash in the design of monitoring systems (Fraden 2013; Hać 2013; Wang 2010; Wiersma 2004). The first assumes the location stability of environmental conditions, whose measurement is the purpose of constructing the system. This means that the sources of potential contaminations are known and invariable. Therefore, the basic criteria for the distribution of measuring nodes are the location of pollution sources and the ways of their movement in the environment (air, water, or soil). The solution of the design task comes down to the location of the nodes in places with the maximum concentration of harmful factors. Connecting such nodes can be a complex task, which is a consequence of the territorial heterogeneity of the location of the nodes, and the system itself can have heterogeneous communication. Despite these difficulties, due to the small number of components, such systems are relatively cheap to design, construct, and operate. However, they do not provide tracking of the state of the environment, except the previously designated areas.

The second strategy assumes that the occurrence of harmful factors has the same probability throughout the entire monitored area and requires the distribution of measuring nodes in the entire domain. Owing to this, not only stationary sources of pollution are tracked but also routes of harmful substances or contamination resulting from criminal activities. From the point of view of communication design, this task is simpler than the previous one – usually a homogeneous mesh network is built. The assumptions made in both strategies can be an additional source of new attacks or simplification of the implementation.

3.3 HIERARCHY OF SENSOR MONITORING NETWORKS

Most modern monitoring systems have a hierarchical organization, which complementarily affects their design and security. First, hierarchy increases the complexity of relationships between components, as well as the number of parameters and characteristics that describe them. The consequence of this is increasing complexity: memory and time of the design process. On the other hand, however, using methods of hierarchical systems theory (Mesarovic, Macko and Takahara 1970; Smith and Sage 2005), the design process can be divided into simpler, relatively independent stages with an acceptable level of both of the above complexities. For example, when designing a hierarchical monitoring network, the stages of creating the core and access networks are considered as two independent design processes. Unfortunately, the division of the task into stages deteriorates the quality of the project and the created solutions are not optimal, which increases the likelihood of a serious failure (Levitin 2006; Wasson 2005). Despite the high complexity, one-step design allows to obtain the optimal solution of the task that meets the constantly increasing expectations of users. However, it requires the use of new, often poorly recognized methods and means of mathematical modeling, as well as high-performance processing techniques. Generally, design (also hierarchical systems) is based on principles for constructing a set of system elements and connecting them with relationships. There is no universal method useful for designing any objects. On the other hand, the design of a specific system can be accomplished with similar results using a number of methods that differ in the techniques of analysis and synthesis used (McCabe 2007; Mendez-Rangel and Lozano-Garzon 2012; Slyke 2001; Spohn 2002).

Breaking down the monitoring system into levels enables independent protection of each layer. Owing to the variety of protective mechanisms, the level of security can be significantly improved. On the other hand, interlayer communication at logical (interprotocol interfaces) and physical (communication links) levels can be a place where dangerous security gaps appear. In both cases, the physical and logical organization of the measurement network will play an important role in ensuring security.

To propose an effective solution to the task of topological design of sensor networks, let's consider their organization first. These networks are derived from cellular wireless networks that are currently built as multilevel hierarchical architectures, consisting of three to five communication levels (Freeman 2005; Hać 2013; Illyas and Mahgoub 2012; Santi 2005). In addition to hierarchy, these networks are characterized by heterogeneity, which results from the diversity of requirements set on traffic handling at each level of the hierarchy. Although theoretically it is possible to imagine the use of a flat homogeneous network for this purpose, such solution is currently not justified. Heterogeneity of communication has many advantages: it provides a wide range of available ways of integrating the user with the network, as well as the possibility of building multiprotocol multimedia communication systems at its higher levels. Owing to this, the use of environmental monitoring networks can be much wider. If no serious weather events are expected, it can be used to provide alternative connectivity, especially in areas with poor telecommunications infrastructure.

Hierarchy places environment monitoring networks to the main direction of the development of universal information systems (Hennessy and Patterson 2002; May 2004; Perahia and Stacey 2008; Raychaudhuri and Gerla 2011; Tannenbaum 2006). Because of the wide spread of hierarchical systems, the following are usually indicated (Hajder, Dymora and Mazurek 2002; Macko and Takahara 1970; Mesarovic, Smith and Sage 2005; Tannenbaum 2006):

- Easier analysis and synthesis of complex systems divided into smaller components;
- High specialization of functional blocks;
- Simplification of exploitation, maintenance, and services, especially in the area of information security.

The most common features of hierarchical systems are (Smith and Sage 2005) sequential deployment of functional subsystems while maintaining their vertical dependence; superiority of higher levels over lower levels; the use of feedback, guaranteeing the effect of lower levels on higher levels. The general structure of

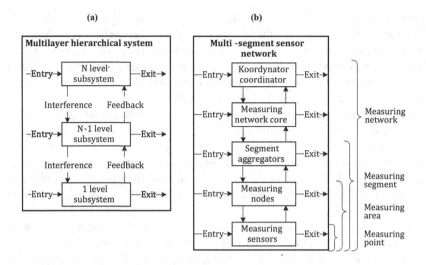

FIGURE 3.3 Layers of the hierarchical sensor monitoring network.

multilayered hierarchical systems is shown in Figure 3.3. Figure 3.3a shows the interrelationship of subsequent layers and how is the user involved in their functioning. Using the generalized model in Figure 3.3b, a hierarchical model was proposed, reflecting the actual sensor network. The hierarchy consists of five levels, the lowest of which groups measurement sensors, while the highest contains a central management node, the network coordinator.

Elements of other hierarchy levels include measuring nodes, segment aggregators, and the core of the measuring network. An example of a network operating in accordance with the above model is shown in Figure 3.4.

In the lowest level of the model, there are distributed measurement sensors, most often connected with measurement node using wired methods. Using RFID technology enables wireless integration of sensors with network; however, due to the power consumed by the sensor during measurement and communication, this solution is used less often. The location of the measuring device is called the measuring point. The number of sensors connected to a single node is determined by its architecture and usually does not exceed several sensors. An additional limitation on the number of sensors may also be the power consumption. The area where sensors connected to the same node are located is called the measurement area. Nodes are combined into measurement segments using segment aggregators. In addition to greater computing power, these devices have a wide set of external interfaces to build the network core. A network coordinator is an extended unit whose primary task is to provide external communication for the entire measuring network. The task of the coordinator may also be to integrate aggregators within the core of the measurement network.

When analyzing the organization presented in Figure 3.4, a wide spectrum of threats to information security of the environmental monitoring network can be distinguished. These include attacks on the availability of nodes, system

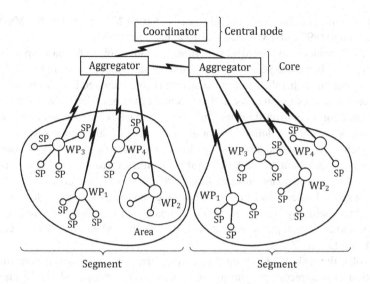

FIGURE 3.4 Hierarchical sensor network architecture. Designations: WP – measuring node, SP – measuring sensor.

compromise by unauthorized users, as well as various attacks on the communication integrity.

3.4 THE TASK OF HIERARCHY DESIGN

According to the theory of hierarchical systems (Dandamudi 2003; Mirkin, Rzhetsky and Roberts 1997), all the levels in the hierarchy can be in the form of layer, stratum, or echelon. The distinction between three different layered organizations results from the search for a compromise between the simplicity of description and the complexity of interrelationships. Real systems are usually a complex combination of the above basic types, and the way the hierarchy is presented depends on the system architecture.

Stratum is a level with its own description of functioning, owing to which, by analyzing the system, we can limit the analysis to the layer we are interested in. The system model, whose levels are in the form of stratum, is called the stratified model, and the system itself is called the stratified system. They are characterized by high independence of stratums, owing to which it becomes possible to precisely define its properties, which are the superposition of the functions of individual stratums. In the first approximation, wireless monitoring networks can be treated as stratified systems; however, in real systems, there is no absolute independence of stratums, and the description of their functioning must take into account all the stratums used, as well as the relationships between them.

The synthesis and analysis of systems presented in the form of multilevel organizations with a hierarchical structure is one of the known directions in the study of large-scale and complex systems, such as EMSs. Numerous researchers have dealt with general issues in this area, including Mesarovic (1970), Murtagh (1992), Saaty (1997), and Willett and El-Hamdouchi (1989).

For the considered hierarchical structures, it is advisable to formally specify the number of levels and the selection of elements of individual layers and methods of their connection so that the resulting structure is characterized by minimal design, construction and operation costs, and maximum efficiency. Until now, for this purpose, based on a compact description, a set of acceptable structures and criteria for their assessment was defined. Although this approach allowed to solve the task of hierarchy selection, the solution was limited to the area of a specific, compact description. In addition, the task of hierarchy synthesis was performed only at the qualitative level, and quantitative models were either not considered at all or were only of a special nature. To formally solve the given task, methods of determining the optimal hierarchy were used so far, primarily in management, control, and bioinformatics. This topic was dealt with by Mirkin and Koonin (2002), Gubko (2006), and Levin (2007).

To solve the task of searching the optimal hierarchy, we use our own implementation of the integer programming method (Bronsztejn et al. 2004; Hamdy 2005; Schrijver 1998). To this end, let's consider a set of graphs G of acceptable hierarchies $G = \{G_1, \cdots, G_n\}$, where $i = 1, \cdots, n$. Suppose we are looking for a hierarchy described by graph $G^* \in G$, characterized by the maximum value of the evaluation of the selected property or its set. For this purpose, we will analyze the vector Ψ of partial graph performance indicators $G_{i_1} \in G_i$, defined as:

$$\Psi(G_{i_1}) = \left(\psi_1(G_{i_1}), \ldots, \psi_k(G_{i_k}), \ldots, \psi_m(G_{i_m})\right) \tag{3.1}$$

Then, the objective function will have the form:

$$\max_{G^* \in \{G_i\}} \psi_k(G^*), \quad \forall k = 1, \cdots, m \tag{3.2}$$

It is also possible to search for a hierarchy that is most similar to predetermined pattern structure or set. As $G_C = (V_C, E_C)$, let's denote the graph of the target (reference) hierarchy, where $G_C \in \{G_i\}$. To determine the distance of the chosen graph from the graph of the target hierarchy, we introduce the proximity function ρ. For two arbitrary graphs G_{i_1} and G_{i_2} belonging to the set of permissible hierarchies (i.e. $G_{i_1}, G_{i_2} \in \{G_i\}$), function $\rho(G_{i_1}, G_{i_2})$ determines the closeness between them. Thus, the objective function will have the form:

$$\min_{G^* \in \{G_j\}} \rho(G^*, G_C) \tag{3.3}$$

Using the integrated definition of tasks equations (3.2) and (3.3), for $\psi_k(G^*) \geq r_k$, $\forall k = 1, \cdots, m$, the new objective function can be written as:

$$\min_{G^* \in \{G_i\}} \rho(G^*, G_C) \tag{3.4}$$

where r_k is the limits on the value of the rated properties.

The above optimization tasks correspond to complex models of integer or mixed integer programming, which can be solved with various methods, from naïve approaches to artificial intelligence methods (Hamdy 2005; Schrijver 1998; Sinha 2006).

The task of determining the number and types of network levels can also be presented as a multilevel distribution task, and to obtain an approximate result, a method based on linear relaxation and solving a dual task with local improvement of the result can be used. If an exact solution is required, the branches and borders method can be used. When additional restrictions are imposed on the structure of the designed network, the following methods should be applied: searching for a spanning tree with a limited radius; building an optimal Steiner tree with limited path lengths and a limited number of Steiner points; and searching for a rectangular Steiner tree with equal length paths (Gross and Yellen 2004; Mao and Fidan 2009; Santi 2005). Unfortunately, all the above methods belong to the group of NP-complete algorithms and are useless for solving a task quickly. Another group of methods useful for determining an effective hierarchy is hierarchical clustering, in particular those derived from it – agglomeration method (Jain, Murty and Flynn 1999; Jardine 1971) and hierarchical method with intersection of clusters (Levin 2007).

3.5   MONITORING SYSTEM AS A COMPLEX SYSTEM

3.5.1   FUNCTIONAL COMPONENTS OF THE SYSTEM

The concept of a monitoring system is a set of elements that form a structure designed to collect and process information about the state of the surrounding environment. These elements include monitoring facilities and entities, its instrumentations, and a set of monitoring indicators and monitoring activities. The interrelationships between these elements are shown in Figure 3.5.

Generally, *the monitoring objects* are complex systems and phenomena. The common feature of all monitoring objects is the high dynamics of changes occurring in them, only in this case it is appropriate to monitor them. Objects whose

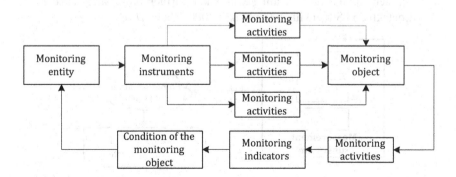

FIGURE 3.5   Functional elements of the monitoring system and their mutual relations.

behavior has static character can be most often observed by classical methods. *The monitoring entities* usually have monitoring functions, that is, organizations, organizational structure, and people who perform above tasks. The entity is not only making monitoring disposition but is also interested in their results. *A set of monitoring indicators* is a set of measured values, knowledge of which provides a comprehensive description of a state of the environment, in particular data on its quantitative and qualitative changes. *Instrumentation of monitoring* creates a set of hardware and software measures necessary to perform measurements, their statistical processing, forecasting, as well as to inform and warn the public about the state of the environment and potential threats. It is used in its activities by monitoring entities. *Monitoring activities* are a set of functional procedures, including collection and processing of information, its visualization, as well as preparation of necessary actions that are a response to the state of the environment, such as changes in the operation of the monitoring system itself.

We distinguish three basic types of activities performed in monitoring systems:

(a) Organization and implementation of monitoring, for which the measuring subsystem is used;
(b) Collecting measurement results performed by the communications subsystem;
(c) Processing of measurement data together with the formulation of recommendations concerning their use of the information and the analytical subsystem.

The interconnectedness between the activities and the components implementing them is shown in Figure 3.6.

Monitoring as a sequence of interrelated activities can be divided into three subsequent stages:

(a) Preparation to the legal and normative organization of monitoring (Bac and Rojek 1981), (Kożuchowski 2006);
(b) Execution, during which measurements are carried out and their results are sent to the node dealing with their further processing (Hajder, Loutskii and Stręciwilk 2002; Makki 2008; Zheng 2009);

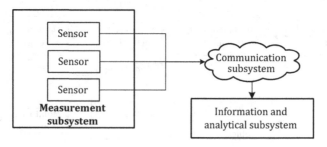

FIGURE 3.6 Interconnection of monitoring components.

(c) Analytical and decision-making, when the results of monitoring are processed and then used in the management process (Grocki, Mokwa and
Radczuk 2001; Hajder, Loutskii and Stręciwilk 2002).

3.5.2 ENVIRONMENTAL MONITORING OBJECTS

Because the monitored environment is a complex system, the proposed theory
contributes to the development of the theory of such systems. A complex system
is an object consisting of multiple elements, each of which we can consider as a
system. Generally, these elements are bound together in a single integral, or are
linked by corresponding functional relationships. At any point, elements of any
complex system are in one of the possible states. The transition between them
is made under the influence of internal or external factors. The dynamics of the
behavior of a complex system manifests itself in the fact that the state of an element and its output signals, at any time, are determined by its previous states and
by the input signals from other system elements. In the theory of complex systems,
the term external environment is a set of objects that are not a part of a given system, and interaction with which is considered in the process of its study. Elements
of complex systems function in mutual connection: the properties of each element
depend on the conditions set by other elements of this system. The properties of
the complex system are determined not only by the properties of its components
but also by the nature of the interaction between them (Shidlovskiy 2006).

The basic method of studying complex systems described by theory is mathematical modeling. To deal with this, it is necessary to formalize the processes
of system functioning, that is, to present it in the form of a sequence of specific
events, phenomena, or procedures, and then create its mathematical description.
According to modeling theory, to formalize the representation of any object O,
first it is required to specify all its attributes, thus creating a static object model.
Then, the process of describing Q changes their value in time, which is the result
of various factors and creates a model of object behavior under given conditions.
In further considerations we will assume that K is the identifier of the object contained in the classification space; A is the description of the invariant K attributes
of the object; and V is the description of the properties, relations, and functions
determining the behavior of the examined object. Because this model is static, it
always describes the state of the system at some point t. Considering the above
findings, the static object model O can be written as:

$$O \rightarrow \left(K, A, V, t \right) \tag{3.5}$$

The process of object state changes over time under the influence of a set of
internal and external factors, and is called object behavior. This process can be
described by the following expression:

$$Q \rightarrow (K, G, F, T) \tag{3.6}$$

where $F-$ is the database of space factors that affects the behavior of the object.

As noted earlier, from the point of view of the behavior of the object, time plays a crucial role, time T is beyond the set of F factors. In turn G, is a set of all object attributes, divided into two subsets: A is a subset of invariant attributes; X is a subset of parametric attributes that change over time under the influence of internal or external factors contained in the set F. At the same time, $G = A \cup X$. The subset A includes such attributes as the name of the object, its identification number, geographical location, etc. In turn, the subset X contains characteristics that are the parameters of the object, which are functions of time and factors from the set F affecting the object O.

The set of values of all object attributes at the time T is called the state of the given object. The set of attributes $(A_1, A_2, \ldots, A_s, X_1, X_2, \ldots, X_n, t)$ creates the state space of the object O, and the set of values of these variables is called its state coordinates. The sequential change in the state of monitoring objects, expressed by means of monitoring indicators, is called the monitoring process. According to the introduced markings:

$$Q = f(K, A, X, T) \tag{3.7}$$

is a mathematical description of the monitoring object state change process. From the viewpoint of information security, the process of Q monitoring object changes over time and may interfere with attacks on each of the arguments to the f function, which determines a wide range of vulnerabilities.

For the functioning of the monitoring system to provide reliable data on the state of the environment with minimal expenditure on its implementation, it is necessary to introduce hierarchically linked levels of information generalization on monitoring objects. If objects are subject to generalization and are also the result, we can distinguish three basic types described by argument K from equation (3.7). These are the objects of observation, generalization, and monitoring, and relations between them are shown in Figure 3.7.

Observation objects are objects that are subject to continuous tracking of their selected characteristics, which is performed by measuring them directly. *The object of generalization* will be the set of observed objects, grouped using thematic, spatial, or temporal criteria based on the analysis of the state of the environment and forecasting its changes. While observation objects are described by parameters whose significance is determined by measurement, generalization objects are described by calculation parameters determined based on mathematical

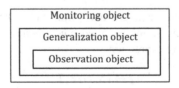

FIGURE 3.7   Relationships between classes of objects.

or statistical formulas. *Monitoring objects* are complex system objects, whose condition is described by means of integral assessments, which allow to present overall quantitative and qualitative changes in the state of the examined system. Therefore, we can write that $K = \{K_o, K_u, K_m\}$, where K_o is a subset of observation objects; K_u is a subset of generalization objects; and K_m is a subset of monitoring objects.

In the description based on the equation , A denotes the attributes of the objects and their most important properties. At the same time, the elements of the set A of the object attributes are the sums of subsets of the objects, that is, $A = \left\{ \cup A_{K_o}, \cup A_{K_u}, \cup A_{K_m} \right\}$. In turn, X describes parametric properties characterizing the state of the object, which are determined by internal or external factors. As in the case of attributes, we assume that $X = \left\{ \cup X_o(t), \cup X_u(t), \cup X_m(t) \right\}$. The argument T in equation defines the periodicity of recording the dynamics of changes in the state of objects, that is, obtaining measurement results, their generalization, and obtaining monitoring results, as well as the moment t_0 to start the observation process. This argument has the form: $T = \{T_o, T_u, T_m, t_0\}$. Using the above record, you can examine the system's vulnerability to information security threats. Each subset of objects should be subject to independent analysis.

3.6 OPTIMIZATION OF MONITORING SYSTEM INSTRUMENTATION

To present the solution to the optimization task, let's consider an EMS. The key feature of the system is the continuity of its work. It should provide reliable information on the state of the environment without interruption at acceptable cost. Therefore, the monitoring system should be based on a unified set of multifunctional, maintenance-free sensors with high reliability, and the connection network should be redundant.

The above conditions have decided to develop their own methods of selecting EMS instruments and connecting network design. To design the instruments, a two-stage approach was proposed: in the first step, the range of sensors used is selected, and in the second, the criteria are determined on the basis of which the optimization of the set of measuring means will be performed. The preselection of components is a typical subtask of a complex system design, consisting of determining a set of specific devices and comparing it with alternative solutions. The comparison stage is usually based on classical qualitative methods, expert systems, and now more and more often on artificial intelligence. The step results are used further during the optimization of the set of devices to improve the EMS operating parameters. From the viewpoint of operational efficiency, it is appropriate to consider the characteristics describing the number of measurements performed by a specific set of sensors per unit of time. Considering the periodicity of environmental phenomena, the quantity should relate to a relatively long time, minimum quarter, preferably a year. It is possible to ensure maximum loads on the sensor or their sets during the implementation of a given monitoring program or to strive to minimize the unit costs of measurement.

The data source for the selection and optimization step are the list a_i specifying the measured environmental parameters $(i = 1,\ldots,n)$ and the list b_j sensor types used to measure the parameters specified by the list a_i $(j = 1,\ldots,m)$ of the measured parameters. Because the sensors used are multifunctional, they can measure many different parameters and $n \gg m$. Considering that the operating conditions of measuring equipment are different for different seasons, let's assume that the accounting period will be one year. Let τ_{ij} denote the time necessary for the j-th sensor to perform the i-th parameter measurement, $\tau_{ij} \in \mathbb{Q}$, and R_i is the number of i-th parameter measurements in the accounting cycle, $R_i \in \mathbb{N}$. Then, as T_j we mean the total operating time of the j-th type sensor in the billing period, which is $T_j = \sum_{i=1}^{n} R_i \tau_{ij}$. In turn, the total time T of the set of sensor types during the year is $T = \sum_{i=1}^{n}\sum_{j=1}^{m} R_i \tau_{ij}$, where $T_j \leq T$. In further considerations, time τ_{ij} plays a key role. Its value can be obtained by experimental methods or expert evaluation based on fuzzy set theory.

Consider the first of the system optimization criteria, which is *maximizing the load* of the measurement sensors. The load of the j-th sensor by obtaining the value of the i-th ecosystem parameter will be determined by the time T_{ij} necessary to make the measurements. Time T_{ij} is part of the total time T_i. In the solution of optimization procedure, it was assumed that the time of measurement of the i-th parameter by the j-th sensor is equal to the number of measurements in the billing cycle, that is, $T_{ij} = R_i$. Then, we can describe the function Z target as:

$$Z = \sum_{i=1}^{n} \sum_{j=1}^{m} T_{ij}/\tau_{ij} \to \max \qquad (3.8)$$

subject to the following restrictions $\sum_{i}^{n} T_{ij} \leq T_j,\ \forall i = 1,\ldots,n;\ \sum_{j}^{m}\left(T_{ij} / \tau_{ij}\right) \leq R_i$, $\forall j = 1,\ldots,m$. The optimization equation (3.8) guarantees the maximization of the number of measurements performed by a specific sensor during the billing period.

According to the second approach, we will be looking for EMS architecture that ensures *minimization of measurement costs*. It is an economic and technical criterion as it allows to obtain the minimum cost of measurement in the accounting cycle (quarterly, annually, etc.). The basic cost carriers are annual C_j cost of reconstruction of the j-th measurement sensor together with the part of the cost of restoring autonomous power supply and communication devices attributable to it; annual C_{wz} cost to restore the management node; annual C_p cost of renting the space necessary for the location of sensors and the management node; annual C_e operating costs of the system, including, among others, lease payments for communication bands and power costs for the management node; C_0 personnel costs reflecting employee salary.

Unit measurement costs can be determined in two ways. First, let's consider the average cost of the measurement, which we determine based on the following expression:

$$C_{\text{avg}} = \frac{\sum_{j=1}^{m} C_j + C_{wz} + C_p + C_e + C_o}{\sum_{i=1}^{n} R_i} \tag{3.9}$$

If the approximation of the costs offered by the equation (3.9) is insufficient, the optimization process can use the cost C_{ij} for measuring the i-th parameter by the j-th measuring device. This cost can be determined on the basis of the expression: $C_{ij} = \left(C_j + C_{wz}^j + C_p^j + C_e^j + C_o^j \right) / R_{ij}$, where $C_{wz}^j, C_p^j, C_e^j, C_o^j$ are the costs specified for the equation (3.9) converted for the j-th measurement sensor; R_{ij} is the number of measurements of the i-parameter by the j-ty sensor. The objective function Z for the second considered approach has the form:

$$Z = \sum_{i=1}^{n} \sum_{j=1}^{m} C_{ij} \rightarrow \min \tag{3.10}$$

including the following restrictions: $\sum_{i}^{n} R_{ij}\tau_{ij}, \quad \forall i = 1,\dots,n; \quad \sum_{j}^{m} n_{ij} \leq R_i,$ $\forall j = 1,\dots,m.$

The methodology described was used to design various types of EMS. It favors equipping systems with unified sets of equipment and measurement methodologies. It allows to obtain reliable results, thus increasing the quality of EMS functioning.

3.7   SYNTHESIS OF TOPOLOGICAL ARCHITECTURE

Now consider the synthesis of the topological architecture of a sensor measuring network. The design process is presented in the form of a block diagram in Figure 3.8. The design begins with the introduction of initial requirements for EMS. The list of requirements includes such parameters as network life cycle, range, level of information timeliness, data transmission speed, and level of network reliability. Defining parameters can have varying degrees of detail determining the accuracy of the input data.

Actions taken at the stage of defining initial requirements are formally described by the expressions: $W = f(P)$, where: $P = \{p_1,\dots,p_n\}$ – the initial requirements vector; $W = \{w_1,\dots,w_n\}$ – vector of formalized requirements; and $f(p)$ – formalization function; n – number of requirements. The list of initial requirements is defined in the description of the subject of the contract and consists of several initial requirements. The effect of the project activities is to create a vector of formal requirements of a similar size.

The next step in the procedure in Figure 3.8 is choosing the components database. The rapid development of EMS technology is primarily due to the dynamic development of microelectronics, which ensured the availability of components with an appropriate technical level. The basic parameters considered by the designers in the process of designing sensor networks include frequency range, transmission speed and range, sensitivity and signal output power, supply voltage

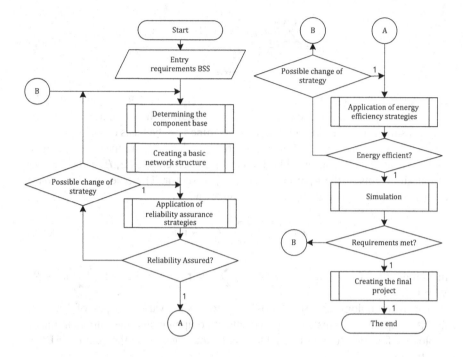

FIGURE 3.8 Procedure for designing a wireless sensor network with static connections.

and power supply in reception, and transmission and sleep modes. The stage of selecting the component base is to define the subset V of the U set of technical devices enough to build the designed sensor network. Theoretically, the best solution is to minimize the size of the V subset. In practice, a small subset V minimizes the cost of maintaining a spare parts warehouse. Let $U = \{u_1,\ldots,u_k\}$, where k is the elemental set of available system components. Then the function defining the component base can be written using the expression: $V = f(W,U)$, where $V \subset U$, which is a subset of the available components forming the EMS, selected by the design support system or by the designer based on the vector analysis of W formal requirements; $f(x)$ is the function to select a subset of elements.

The next stage of the design procedure is to determine the underlying structure of the monitoring network. The core network consists of a minimal set of elements necessary to execute the tasks facing the EMS. The basic requirement of such a structure is to ensure consistency of the designed network. At this stage of EMS design, two subtasks are solved:

a. Sensor network segmentation;
b. Building links between EMS segments and its central node.

The function of creating a basic structure can be described by the formula: $(D,E) = f(W,V)$, where D is the connection vector between network nodes;

E is the set of intermediate (transit) nodes connected to the network to ensure connection between clusters; and $f(x)$ is the function for creating a basic network structure.

In the next step of the design procedure, an attempt is made to ensure network reliability. One of the main features of EMS is the level of network reliability. In this context, the terms reliability and fault tolerance are treated as synonyms. According to the classic definition, reliability determines the probability of system failure under the influence of external factors. The parameter derived from graph theory that perfectly illustrates reliability is graph coherence. Graph $G(V,E)$ is a mathematical object consisting of n-element set of V vertices $V = \{v_1,...,v_n\}$ and set E of edges $E = V \times V$. In fact, the graph is the binary relation $E = V \times V$ on the set of vertices V. An auxiliary concept used to define consistency is the path to define the concept of *the route*. *The route* in graph G is a finite progression of character edges: (v_1,v_2), (v_2,v_3),...,(v_{n-1},v_n), in which two adjacent edges are adjacent or identical. The vertex v_0 is the beginning vertex of the route, v_n – the end. The above entry shows the route from vertex v_1 to v_n. *The path* in graph G is the route where all edges are different. If the path also has all its vertices different (except when $v_1 = v_n$), then the path is called a path. The road or path is closed if $v_1 = v_n$. A *cycle* is a closed path that has at least one edge, so each loop or pair of multiple edges is a cycle. A graph G is called connected if at least one path exists between any pair of its vertices.

Graph coherence can also be considered from the viewpoint of its components. Let's consider two graphs: $G = (V(G),E(G))$ and $H = (V(H),E(H))$, whose sets of vertices $V(G)$ i $V(H)$ are disjointed. The sum of $G \cup H$ graph will be a graph whose set of vertices is set $V(G) \cup V(H)$ and set of edges $E(G) \cup E(H)$. A graph that can be represented by the sum (in this case) of two graphs is called an inconsistent graph, otherwise the graph is called a connected graph. Of course, a disconnected graph can be represented by the sum of connected graphs called graph components.

The basic method of changing the network's reliability is the reconstruction of its topological structure. The failures result from a violation of the information transmission paths from the sender to the recipient. Such interference is possible, for example, in the event of a failure of one of the transit nodes. Therefore, increasing the number of possible information transmission routes linearly increases the reliability of the system. In this way, the reliability function can be represented by the following expressions $(D^*,E^*) = f(D,E,V)$, where D^* is the vector of connections between nodes supplemented with new paths, E^* is the set of nodes transit taking into account added at a given design stage; and $f(x)$ is the function for adding new tracks. The result of the η_r network reliability check operation can be written as: $\eta_r = g(D^*,E^*,W)$, where $\eta_r \in \{0,1\}$ is the result of the network reliability check, and $g(x)$ is the reliability check function.

Ensuring energy efficiency is a key step for networks fed from internal energy sources. It allows you to maximize your network's uptime. To solve this problem, it is necessary to correctly arrange the transit nodes, which will allow even distribution of the load throughout the network, and ensure a minimal change in the indicators of the average energy consumption of devices. The second way to achieve optimal energy consumption is to set the correct mode of access to EMS resources. Each of the functional and transit nodes can be in one of the four modes: transmission, reception, computations, and sleep. Minimizing node operation times in modes with the highest power consumption is the main way to maximize network uptime. The function of ensuring energy efficiency can be written as: $\left(D^\blacksquare, E^\blacksquare\right) = f\left(D^\blacksquare, E^\blacksquare, V\right)$, where $D^\blacksquare$ is the vector of internode connections, supplemented with new paths; $E^\blacksquare$ is the set of intermediate nodes taking into account changed or added at a given stage; and $f(x)$ reallocation function of transit nodes to ensure energy efficiency. The result of the η_e energy efficiency check operation can be represented as: $\eta_e = g\left(D^\blacksquare, E^\blacksquare, V\right)$, where $\eta_e \in \{0,1\}$ is the result of the grid energy efficiency check, and $g(x)$ is the energy efficiency check function.

The penultimate step in the connection design procedure are simulation tests. During this process, the designed EMS is modeled. In the modeling process, the basic features of the system are estimated, which are the basis for conducting multicriteria EMS analysis. If the parameters specified by mathematical methods meet the requirements defined at the stage of introducing and analyzing initial assumptions, the transition to the stage of generating the final project is initiated. Otherwise, improvements will be made to the project; for this purpose, the individual steps of the design procedure will be carried out again: $C = f\left(D^\blacksquare, E^\blacksquare\right)$, where C is the vector of output parameters of the designed network, and $f(x)$ is the function for determining output parameters. Then we get the result of the multicriteria analysis with respect to the whole vector W: $\eta_c = g(C, W)$, $\eta_c \in \{0,1\}$, where $g(x)$ is the multicriteria analysis function.

The design procedure completes the creation of the final project, consisting of the final configuration of the EMS model.

3.8   SOFTWARE AND HARDWARE ARCHITECTURE

3.8.1   FUNCTIONAL COMPONENTS OF THE SYSTEM

Until now, we assumed that information security is a risk mainly at the stage of communication and processing. The attacks analyzed involved a refusal to process or communication because of attack on accessibility. In addition to these concerns, there may also be a loss of consistency in the monitoring system. Note, however, that the real threat may also be attacks on software components of the system. The EMS represents a complex structure that is formed by interrelated software and hardware components. From a functional viewpoint, it distinguishes six basic types of modules: Measurement module, communication module (remote and local), computing module, forecasting module, archiving module, and information and warning subsystem. Their relationship is shown in Figure 3.9. The absolute need to protect information appears in each of the modules.

The measuring module is in places that have a special impact on the state of the environment, which is reflected in the model describing it. The parameters

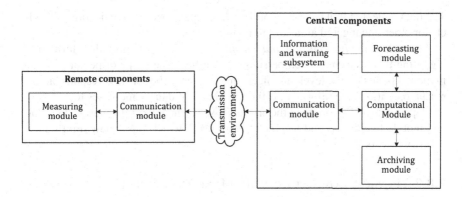

FIGURE 3.9   General architecture of the environmental monitoring system.

used in the model and all other parameters necessary for efficient management (monitoring of communication infrastructure, video monitoring, etc.) can be measured. Unlike most operated systems, the measurement module is preinterpreted in the measurement module, and it is also possible to autonomously change its operating modes. Owing to this, EMS becomes less sensitive to attacks on the transmission environment. Even if it is turned off, measurements continue. Remote components also measure elements owned by residents and other entities located in the monitored area. Although they play a supporting role, they can also be attacked, even more so when they are managed by unprofessional staff. In Eastern Europe, measuring modules are additionally exposed to theft or intentional destruction.

Measuring modules can be connected both with each other (optional) and with the central components of the system (obligatory). For this purpose, local and remote *communication modules* are used, respectively. The channels they create can also be used to transmit other data, usually to provide access to the Internet. The main threats to communication arise from its wireless nature. Connections between system components are exposed to threats typical for this type of communication; therefore, it should use protective methods developed for wireless networks. Methods of protecting confidentiality and ensuring communication integrity and authentication play a special role here.

The task of the *computational module* is to provide the computing power necessary to develop a forecast of environmental changes and the archiving of measurement data, including sharing of data. *The forecasting module*, based on information on the current state of the environment and archival data, prepares forecasts describing its likely changes. The forecast preparation procedure uses a classic hydrological model, verifying its accuracy using artificial intelligence methods. Owing to this, the accuracy of forecasts increases with the system's lifetime. The most important threat of this group of components arises from the operating environment and the need to provide input to the forecasting process.

The task of the *archiving module* is to collect measurement data and forecasts and make them available via the Internet to all interested persons and entities.

The data collected in it can be helpful in the preparation and verification of models for other endangered areas of our country.

The task of the *information and warning subsystem* is to provide information regarding threats to interested persons and entities. The scope of personal information depends on the level of danger: initially, only the services responsible for rescue operations are informed; if the likelihood of danger increases to a certain level, people who live or are in the area are also warned. For information, landline and mobile telephones and the Internet are used, which raises new information security threats.

3.8.2 Physical Components of the Monitoring System

The physical components of the EMS are shown in Figure 3.9. The general architecture of the EMS is shown in Figure 3.10.

System components can be divided into *remote*, also called elevated, and *central*. The basic remote components are *remote measuring system* (RMS) and *aggregation nodes* (AN). The RMS module is intended for the local collection of information about the environment and forwarding it via the aggregating node to the *central management node* (CMN) and *redundant management nodes* (RMN). Both RMS and AN are located directly in the monitored area. In accordance with current trends, the measuring system is based on a hierarchical sensor network with a star or lattice architecture. In the terminology of sensor networks, RMS will be served by data collectors, and the aggregating node is called the data aggregator.

AN, referred to as aggregators, are highly specialized computers used to collect and preprocess information sent by data collectors. In addition, using traditional ICT networks, they transmit the collected information to the places of

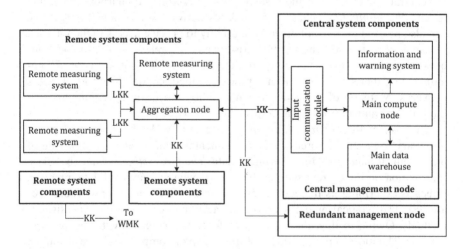

FIGURE 3.10 Physical components of the environmental risk monitoring system. Designations: LKK – local communication channel; KK – communication channel; WMK – input communication module.

their collection and processing. Aggregators can also be equipped with their own measuring sensors acting as both a collector and a data aggregator. Due to the specificity of functioning and the special role played in the system, aggregators are powered from sources ensuring full operational autonomy. The functional structure of the data aggregator is presented in Figure 3.11.

First, the aggregator collects data from RMS. The aggregator's external communication is performed via wireless technologies (Wi-Fi, LMDS, WiMAX, GPRS, VSAT, LTE) using the TCP/IP protocol stack. Wired Ethernet and DSL interfaces are also available. The choice of a technology is made based on its availability. First, the use of Wi-Fi networks working in the 802.11 standard can be considered. They are the cheapest and provide sufficient communication parameters. Because the proposed aggregators are equipped with several alternative external interfaces, the selection of a specific one will be made after its installation.

If the normal functioning of CMN is disturbed, its role is taken over by RMN, which temporarily obtains the status of a central node. If the correct functioning of the original CMN is restored, it is initially connected to the system as a redundant node, and after the so-called *grace time*, reducing the possibility of re-disconnection, again becomes CMN. The procedure of connecting any redundant node begins with data synchronization; however, to limit the size of inter-node communication, it is carried out from the last checkpoint available in both nodes. Checkpoints are created periodically when monitored threats are minimal. The triggering factor can be the time or level of changes made to the archive. Communication relations between the aggregating node and CMN and RMN nodes are shown in Figure 3.12.

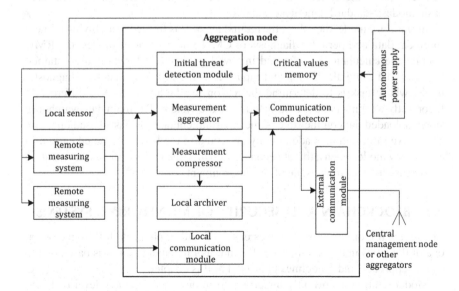

FIGURE 3.11   Functional structure of the remote aggregation node.

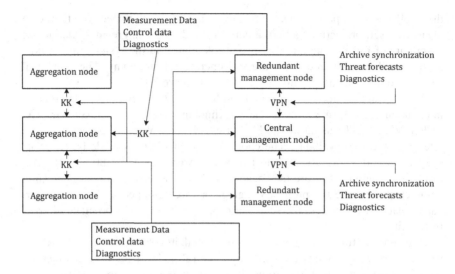

FIGURE 3.12 Information interoperability of aggregation nodes and the central and redundant management node.

Connection of CMN with redundant nodes is performed via the Internet, in particular based on VPN channels. These channels, in addition to synchronization of archives, are used for diagnostics and sending forecasts of threats generated additionally in redundant nodes.

The level of cooperation of redundant nodes with CMN depends on the mode of operation of the system. In the offered version, the system operates in three basic modes: standard, emergency, and disaster.

In *standard* mode, the role of the redundant node is limited to archiving of measurement data and periodic diagnosis of CMN status. In other modes, the RMN doubles the calculations performed in CMN. If different forecasts are obtained based on the same algorithms and input data, the system switches to diagnostic mode, whose task is to determine the reason for the appearance of differences. Incorrectly working node is eliminated from the system. The main data warehouse stores uploaded data on the state of the environment and forecasts based on them. The above information is additionally protected in the RMN. As noted earlier, in emergency mode, when a threat becomes probable, the RMN expands its operation to forecast changes in the state of the environment.

3.9 BLOCKCHAIN AND SECURITY OF MONITORING SYSTEMS

The regional monitoring system was designed and implemented in Boguchwala, one of the cities in the south-east of Poland. Its implementation was based on the classic approach and guidelines presented in this chapter.

Modernization is currently underway to improve the security level of EMS, especially in the area of using IoT components in it. Security is widely believed to be a key issue in the area of IoT. The use of wired and wireless communication channels in IoT creates virtually unlimited possibilities of interference by cybercriminals. One of the basic methods to improve security is to use digital signatures to confirm the authenticity of the code used. The digital signature guarantees its immutability, which allows you to quickly and confidently identify the code sent by criminals. If at the application or microprogram level, force all system components to play only the signed code, as a result of which its penetration into the EMS will be very difficult. This solution additionally requires limitation and continuous control of server resources, especially in the area of file access control and computing resources. If the vulnerabilities in the IoT devices are from the manufacturer, it should be expected that they will be removed soon. However, if they result from the actions of an EMS-enabled programmer, their removal will rest with the management staff.

In previous paragraphs of this chapter, the threats resulting from EMS internode communication have been emphasized. These problems are even more important due to the use of IoT elements in the system. Therefore, another area of activities undertaken as part of investment modernization is the development of proprietary software for information exchange between EMS components. The solutions prepared should additionally ensure the possibility of temporary storage of information in EMS nodes and simplicity of extending the EMS to new nodes. These activities should not dramatically affect the operation of an existing network. Due to low and dense buildings, the use of Wi-Fi technology in Boguchwala to implement internode communication has been limited. Owing to this, counteracting violent attacks and swapping nodes or attacks on accessibility is of secondary importance. However, a return to this communication technology is not excluded if the share of IoT components in the whole system is expanded. Therefore, to minimize the possibility of the data being taken over by cybercriminals, a blockchain-based protocol for data exchange between Wi-Fi network nodes has been developed and is currently being implemented. On each node of the system, copies of blockchain containing information are stored independently. Each newly created block stores information about ordered records and headers. The created block is checked by other network nodes, and then, if everyone confirms the addition of the block, it connects to the end of the chain. No more changes can then be made to it. The database is automatically updated on all devices connected to the system. The software necessary to implement the above activities was prepared using the Python language.

The second improvement currently being introduced to the functioning system is the construction of a distributed data repository. At the moment, it will be designed to store the most valuable information contained in the system. These are forecasts prepared based on data obtained from measuring and machine learning nodes as well as their verification. These data are the basis for the operation of machine learning algorithms and the accuracy of forecasts formulated in the future depends on their availability. Selected components of the information system of the city of Boguchwala and the university's research team will be used to disperse the data. In addition to scattering, an encryption algorithm using a hash function and an electronic digital signature based on two public and private keys are an obstacle to falsifying valuable data. A public key is needed to verify the signature itself and a private key is used to create the digital signature. The hash function ensures that all recorded data remains unchanged.

3.10 SUMMARY AND FURTHER WORK

This chapter discussed the operational security of regional EMS. Regional phenomena are characterized by high intensity and short duration, and their detection is possible immediately before their occurrence. Therefore, the use of solutions characteristic for state threat monitoring systems to monitor them does not bring expected results. Financial conditions are another significant limitation in the construction of regional EMS. Local governments will be the owners and operators of such systems, who will try to minimize fixed expenses, which are operating costs. Developed, tested, and implemented design algorithms based on multicriteria optimization meet these requirements.

However, the operators' biggest concerns are security restrictions in the processing of personal data resulting from the Council of Europe regulation. Unfortunately, in most cases, the interpretation of applicable regulations is assured and redundant and significantly exceeds the necessary scope. In the EMS in use, residents who register personally will be informed of potential security threats. Therefore, a blockchain-based authentication system is being developed. Other works, the progress of which allows obtaining interesting results, include use of blockchain for forecasting threats, improving security on a wireless network, and organization of network traffic, ensuring the lifetime of the monitoring system. Specific effects of the described works are expected in the coming months.

REFERENCES

Acevedo, M. F. 2013. *Real-Time Environmental Monitoring: Sensor and Systems*. I. Boca Raton, FL: CRC Press.

Bac, S., and M. Rojek. 1981. *Meteorologia i klimatologia*. Warszawa: Państwowe Wydawnictwa Naukowe.

Bronsztejn, I. N., K. A. Siemiendiajew, G. Musiol, and H. Muhlig. 2004. *Nowoczesne kompendium matematyki*. Warszawa: Wydawnictwa Naukowe PWN.

Canter, L. W. 2018. *Environmental Impact of Water Resource Projects*. Boca Raton, FL: CRC Press.

Dandamudi, S. 2003. *Hierarchical Scheduling in Parallel and Cluster Systems*. New York: Springer.

Davis, L. 2009. *Natural Disasters*. II. New York: Facts On File.

Fraden, J. 2013. *Handbook of Modern Sensors. Physics, Designs, and Applications*. 4th. New York: Springer Science.

Freeman, R. L. 2005. *Fundamentals of Telecommunications*. II. Hoboken, NJ: John Wiley & Sons.

Grocki, R., M. Mokwa, and L. Radczuk. 2001. *Organizacja i wdrażanie lokalnych systemów ostrzeżeń powodziowych*. Wrocław: Biuro Koordynacji Projektu Banku światowego.

Gross, J. L., and J. Yellen. 2004. *Handbook of Graph Theory.* London: CRC Press.

Gubko, M. V. 2006. Mathematical *Models* of *Optimization* of *Hierarchical Structures.* Moscow: Lenand.

Gunn, A. M. 2008. *Encyclopedia of Disasters: Environmental Catastrophes and Human Tragedies.* Westport, CT: Greenwood Press.

Hać, A. 2013. Wireless *Sensor Network Designs.* 2nd. Hoboken, NJ: John Wiley & Sons Inc.

Hajder, M., P. Dymora, and M. Mazurek. 2002. "Projektowanie topologii transparentnych sieci optycznych." In *Konferencja Polski Internet Optyczny: Technologie, usługi i aplikacje.* Poznań: Instytut Informatyki Politechniki Poznańskiej. 183–195.

Hajder, M., H. Loutskii, and W. Stręciwilk. 2002. In *Informatyka. Wirtualna podróż w świat systemów i sieci komputerowych.* I. Edited by M. Hajder. Rzeszów: Wydawnictwo Wyższej Szkoły Informatyki i Zarządzania.

Hamdy, T. A. 2005. *Operations Research: An Introduction.* Upper Saddle River, NJ: Prentice Hall.

Hennessy, J. L., and D. A. Patterson. 2002. *Computer Architecture a Quantitative Approach.* I. San Francisco, CA: Morgan Kaufmann.

Illyas, M., and I. Mahgoub. 2012. *Handbook of Sensor Networks: Compact Wireless and Wired Sensing Systems.* 2nd. Boca Raton, FL: CRC Press.

Jain, A. K., N. Murty, and P. J. Flynn. 1999. "Data Clustering: A Review." *ACM Computing Surveys* 31 (3): 264–323.

Jardine, N. 1971. *Mathematical Taxonomy.* London: John Wiley & Sons.

Kożuchowski, K., ed. 2006. *Meteorologia i klimatologia.* Warszawa: Wydawnictwa Naukowe PWN.

Lee, K., D. Kim, and D. Ha. 2015. "On security and privacy issues of fog computing supported Internet of Things environment." 2015 *International Conference on the Network of the Future, NOF 2015.*

Lei, M., Y. Yang, X. Niu, Y. Yang, and J. Heo. 2017. "An Overview of General Theory of Security." *China Communications* 14 (7): 1–10.

Levin, M. S. 2007. "Towards Hierarchical Clustering." In *Computer Science – Theory and Applications: Second International Symposium on Computer Science in Russia.* Edited by V. Diekert, M.V. Volkov, and A. Voronkov. New York: Springer. 205–215.

Levitin, A. 2006. *Introduction to the Design and Analysis of Algorithms.* Phenix: Addison Wesley.

Makki, K., ed. 2008. *Sensor and Ad Hoc Networks. Theoretical and Algorithmic Aspects.* New York: Springer.

Manfreda, S., M. McCabe, and P. Miller. 2018. "On the Use of Unmanned Aerial Systems for Environmental Monitoring." *Remote Sensing* 10: 641.Mao, G., and B. Fidan. 2009. *Localization Algorithms and Strategies for Wireless Sensor Networks.* Hershey, PA: Information Science Publishing.

May, E. 2004. *Wireless Communications & Networks.* 2. Upper Saddle River, NJ: Prentice Hall.

McCabe, J. D. 2007. *Network Analysis, Architecture and Design.* 3. New York: Morgan Kaufmann.

Mendez-Rangel, J., and C. Lozano-Garzon. 2012. "A network design methodology proposal for E-health in rural areas of developing countries." *6th Euro American Conference on Telematics and Information Systems.* Valencia. 1–7.

Mesarovic, M. D. 1970. "Multilevel Systems and Concepts in Process Control." *IEEE Journals & Magazines* 58 (1): 111–125.

Mesarovic, M. D., D. Macko, and Y. Takahara. 1970. *Theory of Hierarchical, Multilevel Systems.* New York: Academic Press.

Mirkin, B., A. Rzhetsky, and F. S. Roberts. 1997. *Mathematical Hierarchies and Biology*. New York: Amer Mathematical Society.

Mirkin, B., and E. Koonin. 2002. "A Top-down Method for Building Genome Classification Trees." *DIMACS Series* 61: 97–112.

Murtagh, F. 1992. "Parallel Algorithms for Hierarchical Clustering and Cluster Validity." *IEEE Journals & Magazines* 14 (10): 1056–1057.

Patnaik, P. 2010. *Handbook of Environmental Analysis: Chemical Pollutants in Air, Water, Soil, and Solid Wastes*. 2nd. Boca Raton, FL: CRC Press.

Perahia, E., and R. Stacey. 2008. *Next Generation Wireless LANs: Throughput, Robustness, and Reliability in 802.11n*. Cambridge: Cambridge University Press.

Raychaudhuri, D., and M. Gerla. 2011. *Emerging Wireless Technologies and the Future Mobile Internet*. Cambridge: Cambridge University Press.

Ripka, P., and A. Tipek. 2007. *Modern Sensors Handbook*. Chippenham: Antony Rowe Ltd.

Saaty, T. L. 1997. "The Analytic Hierarchy and Analytic Network Processes for the Measurement of Intangible Criteria and for Decision-Making." In *Multiple Criteria Decision Analysis*. Edited by J. Figueira. New York: Springer. 345–407.

Santi, P. 2005. *Topology Control in Wireless Ad Hoc and Sensor Networks*. Chichester: John Wiley & Sons.

Schrijver, A. 1998. *Theory of Linear and Integer Programming*. Chichester: John Wiley & Sons.

Shidlovskiy, S. V. 2006. *Automated control. Reconfigurable structures*. I. Tomsk: Tomsk State University.

Sinha, S. M. 2006. *Mathematical Programming: Theory and Methods*. Delhi: Elsevier Science.

Slyke, R. V. 2001. Network *Planning* and *Design*. 10. ftp://ftp.shore.net/members/ws/Support/BDC/nd.pdf.

Smith, N. J., and A. P. Sage. 2005. *An Introduction to Hierarchical Systems Theory*. Dallas, TX: Information and Control Sciences Center, SMU Institute of Technology.

Spohn, D. L. 2002. *Data Network Design*. New York: McGraw-Hill.

Such, J., and M. Szcześniak. 2007. Filozofia nauki. Poznań: Wydawnictwo: Naukowe UAM.

Tannenbaum, A. S. 2006. *Strukturalna organizacja systemów komputerowych*. V. Gliwice: Helion.

Wang, B. 2010. *Coverage Control in Sensor Networks*. London: Springer-Verlag.

Wasson, C. S. 2005. *System Analysis, Design, and Development: Concepts, Principles, and Practices*. Hoboken, NJ: Wiley-Interscience.

Wiersma, G. B., ed. 2004. *Environmental Monitoring*. Boca Raton, FL: CRC Press.

Willett, P., and A. El-Hamdouchi. 1989. "Comparison of Hierarchic Agglomerative Clustering Methods for Document Retrieval." *Computer Journal* 31: 220–227.

Zheng, J., ed. 2009. *Wireless Sensor Networka*. Hoboken, NJ: Institute of Electrical and Electronics Engineers.

Zhu, Y., J. Song, and F. Dong. 2011. "Applications of Wireless Sensor Network in the Agricultureenvironment Monitoring." *Procedia Engineering* 16, 608–614.

4 IoT-Enabled Surveillance System to Provide Secured Gait Signatures Using Deep Learning

Anubha Parashar
Manipal University Jaipur

Apoorva Parashar
Maharshi Dayanand University

Rajveer Singh Shekhawat
Manipal University Jaipur

Vidyadhar Aski
Manipal University Jaipur

CONTENTS

4.1   INTRODUCTION

This chapter focuses on discrete testing using an accelerometer for the identification of biometric gait based on a smartphone. Having an authentication mechanism makes it possible for a cell phone to identify its user on the basis of how it functions. This strategy has two big advantages. First, mobile acceleration sensors can detect gait, which are already built in. There are also no additional engineering costs associated with the introduction of this program. Second, defining a gait during research does not require specific user feedback.

This chapter takes a different proposal for gait authorization method (GAM). The aforementioned factors ensure the excessive usage of a biometric gait recognition using an accelerometer-based system, which does not require additional interaction time. This overcomes problems with current mobile device authentication methods. In fact, most phones only provide authentication through a personal identification number (PIN). Lately, methods for authenticating graphically have become adaptable [1]. Study findings indicate that most of the mobile device owners do not disable PIN authentication, mostly because of its poor user-friendliness.

The perks of aforementioned approaches are that all the confidential details which are required for verification are connected to the subject matter without any interruptions; thus, it cannot be passed onto anyone easily. In a biometric method, the approach is different. While enrolling, the subject is required to apply his biometric feature to the sensor which is stored for comparing it later on [2,3].

Initially, these were bulky devices that only allowed telephone calls to be made or short messages to be written (SMS) and were mainly used by businessmen. In reality, almost everybody uses smartphones or laptops. They are consistently appending features so the accessible applications are increasing figuratively. Consequently, the volume and processed data range has also increased. Any fraudster, who can ingress entries of calendar, log details, social networking accounts, emails, etc., can mimic to own the phone and cause harm. Thus, the authentication mechanism for the phone and the data stored therein must be secure. Most mobile devices do not need to insert a PIN after a stand-by process, which allows an attacker to easily access data when it is reactivated [4].

Major contributions of this chapter are:

- Specific sensors were used for collecting and analyzing the database; thus making a point about inbuilt sensor's accuracy in the smartphones.
- The four algorithms used here (SVM, HMM, KNN, and DNN) have not been used in the literature yet. Hence, we gave the utmost priority to extraction of gait cycles and algorithms based on template classification.
- The testing database that we used was vigorously tested and is large unlike other testing databases, in which data collection is done on a single day and the database itself consists of the data of subjects walking on a flat ground.

All the features for deploying a gait recognition system based on the accelerometer on the mobile handsets have been answered by this research. The application design for authorization has been submitted. This chapter focuses on forming a satisfactory gait-based identification approach and appropriate feature recognition. To evaluate constructive results, distinct databases that focus on distinct rules are used. To compare these outcomes, various approaches are compared in every database.

This chapter is organized in seven sections. Section 4.1 consists of the introduction. Section 4.2 consists of literature review and details about the techniques of authentication system. Section 4.3 presents the proposed methodology, dataset used, components of the proposed methodology, and user interface details. Section 4.4 gives technical details about Modules of Gait Authentication System. Section 4.5 gives the implementation details, which include how the user will get registered, authorization steps, preprocessing steps, and feature extraction techniques. Section 4.6 discusses the techniques used and the results obtained. Section 4.7 provides the conclusions of the implemented work.

4.2 LITERATURE REVIEW

4.2.1 KNOWLEDGE-BASED AUTHENTICATION

Each application requires a specific format for a strong password, such as being-caps, numeric, and special characters. Memorizing this is a hectic job. The outcome of such a system is that people create easy passwords that they can easily recall like names, date of births, etc. Consequence of this is increase in dictionary attacks as the search space of the attacker is limited [5].

4.2.2 GRAPHICAL AUTHENTICATION METHODS

There are two objectives of using an illustrative authentication technique. Next, an excessive amount of security is achieved as either an image or a pattern is the key needed for authentication. It is harder to write than a standard password or to transfer it to someone else. Second, it is presumed that recollecting secret is easy as the images are easier to recall than the words under the so-called image dominance effect [6].

4.2.3 BIOMETRICS

Biometry may be utilized in authenticating people based on their physiological features and behaviors. Fingerprints, iris, face speech, keystroke, and gait recognition are some of the modes used for obtaining biometric information. Such authentication methods can be combined with the growing computational capacity of mobile devices. Different biometrics techniques are shown in Figure 4.1.

4.2.3.1 Facial-Based Recognition

For front cameras being built into cell phones, facial recognition is a very simple form of cell authentication. In Ref. [7], a database obtained through a cell phone and a laptop [17] was used to test various face and speaker recognition

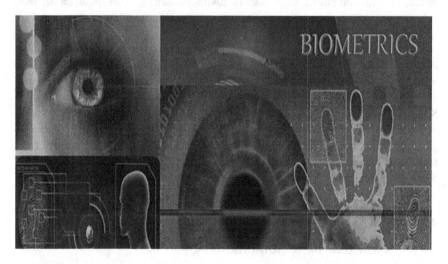

FIGURE 4.1 Biometrics.

systems. Eighteen devices were analyzed from nine institutes. The best results were obtained by combining the results of a local histogram's binary pattern to local step quantization histogram. Then, score normalization was applied to get the best result, which was half of the total error rate (HTER) of 10.91.

4.2.3.2　Audio Recognition

The identification of speakers on cellular devices is seen either as a never-ending or a blunt authority. During ongoing phone calls, speech is evaluated for continuous authentication, which is done in hindsight [8–10]. Example of an audio recognition system is shown in Figure 4.2.

4.2.3.3　Minutiae-Based Identification

Fingerprints can be realized on cellular devices using one of the following ways – one alternative is to use a devoted sensor. Example of minutiae-based recognition is presented in Figure 4.3.

4.2.3.4　Biometrics Gait Identification

This is a method of recognizing people with the way they walk. R. P. Trommel [11,12] identified walking individuals' subject-specific behaviors. Thomas Wolf [18] organized gait identification into three categories which were built around sensors used to record the gait. Illustration of gait identification is shown in Figure 4.4.

4.2.3.4.1　Gait Biometrics

A way to walk or step on foot is known as gait [13]. Walking pattern of humans is composed of several repetitive loops of gait. Growing gait process involves two stages. Figure 4.5 provides a graphical portrayal of a single gait period and the vertical accelerations as measured.

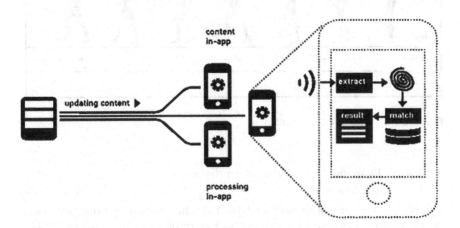

FIGURE 4.2　Audio-based recognition.

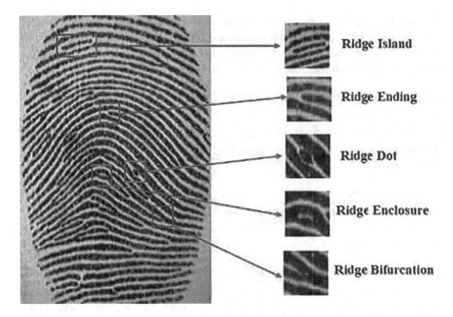

FIGURE 4.3 Minutiae-based recognition.

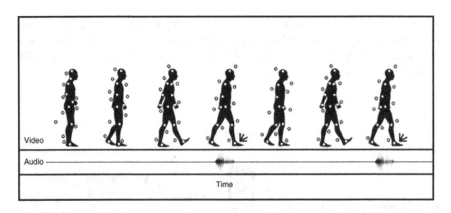

FIGURE 4.4 Gait-based recognition.

4.2.3.4.2 *Biometric Gait Identification Based on Machine Vision*

This includes detection of diseases, analysis of athletic results, monitoring, machine–man interfaces, videoconferencing, and storing of images based on content [14]. Although they follow distinct objectives, they have a lot of general handling steps, such as context segmentation, identification of body joints, etc. (Figure 4.6).

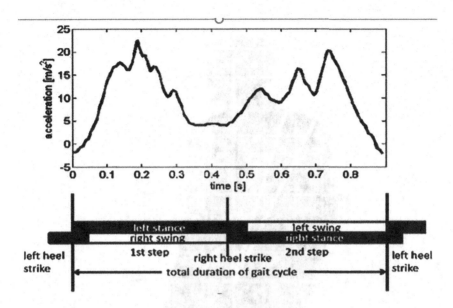

FIGURE 4.5   Graphical representations of one gait period.

In [15], a checklist is provided by the Institute of Forensic Medicine in Denmark to recognize gait patterns and to illustrate this on the actual instance of a bank thief. Imprisonment of a burglar over unusual walk was reported by BBC [16]. This indicates a good implementation of gait recognition in monitoring based on vision as an experienced podiatrist did in this case. The monitoring scenario is distinct from many other recognition scenarios in which a person is involved in the tracking, as shown in Figure 4.7.

Michal Balazia [17] categorized the outlook of feature extortion into holistic approaches based on models. He included a thorough analysis of vision-based gait recognition. No specific model is considered in these types of methods, and the attributes are based primarily on the obtained outlines. The attributes of the figures are focused on the mean length between the center of the outline and the pixels in the foreground (Figure 4.8). In addition, the outline is divided into angular sectors [18].

Model-based methods presume an underlying mechanism constructed on human walk. Such approaches are typically hard, but they have the perks of becoming more resilient to noise, difference in clothing, and viewpoint. The attributes that are taken out include fixed specifications (such as separation in a feet or height) [19] or ellipse parameters (in which the ellipse refers to different regions like lower part of leg or a person's head) in the body [20]. Although the given attributes permit a close-packed description of a gait, these are susceptible to inaccuracy that restrain the proper extortion of the needed guidelines.

The HumanID hassle of challenges by gait was scripted, wherein the problem was multiple experiments, a baseline algorithm, and 122 volunteers [21].

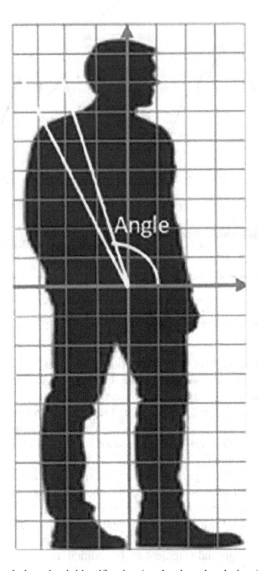

FIGURE 4.6   Angle-based gait identification (marker-based technique).

Maximum function of surveillance-based gait identification was conducted [22]. They released a broad database that included data on 118 subjects [23].

4.2.3.4.3   3D-Video

Work on gait recognition has begun in the last few years using data extracted from 3D videos. Pau et al. [24] proposed an exemplary technique with the help of 3D point cylinders for lower body parts, such as shins and thighs. Center of the hip is at the root of the global scheme of coordinates, as shown in Figure 4.9. Extracting gait kinematics is built by the angles derived from the orientations of the cylinder.

FIGURE 4.7 Silhouette-based gait identification (marker-less-based technique).

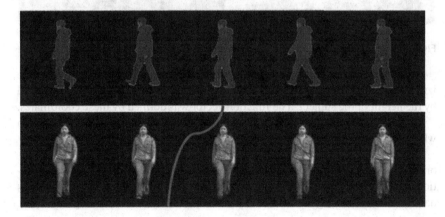

FIGURE 4.8 Skeleton-based gait identification.

4.2.3.4.4 Surveillance and Secured Monitoring

In [25], use of gait monitoring has been assessed. A motion map was developed using a frame differentiation approach. A hair-based template matching system was applied to locate shoulders, knees, and ankles, and the outcome was updated with the help of anthropometric and gait kinematic information.

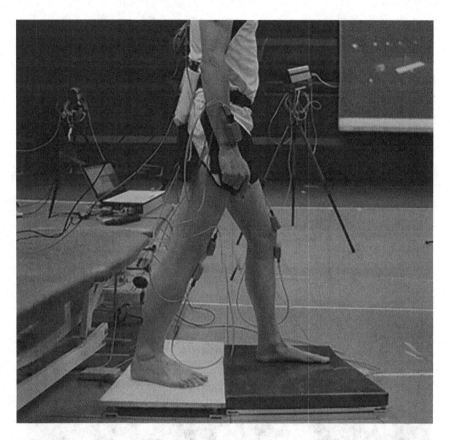

FIGURE 4.9   3D-based model of human gait.

The joint coordinates were measured in the metric system originating in the left ankle (images of frontal video are taken into account) and were normalized by the height of the subject. The function vector was defined by these coordinates.

Ronald B. Postuma [26] concentrated on vigorous authorization of videos in which the frame rate was low and available for CCTV.

Daniele Ravı [27]'s key emphasis was on the creation of a framework for calibrating gait recognition. Advancement was done using images with poor resolution, surveillance videos will combine gait and face recognition. Human studies have stated that mixing enhances the efficiency of body or facial recognition [28].

4.2.3.4.5   Covariates Influence

Trung et al. [29] tested the effect of time on gait identification based on vision. Other variables (e.g., surroundings and attire) remained constant to collect data over 9 months in quad sittings.

On comparing the probe data collected when the subjects wore regular clothing on similar day comparisons received via volunteers who wore overalls, the identification output decreased approximately to 40% in Figure 4.10.

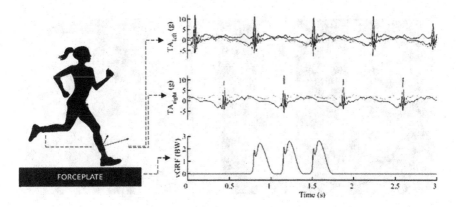

FIGURE 4.10 Variations due to covariates.

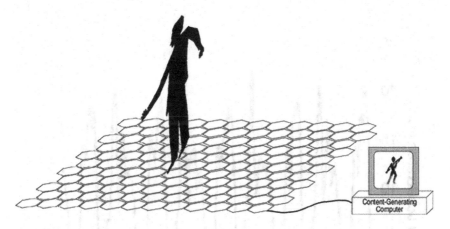

FIGURE 4.11 Pressure sensors on floor.

T. T. Ngo [30] introduced an outline function and used various codes of comparison such as Hidden Markov Mode.

4.2.3.4.6 Gait Identification Based on Sensors on Floor

Biometrics based on floor sensors have a lot of uses like smart homes and access control, and they can be applied to analyze gait clinically [31] along with recording videos. These sensors can directly be installed on mattress or floors. Gait overview can be found in [32], and this technology can be found with the use of key "Footstep Recognition" in Figure 4.11.

4.2.3.4.7 Gait Recognition Based on Wearable Sensors

The most recent technique is a wearable-sensor-based approach shown in Figure 4.12. We have overviewed several applications with the help of accelerometers for collecting data, along with the described Genuine Match Rate (GMR), CCR, and EER (Figure 4.13).

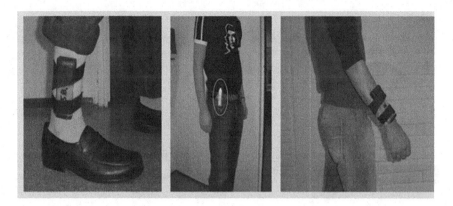

FIGURE 4.12   Wearable sensors.

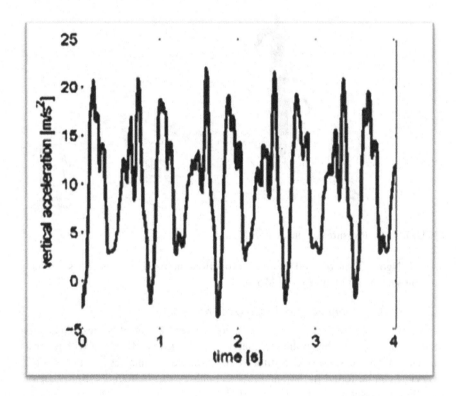

FIGURE 4.13   Acceleration.

Extraction of appropriate clinical information from the sensor was the main aim. Second, the experiment helped in distinguishing people who had Parkinson's disease from those who did not. Nonetheless, whether subjects could be identified with the help of their gait was analyzed [33,34].

4.3   PROPOSED METHODOLOGY

GAM has been developed to provide evaluation and mounting of distinct authorization techniques on cell phones. GAM is an Android application that produces a framework on cellular devices. Figure 4.14 presents the architecture of the system.

4.3.1   DATASET

Particulars of a module are stored in a database (e.g., level of priority and status of activation), enrolled customers (e.g., volunteer ID, time of enrolment), count (in calculating total authorizations) because of space constraints the links of cited data are stored in distinct files. Cited data can be stored on an external server rather than a cell phone. As the biometric data keeps changing, methods of encryption with the help of cryptography and encryption are not suitable. Hence, there is a requirement for techniques that protect templates, such as in [35].

4.3.2   MODULES IN GAM

The entire code for the authorization method is available in a different application package in the GAM module. Modules are of two different types: background and foreground modules. No user interactions are needed for background sections as they are run in hindsight. Users interact with the authorization process in the foreground.

4.3.3   USER INTERFACE

Access to available modules is granted by user interface. (Figures 4.15 and 4.16).

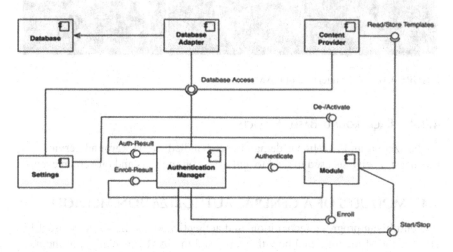

FIGURE 4.14   GAM architecture.

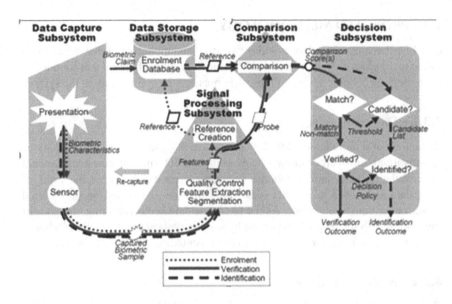

FIGURE 4.15   Modules of GAM.

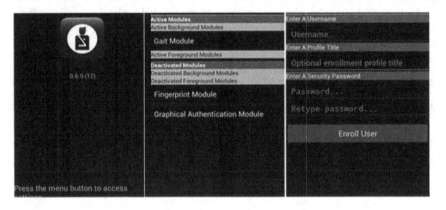

FIGURE 4.16   User interface of GAM.

4.3.4 BACKGROUND-BASED SERVICES

Authorization and enrolment demands are handled by background services. A broadcast receiver is used to tend to authorization requests and on-screen events.

4.4 MODULES OF A GENERAL AUTHORIZATION METHOD

The enrolment progress, recognition, and authentication is shown in Figure 4.15. Description of modules and how they are used in GAM is explained in the next section.

4.4.1 Data Capture Module

This is a system that helps to capture the data of the users. Volunteers are requested so that we can obtain features based on accelerometer biometrics.

4.4.2 Signal Processing Module

Several factors such as low ciphering proportions of a cellular device are considered at the time of constructing an authorization system like GAM because it is performed on the cellphone.

4.4.3 Data Storage Module

Database and citation of enrolled users and their information (e.g., their ID and creation date) is stored in this subsystem. SQLite database is the ground basis of GAM as very few people own the same type of phone so entries in the database will be very few.

4.4.4 Comparison-Based Module

Results are compared to all the available citations in the recognition mode. These comparisons are known as 1:n, and can be the outcome instead of a tally score for authorizing techniques such as SVMs.

4.4.5 Decision-Based Module

The authorization settlement is made based on the calculated contrast score(s). In certain schemes, this just means matching scores with the program-specific entrance.

4.5 IMPLEMENTATION OF GAM

The biometric devices can be operated on three different modes: validation, enrolment, and recognition. Validation and recognition can be clubbed by the term authorization (Figure 4.17). This section presents a brief discussion about the implementation process of different GAM modes.

4.5.1 Registration Module

The user can enroll themselves in an operative section upon creating their accounts.

4.5.2 Authorization

The authorization requests are vaulted to the screensaver as GAM is built on a cellular device (Figure 4.17). The screensaver starts operating as soon the phone is not being used. GAM starts the authorization process when the user disables the screensaver.

Context sections do not require any interaction with users. Modules consisting gait data require never-ending authorization. Therefore, it can be concluded that upon turning on the screensaver, the sections start accumulating data.

4.5.3 Preprocessing

Prior to module distribution, the data is preprocessed with the help of the procedure shown in Figure 4.18.

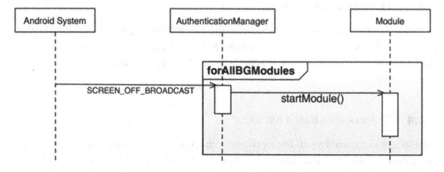

FIGURE 4.17   Authorization procedure executed in GAM.

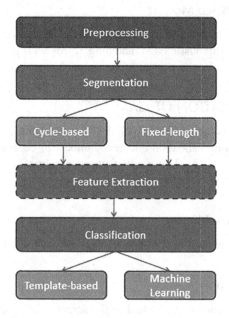

FIGURE 4.18   Preprocessing procedure executed in GAM.

4.5.3.1 Feature Extraction

To process relevant data, it is of utmost importance to classify certain frag-ments of the data that are obtained when the volunteer walks. In general, this is achieved with the help of identifying behaviors set out in [25] (Figures 4.19 and 4.20).

There are some sections in the data where the volunteer is not moving. The stroll intervals in which the volunteer is walking are extracted as those bits con-tain all the information that is required.

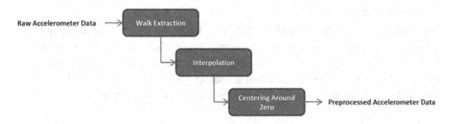

FIGURE 4.19   Preprocessing procedure executed in GAM.

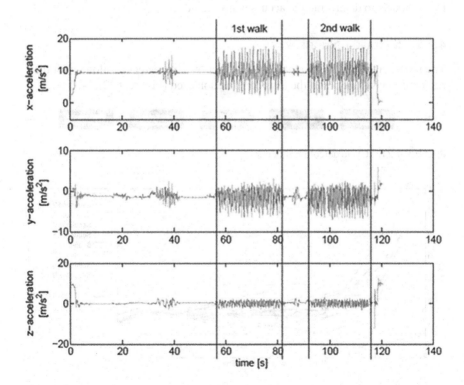

FIGURE 4.20   Accelerometer data of x, y, z axis.

4.5.3.2 Zero Center

The data is focused around zero to decrease the effect of this occurrence. To achieve this, the mean values of strolling acceleration (μ) is subtracted from the data values: sá(t)=sa(t)$-\mu$a, a ácido $\{x, y, z\}$.

4.5.3.3 Segmentation

After preprocessing, the data is segmented by gait cycles and extraction is not compulsory at this step. Unprocessed data is acquired from gait cycles, and then features are drawn out from these components, as shown in Figure 4.21.

4.5.4 Classification

For classification, KNN, SVM, HMM, and DNN were used and were trained with half of the tested and traces. Frequently, an enrolled user is not identified by their cell phones and an unknown volunteer is proceeding in unapproved access. Forty-eight tests were conducted and a subject was marked as authentic in each one of them. Volunteers should be identified by the application as genuine users.

4.6 DISCUSSION AND RESULT

We considered ten variant sequences to evaluate the HMM, KNN, SVM, and DNN models to determine the accuracy training data.

4.6.1 K NEAREST NEIGHBOR

The model stored in KNN algorithms is a set of feature vectors along with their respective class labels and the distance function used (Figure 4.22).

FIGURE 4.21   Cycle detection process.

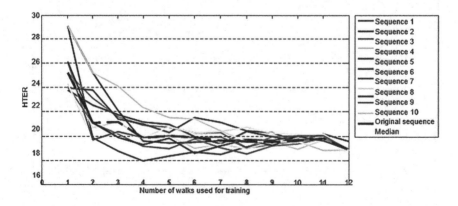

FIGURE 4.22   Accuracy obtained from KNN model.

4.6.2 HIDDEN MARKOV MODEL

Hidden Markov model provides a graph with probability of transitioning provides a perfect description of Markov chains (Figure 4.23).

4.6.3 SUPPORT VECTOR MACHINE

These are supervised learning algorithms that are used for analyzing regression and classifying data. To evaluate SVM in this database, data were inserted in the sampling rates below 200 Hz and then features were extracted. Results of SVM are shown in Figure 4.24.

4.6.4 DEEP NEURAL NETWORK

Data flows from the input layer to the output layer in a DNN. In the database, separate testing was conducted to analyze the single feature capabilities. Walk samples were collected on day 1 to train, and on day 2 the recorded dataset was used for testing (Table 4.1). Results obtained from DNN are shown in Figures 4.25 and 4.26.

4.7 CONCLUSIONS AND FUTURE WORK

This chapter presented and assessed a novel procedure for gait identification based on accelerometer. Direct implementation framework is the authorization on cellular devices, and the focal query was whether gait recognition built on accelerometer is useful for authorizing people via a mobile application? As a solution to this query, the evolved techniques were appraised on the grounds of three distinct databases collected by gait with the help of two separate cell phones.

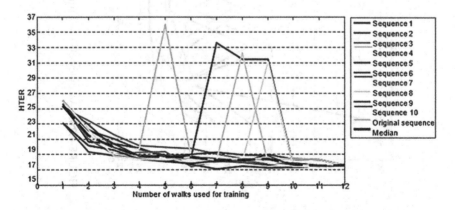

FIGURE 4.23 Accuracy obtained from SVM model.

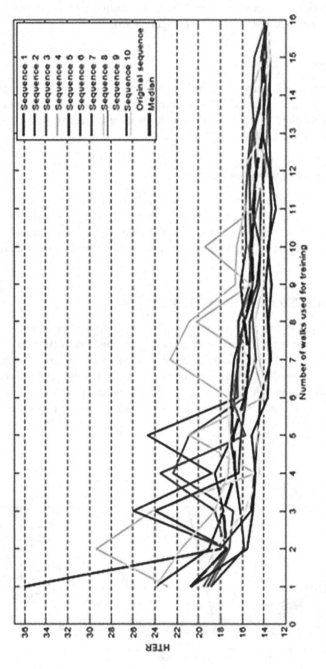

FIGURE 4.24 Accuracy obtained from SVM model.

TABLE 4.1
Result Attained from Different Models

Features	KNN (%)	SVM (%)	HMM (%)	DNN (%)	Fusion (%)
Maximum	97.8	97.6	95.5	99.3	82.2
Minimum	98.2	96.4	96.5	99.3	77.2
Mean (M)	98.9	98.4	98.3	99.6	89.4
Standard deviation	97.6	97.4	98.9	99.6	81.6

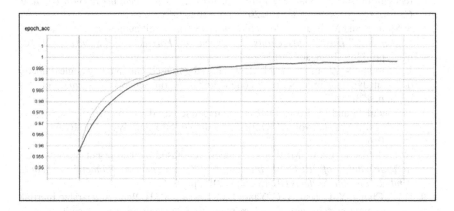

FIGURE 4.25   Accuracy obtained from DNN model.

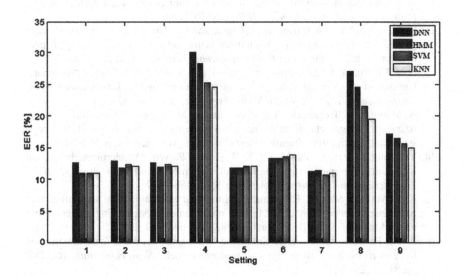

FIGURE 4.26   Comparison of accuracy obtained from different models.

REFERENCES

1. M. J. Marín-Jiménez, F. M. Castro, N. Guil, F. de la Torre, and R. Medina-Carnicer, "Deep multi-task learning for gait-based biometrics," in *Proceedings of International Conference on Image Processing (ICIP)*, Sep. 2017, pp. 106–110.
2. F. M. Castro, M. J. Marín-Jiménez, and N. Guil, "Multimodal features fusion for gait, gender and shoes recognition," *Machine Vision and Applications*, vol. 27, no. 8, pp. 1213–1228, Nov. 2016.
3. Y. Zhong and Y. Deng, "Sensor orientation invariant mobile gait biometrics," in *Proceedings of IEEE International Joint Conference on Biometrics*, Sep./Oct. 2014, pp. 1–8.
4. Z. Wei, W. Qinghui, D. Muqing, and L. Yiqi, "A new inertial sensor-based gait recognition method via deterministic learning," in *Proceedings of 34th Chinese Control Conference (CCC)*, Jul. 2015, pp. 3908–3913.
5. S. Sprager and M. B. Juric, "An efficient HOS-based gait authentication of accelerometer data," *IEEE Transactions on Information Forensics and Security*, vol. 10, no. 7, pp. 1486–1498, Jul. 2015.
6. Y. Zhao and S. Zhou, "Wearable device-based gait recognition using angle embedded gait dynamic images and a convolutional neural network," *Sensors*, vol. 17, no. 3, p. 478, 2017.
7. M. Gadaleta and M. Rossi, "IDNet: smartphone-based gait recognition with convolutional neural networks." [Online]. Available: https://arxiv.org/abs/1606.03238, Oct. 2016.
8. A. H. Johnston and G. M. Weiss, "Smartwatch-based biometric gait recognition," in *Proceedings of IEEE 7th International Conference on Biometrics Theory, Applications and Systems (BTAS)*, Sep. 2015, pp. 1–6.
9. H.-C. Chang, Y.-L. Hsu, S.-C. Yang, J.-C. Lin, and Z.-H. Wu, "A wearable inertial measurement system with complementary filter for gait analysis of patients with stroke or Parkinson's disease," *IEEE Access*, vol. 4, pp. 8442–8453, 2016.
10. M. Alotaibi and A. Mahmood, "Improved gait recognition based on specialized deep convolutional neural networks," in *Proceedings of IEEE Applied Imagery Pattern Recognition Workshop (AIPR)*, Oct. 2015, pp. 1–7.
11. R. P. Trommel, R. I. A. Harmanny, L. Cifola, and J. N. Driessen, "Multi-target human gait classification using deep convolutional neural networks on micro-Doppler spectrograms," in *2016 European Radar Conference*, London, 2016, pp. 81–84.
12. F. M. Castro, M. J. Marín-Jiménez, N. Guil, and N. Pérez de la Blanca, "Automatic learning of gait signatures for people identification," in *14th International Work-Conference on Artificial Neural Networks*, Cádiz, 2017, pp. 257–270.
13. A. Mannini, D. Trojaniello, A. Cereatti, and A. M. Sabatini, "A machine learning framework for gait classification using inertial sensors: application to elderly, post-stroke and huntington's disease patients," *Sensors*, vol. 16, no. 1, p. 134, 2016.
14. Y. Zhang, G. Pan, K. Jia, M. Lu, Y. Wang, and Z. Wu, "Accelerometerbased gait recognition by sparse representation of signature points with clusters," *IEEE Transactions on Cybernetics*, vol. 45, no. 9, pp. 1864–1875, 2015.
15. K.-R. Mun, G. Song, S. Chun, and J. Kim, "Gait estimation from anatomical foot parameters measured by a foot feature measurement system using a deep neural network model," *Scientific Reports*, vol. 8, no. 1, pp. 1–10, 2018.
16. S. Yu, H. Chen, Q. Wang, L. Shen, and Y. Huang. "Invariant feature extraction for gait recognition using only one uniform model," *Neurocomputing*, vol. 239, pp. 81–93, 2017.

17. M. Balazia and P. Sojka. "Learning robust features for gait recognition by maximum margin criterion," In *Proceedings of the 23rd International Conference on Pattern Recognition*, pp. 901–906, 2016.

18. T. Wolf, M. Babaee, and G. Rigoll. "Multi-view gait recognition using 3D convolutional neural networks," in *Proceedings of the International Conference on Image Processing*, pp. 4165–4169, 2016.

19. N. Y. Hammerla, S. Halloran, and T. Ploetz. Deep, convolutional, and recurrent models for human activity recognition using wearables. arXiv preprint arXiv:1604.08880, 2016.

20. C. C. Charalambous and A. A. Bharath. A data augmentation methodology for training machine/deep learning gait recognition algorithms. arXiv preprint arXiv:1610.07570, 2016

21. A. Parashar, and A. Parashar, "Identification of gait data using machine learning technique to categories human locomotion," in *Proceedings of the 26th International Conference on World Wide Web*, pp. 351–360, 2017.

22. D. Liu, M. Ye, X. Li, F. Zhang, and L. Lin, "Memory-based gait recognition," In *BMVC*, 2016.

23. M. Pau, S. Caggiari, A. Mura, F. Corona, B. Leban, G. Coghe, L. Lorefice, M. G. Marrosu, and E. Cocco, "Clinical assessment of gait in individuals with multiple sclerosis using wearable inertial sensors: comparison with patient-based measure," *Multiple Sclerosis and Related Disorders*, vol. 10, pp. 187–191, 2016.

24. M. Pau, G. Coghe, C. Atzeni, F. Corona, G. Pilloni, M. G. Marrosu, E. Cocco, and M. Galli, "Novel characterization of gait impairments in people with multiple sclerosis by means of the gait profile score," *Journal of the Neurological Sciences*, vol. 345, no. 1–2, pp. 159–163, 2014.

25. R. Postuma, W. Poewe, I. Litvan, S. Lewis, A. Lang, G. Halliday, C. Goetz, P. Chan, E. Slow, K. Seppi, et al., "Validation of the MDS clinical diagnostic criteria for Parkinson's disease," *Movement Disorders*, vol. 33, no. 10, pp. 1601–1608, 2018.

26. A. Parashar and D. Goyal "Clustering gait data using different machine learning techniques and finding the best technique," *Communications in Computer and Information Science*, vol. 628, pp. 426–433.

27. A. Parashar and A. Parashar, "Identification of gait data using machine learning technique to categories human locomotion," in *Proceedings of the 10th International Conference on Security of Information and Networks*, Oct. 2017, pp. 229–234.

28. S. Sprager and M. B. Juric, "Inertial sensor-based gait recognition: a review," *Sensors*, vol. 15, no. 9, pp. 22089–22127, 2015.

29. N. Trung, Y. Makihara, H. Nagahara, R. Sagawa, Y. Mukaigawa, and Y. Yagi, "Phase registration in a gallery improving gait authentication," in *IJCB*, 2011.

30. T. T. Ngo, Y. Makihara, H. Nagahara, Y. Mukaigawa, and Y. Yagi, "Orientation-compensative signal registration for owner authentication using an accelerometer," *IEICE Transactions on Information and Systems*, vol. 97, no. 3, pp. 541–553, 2014.

31. T. Ngo, Y. Makihara, H. Nagahara, Y. Mukaigawa, and Y. Yagi, "The largest inertial sensor-based gait database and performance evaluation of gait-based personal authentication," *Pattern Recognition*, vol. 47, no. 1, pp. 228–237, 2014.

32. A. Parashar, A. Parashar, S. Goyal, "Push recovery for humanoid robot in dynamic environment and classifying the data using k-mean," *International Journal of Interactive Multimedia and Artificial Intelligence*, vol. 4, no. 2, pp. 29–34, 2016.

33. L. Wu, J. Yang, M. Zhou, Y. Chen, and Q. Wang, "LVID: a multimodal biometrics authentication system on smartphones," *IEEE T-IFS*, 2019.

34. A. H. Johnston and G. M. Weiss, "Smartwatch-based biometric gait recognition," in *BTAS*, 2015, pp. 1–6.

35. S. Yao, S. Hu, Y. Zhao, A. Zhang, and T. Abdelzaher, "Deepsense: a unified deep learning framework for time-series mobile sensing data processing," in *WWW*, 2017, pp. 351–360.

5 IoT-Based Advanced Neonatal Incubator

K. Adalarasu, P. Harini, and B. Tharunika
SASTRA Deemed to be University

M. Jagannath
Vellore Institute of Technology (VIT) Chennai

CONTENTS

5.1   INTRODUCTION

Preterm babies are born in advance of 37 weeks of pregnancy. They are grouped according to their gestational age, as shown in Table 5.1. According to the World Health Organization (WHO), nearly 15 million babies are born too early and approximately 1 million neonates die due to preterm complications. Based on a survey conducted in 2015, India has the greatest number of preterm births. Preterm babies have many short-term and long-term complications. Short-term complications include temperature control problems, respiratory problems, metabolism problems, and immune system problems. Long-term problems include impaired learning, chronic health issues, and hearing and visual problems.

Premature babies have breathing problems due to an immature respiratory system, as their lungs lack surfactant (which allow the lung to expand) causing respiratory distress. Therefore, the respiratory rate and heart rate of preterm babies are monitored continuously in an incubator. An incubator is a device used to screen and control natural conditions reasonable for an infant. It is utilized in preterm births

TABLE 5.1

Grouping of Preterm Babies

Gestational Age (weeks)	Group
>28	Extremely preterm
28–32	Very preterm
32–37	Moderate-to-late preterm

for typical infants. To overcome these complications, preterm babies are placed in a neonatal incubator where they are kept warm to maintain their body temperature [1]. Sensors are attached to the body surface of the neonate for continuous monitoring of heart rate, pulse rate, breathing rate, blood pressure, and temperature. A ventilator is provided in an incubator for babies with respiratory problems. Modern incubators are highly specialized devices. This review discusses the features of neonatal incubators.

This chapter addresses the critical evaluation of neonatal incubators that are available in the commercial market. It recognizes and analyzes related literature to assess a design and theoretical approach, thereby offering readers a contemporary understanding of neonatal incubators. Finally, the authors have proposed an Internet of Things (IoT)-based advanced neonatal incubator.

The chapter is organized as follows: Section 5.2 deals with the systems design of the present problem and illustrates the same by means of examples with temperature control and humidity control. Section 5.3 describes the issues related to portable incubator, and Section 5.4 discusses the IoT-based incubator. Section 5.5 recommends the proposed solution in developing an advanced neonatal incubator. Section 5.6 summarizes the conclusions of the chapter.

5.2 SYSTEMS DESIGN

A neonatal incubator is a rigid box-like system that isolates the infant in a controlled environment. It is normally in the form of a trolley enclosed by transparent perspex. This enclosed cover prevents heat and moisture and isolates neonate from their outer environment. The main parameters that need to be regulated in an incubator are temperature and humidity. The infant's body should be maintained at constant temperature ranges from 36.5°C to 37.5°C [2]. Inside the neonatal incubator, the humidity should be maintained at 70%–75% relative humidity. Measurement of temperature by the axillary method has become the accepted neonatal nursing care. Neonates are usually attached with a noncontact-type temperature sensor DHT11 model for screening temperature continuously. A hygrometer is used for measuring the humidity inside the incubator and current sensors are also used. Below the mattress, there is a heater and a fan to control the temperature, as well as a water reservoir used to humidify the air inside the incubator. Air distributors in the incubator distribute air evenly to ensure equal temperature throughout the incubator to prevent spaces of cold or stale air. It also has four port holes to access the neonate without contaminating the infant's environment.

Electrocardiogram (ECG) is another important parameter that is measured in the neonatal incubator. Three electrodes are placed on the baby's chest for continuous ECG monitoring. They also monitor respiratory rate using the same electrode as PMS can measure. Blood pressure and oxygen saturation are also monitored continuously. Neonates are also connected to the feeding tube and infusion pump for feeding and supplying infusion fluids, respectively. The liquid crystal display (LCD) is attached to the neonatal incubator to display parameter values such as temperature, relative humidity, SpO_2, and respiratory rate. An alarm is included in the system to alert when there is any abnormal change in physiological parameters. Additionally, there is a stand to support the incubator. Sometimes the neonatal incubator also needs some supportive care that may include a ventilator and continuous positive airway pressure devices (which pushes a continuous flow of air to the airway to make the tiny air passage in the lung to open).

5.2.1 TEMPERATURE CONTROL

Sinclair [2] analyzed the effect on death and other clinical effects in maintaining body temperature rather than targeting air temperature. The neonatal death rate is reduced by maintaining the abdominal surface temperature (36°C) rather than maintaining the air temperature (31.8°C). This observation suggests the necessity to control the temperature inside the incubator. Heat loss in neonates may be because full-term babies are warm-blooded but preterm babies have no thermoregulation system as they are late-maturing. The major reason for heat loss in neonates is associated with a high proportion of body surface region to weight; high warmth loss by dissipation and inadequacy of subcutaneous fat tissue; inadequately created muscles; and irregulated blood flow.

The WHO suggests that the typical body temperature of term babies ranges from 36.5°C to 37.5°C. In a closed incubator, the temperature is controlled with the help of a heating module that is activated by the difference in the controlled temperature. The controlled temperature can be chosen as either the body temperature of the neonate or the incubator temperature. The contact temperature sensor placed on the body surface gives a measure of skin temperature of the neonate. In the servo control incubator, either of the temperature is measured and used as feedback to turn on or off the heating module based on the set value. A thermistor probe sensor is placed on the skin, ideally in the upper belly, and the radiator cycles to maintain the skin temperature consistently. Both the body temperature and the air temperature is taken as a controlling temperature in an incubator for a servo control incubator. The skin temperature servo control comprises a more stable system, there is a lag in air temperature servo control system. At a point when the skin temperature of the neonate becomes high, the temperature of the incubator diminishes, while the internal heat level becomes invariant. Thus, both the temperatures must be examined together when skin temperature servo control is utilized. The heating coil is the source of heat and needs a higher voltage to operate. The device works on the principle of forced convection. When the temperature falls below the normal temperature, the heater is switched on and hot air is blown into the incubator through the fan. Therefore, the neonates get continuous warm air.

Shaib et al. [3] used two parts in the warming system – phase-changing material (gel pack) and heater. The heater is periodically turned on and off to maintain the neonate body temperature within the range. Heat from the heater is transformed into gel packs. The gel pack preserves heat to provide a continuous warm environment inside the incubator. They also use a microcontroller to control the heater. Zaylaa et al. [4] developed a handy preterm incubator. In the hardware part, they used ATmega328 microcontroller and Arduino micro to assist the microcontroller. They used optical sensor that included integrated pulse oximeter and heart rate sensor. They also placed the thermometer to measure the temperature. The main part of the portable system is batteries; they used a set of four ultrafire rechargeable batteries to achieve 9800 mAh. They constructed a handy incubator using three major biocompatible materials such as silnylon, Mylar sheets, and bamboo fabric. The silnylon is used to make the outer layer because of its light weight and ability to disconnect the system. Mylar sheets are used because of their high tensile strength, stability, reflectivity, aroma barrier property, and electrical protection. Bamboo fabric has antibacterial properties with a breathable nature and great absorbance of water. They used a 3D printer to fabricate a handy incubator. 3D printers work based on fused deposition modeling, are cost-operative, and provide a personalized design. To control the temperature inside the incubator for maintaining the neonate body temperature, they used two main components – cartridge heater riprap and hot/cold chemical wax. The heater riprap converts electrical energy into thermal energy, and the heating probe connected to the gel pack transfers heat into the gel sack. Then, the chemical wax is used to transfer heat to the neonate via conductance. Use of heaters in incubator cause some noise level, which negatively affects the neonate.

Therefore, Nisha and Elahi [5] used an alternative to heater. They developed a low-cost neonatal incubator. They developed the entire chamber using acrylic sheets as it has less density ranging from 1100 to 1200 kg/m^3 and its transportation and assembly is easier and cheaper. This chamber has a smaller and a larger compartment. The larger compartment keeps the neonate in and the smaller compartment consists of a control circuit. They used DHT11, a noncontact-type sensor, for measuring both temperature and humidity inside the chamber. The temperature controlling unit consisted of an incandescent bulb and a 12 V DC fan in the heating unit to reduce noise level, as well as an aluminum vessel containing ice with a 12 V DC fan in the cooling unit. When the body temperature of the neonate falls below 36°C, the bulb glows and the fan associated with the bulb is switched on and blows hot air to the compartment where the neonate is placed. When the temperature goes beyond 37°C, the bulb is switched off and the fan in the cooling unit is switched on and blows cool air into the compartment until the desired temperature is achieved. This system also has a humidity controlling unit; when the relative humidity falls below 70%, this unit starts automatically and is switched off when the relative humidity reaches 75%. These controlling units were controlled by Arduino microcontroller program. A supply of 220 V AC power is provided to this system.

Tisa et al. [6] designed an improved incubator for neonatal care. First, they aimed for a simple on/off temperature control system where the unit is switched on when the temperature falls below the set point and is switched off once the temperature reaches the set point. However, because this simple control did not fulfill the requirement, they developed a controlling system consisting of a pulse width modulation and on/off control. They used a thermistor to measure the temperature. In addition, they included an alarm circuit to ensure the safety of the neonate.

5.2.2 HUMIDITY CONTROL

Abdiche et al. [7] discussed about calculating partial pressure of water vapor using the measurement of humidity and air temperature. The sensor in the incubator gives an adjusted measure comparing to the air temperature at 10 cm over the sleeping pad and not the air temperature where the sensor is set. The relative humidity sensor is kept at a distance of 15 cm from the air temperature sensor. This helps to provide a stable microclimate inside the incubator. This is designed in such a way that it produces and stabilizes the humidity based on the desired value of partial pressure of water vapor. This is considered as the most descriptive index of water loss from neonatal surroundings.

Costa et al. [8] used a newborn incubator that used distilled water in the reservoir. Silver nitrate was applied with distilled water which helped in proliferating the bacteria. The system included a data acquisition system using a microcontroller, and a SHT11 sensor was used to measure the moisture content of air and the temperature of the incubator. The moisture control system was established using a stepper motor to maintain the moisture content inside the incubator, that is, prescribed a band of comfort (40%–60%).

Silverman and Blanc [9] reported that the clinical course of the infants from the time of admission to 72 hours of age was essentially the same in most respects in the two groups but differs significantly in certain particulars. Among the infants who were in 32%–60% relative humidity, 49% exhibited retraction at some time during their course, and 48% retracted in 80%–90% relative humidity. The degree of retraction as judged by the "retraction scores" was essentially the same in the two groups when they were compared at equivalent ages.

Harpin and Rutter [10] suggested that enhancing the humidity inside incubators improves the control of newborn babies body temperature to the extent, and indeed, even on the principal day of life. On the other hand, newborn babies kept in a dry incubator and secured with a plastic air pocket cover or a slight layer of delicate paraffin are ordinarily hypothermic during the initial days, regardless of the utilization of the most extreme air temperature setting.

5.3   PORTABLE INCUBATORS

The invention of portable incubators has emerged into the field of medical science [11]. The improvement of a portable incubator is still in progress. Recently developed incubators are mostly based on microcontrollers. Vignesh et al. [12] used hardware components such as pulse sensor, MQ-6 gas sensor, DHT11 (temperature and humidity), light sensor, GSM module, and PIC microcontroller. The software components used include PIC microcontroller program (embedded C program) and Cloud: Ubidots.

Shaib et al. [3] developed an advanced potable preterm baby incubator. They extracted features such as temperature, heart rate, and oxygen saturation. The ceaseless warm condition was provided through gel packs that save and change the warmth through conduction to give the required newborn child the perfect temperature. They use 3D-printed incubator parts and introduced Wi-Fi telemedicine to the system. They developed this incubator to make it reasonable, practical, tolerant, well-disposed, and meet the wellbeing prerequisites for preterm babies, particularly in developing nations. The parameters considered included temperature, humidity, light, and oxygenation. The sensors used included DS18B20, DHT11, light-dependent resistor, and MQ-7 for measuring temperature, humidity, light intensity, and carbon monoxide. This system includes a microcontroller Arduino mega2560 that provides programmed control of temperature for the newborn child utilizing closed loop control mechanism. Similarly, it controls the warm water supply as indicated by relative dampness in the newborn child chamber. The control of moisture in the chamber is significant for reducing the loss of warmth from the baby's body. Similarly, controlling light provides appropriate development to newborn child. The CO sensor is advantageous for a newborn child to shield from a number of breath sicknesses. The GSM system can be utilized to decrease the commotion made by the precautions taken during close observation [13]. The design of the neonatal incubator includes raspberry pi model, cooling system, acrylic covering, load sensor, camera, heating system, incandescent bulb, water storage, water heater, and water inlet. This system is mainly used in rural areas and in ambulances at hospitals.

The essential consideration can be given inside a brief period during basic condition which reduces mortality among newborn children. This neonatal incubator has advantages over the traditionally used incubators. It is practical and the utilization of acrylic sheets makes the entire incubator rigid, light, and simple to clean and handle. The incubator sends information consistently to the back end. The information includes the sensor parameters measured inside the incubator, the state of the infant as controlled by the cameras, and the automation exercises being performed by the man-made consciousness running in the PC in the incubator. The doctors and nurses can monitor the activities of the incubator and can override the actions of artificial intelligence in the incubator used for automation by setting their own temperature and humidity levels through a mobile or a web application. This makes it simpler for specialists and attendants to deal with various incubators as the status of the considerable number of incubators would be accessible in one spot.

5.4 IOT-BASED NEONATAL INCUBATORS

Neonatal incubator is a closed system used to help premature infants with sickness. It requires intensive monitoring of various factors such as oxygen level, pulse rate, respiratory rate, temperature level, and humidity level. The conventional approach of neonatal monitoring requires the presence of a nurse or a doctor. Hence, modern technologies use the IoT for uninterrupted monitoring of neonates by doctors even outside the hospitals. If there is any abnormality, the system automatically sends messages to the doctors or nurses or the family members of the neonate.

Shakunthala et al. [14] proposed a method to monitor neonates in incubators using IoT. In this system, several parameters such as temperature, humidity, pulse rate, gas leakage, and light intensity were monitored through IoT. They used DHT11 sensor to measure both temperature and humidity and MQ6 gas sensor to detect toxic gases inside the incubator. This MQ6 sensor is also capable of detecting combustible gases such as methane. The light intensity inside the incubator was detected using LM393 photosensitive light-dependent control sensor module. The pulse rate was measured using pulse sensor and heart rate was measured using pulse rate, indicating the condition of the cardiac system. These sensed data were uploaded in the cloud via Wi-Fi using ESP8266 Node MCU module. The entire system is controlled by Arduino. They used Ubidots service to upload data which is used to store data online via a very simple API. The data is continuously uploaded to the cloud and the message is issued at abnormal conditions to the doctors or family members of the neonate. This system allows doctors to monitor real-time health data regularly.

Sahib et al. [3] developed a system in which the user interface was structured in two different ways – an immediate interface utilizing LCD and a Wi-Fi information transmission by a web server associated directly with the emergency clinic web server and cell phones, which makes this system a telemedicine supporting gadget. Sengeni [15] implemented an IoT-based smart incubator unit in a lab-view environment. They monitored and controlled temperature, humidity, light, and gas leakage inside the incubator. The pulse rate was also monitored using a sensor. They used Arduino and GSM module to upload data into the cloud. If there is any abnormality in neonate's body parameters, the controlling system starts to work and at the same time the system delivers a danger signal to the hospital server or mobile phones of parents by an alarm system and the system's network.

Sowmiya et al. [16] proposed a system to monitor incubator using IoT. They used DHT11 sensor to measure temperature and humidity, MQ6 sensor to monitor gas level, LDR sensor to measure light intensity, M212 to measure pulse rate, and Node MCU. They also used Ubidots to store data in the cloud. The ease of accessing data is ensured by the use of Ubidots. The technologies used are programmable logic controller, wireless transmission, Zigbee, controlled area network bus driver, cloud, bluetooth, and phase-based respiration detection [16].

Programmable Logic Controller: used in incubator system for monitoring and controlling capabilities with multifunctional facilities.

Wireless Transmission: helps in intelligent remote real-time monitoring and networking.

Zigbee: used as low-power wireless mesh network for temperature monitoring and alerting system.

Controlled Area Network: acts as a bus driver for incubator system.

Cloud: used for data storage of monitoring and risk management system for newborn babies.

Bluetooth Wireless Transmission: used in neonatal intensive care units for critical care.

Phase-based Respiration Detection: used for constant monitoring of respiration rate through videoconferencing.

The modern incubator design uses sensor technologies and data communication systems which acquire the data, analyzes data, and stores data in cloud storage. This system incorporates Wi-Fi and infrared equipment that quantify the vital parameters that need to be controlled [17].

According to an estimate, in 2019, roughly 26.66 billion IoT gadgets will be dynamic; by 2025, 75 billion IoT gadgets worldwide will be accessible and remotely associated with the Internet [18]. The overall worldwide spending on the IoT in 2016 was 737 billion dollars and is anticipated to reach 1.29 trillion dollars by 2020. The benefits of of IoT is consistency in monitoring and control of processes, andself-sustainability. Jabbar et al. [19] reported that IoT coordinated into an infant monitoring system can accomplish a fast reaction time and give a more prominent suspicion that all is well and good for the guardians.

5.4.1 ATTACKS IN THE IoT SYSTEM

Various types of attacks in the IoT system have existed since many decades. At its center, the IoT is tied with interfacing and systems administration devices that have not really been associated [20,21]. The effect of each attack can shift significantly, which is contingent upon the biological system, incubator and its condition, accessible insurance level, along others. Recently, there has been developments in botnets among IoT frameworks. A botnet exists when software engineers remotely control web-related devices and apply them for unlawful exercises. An endeavor could have their devices co-picked as an element of a botnet with no data on it. The issue is that various associations need nonstop security answers to track this. Next, is the increase in the number of IoT devices that address extended security vulnerabilities over the endeavor, which is a contemporary challenge for security analysts [22]. The endeavors must guarantee customer data. This is particularly worrisome on the grounds that various pros are using IoT systems supported by their supervisors. Sometimes, private data is undermined; an endeavor's status would persevere through a huge success, which is the one of the reasons that IoT security challenges cannot be disregarded [23]. Hence, patient information privacy needs to be extremely secure while developing IoT-based neonatal incubators.

5.4.2 Requirements of IoT Devices

Some of the common and important characteristics identified during the assessment study include sensors, connectivity, processors, energy efficiency, cost-effectiveness, quality, and reliability and security.

Exploring these difficulties involves mindful planning, field knowledge, and careful execution. Taking into account our study with the existing literatures, we have perceived four basic requirements for methodologies and practices that should be the elements of IoT-based neonatal incubator system development [24].

Edge computing/analytics: The information should be acquired by sensors and processed continuously. This contemplates quick reaction to undetermined change, such as an incubator system reacting to an unexpected change in the vital parameters of a newborn baby or identifying a diagnostic device malfunction [25].

Information ingestion and stream handling: Real-time processes should be done for acquiring information from multiple devices and transforming it for further usage by cloud-based platforms [26].

Device management: We need to guarantee that the IoT devices are safe, impart proficiently, and can be updated with enhanced approaches.

Endeavor integration with business systems: The IoT system experiences should be conveyed to the endeavor system and reference metadata should be obtained to interpret device information.

5.5 PROPOSED INCUBATOR SYSTEM

Based on the review, an advanced neonatal incubator is proposed, as shown in Figure 5.1. The proposed incubator system includes a noncontact-type temperature sensor DHT11 model, humidity sensor, photoplethysmography sensor, respiration sensor, HC12 wireless communication module, rechargeable battery, and microcontroller (Arduino Uno) used for monitoring the health condition of a newborn baby. The proposed neonatal incubator uses the IoT system which is intended to monitor the neonatal baby constantly and send the monitored health parameters to cloud storage for diagnosis. If these parameters surpass the limit level, an alarm signal will be delivered to the neonatal physician or clinical individual or

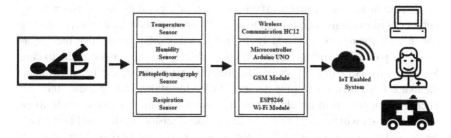

FIGURE 5.1 Proposed IoT-based advanced neonatal incubator system.

ambulance care via mail communication or SMS. The proposed neonatal incubator is run by a rechargeable battery, thus making this system portable.

The major sensor devices used comprise noncontact-type temperature sensor which measures the body temperature, humidity sensor which measures the relative humidity of the environment, photoplethysmography sensor which measures the oxygen saturation level and heart-beat, and respiration sensor which measures the respiration rate data and can acquire more diverse types of data every few seconds. At present, the clinical professional practice is to follow the subjective method rather than to objectively analyze the condition of the baby. To overcome this, we propose that real-time continuous activities should be observed for early diagnosis and detection of the baby with unexpected anomalies to avoid deadly circumstances.

Particularly in neonatal research, enormous datasets emerging from physiology information and different modalities have forced difficulties on information transfer, stockpiling, normalization, representation, investigation, and computation on the web (remote access) or offline mode. In addition, these enormous physiological datasets likewise require huge information storage strategies, for example, distributed computing, and clinical decision support systems incorporated in big data platforms [27]. These platforms that empower information analytics-based decision support systems giving real-time continuous investigation of different physiological information for newborn babies with the capacity to give bits of knowledge and early discovery for several clinical conditions.

High-speed physiological information from clinical devices, for instance, on account of neonatal intensive care, incubators, have limited resource in healthcare. Hence, we believe that the proposed real-time information sharing with clinical decision support system seems to be a vital solution to cope with these challenges. The proposed concept would be a starting point in the intensive care, and the development of these customized clinical devices accessible in expanding numbers to a wide range of buyers would definitely create a new awareness of indigenous incubators with advanced facilities.

5.6 FUTURE RESEARCH TRENDS

With technological improvements and cutting-edge innovation, the issue of the consideration for the neonatal infant has been addressed by an ever-increasing number of individuals. The conventional incubator can just give a relative security condition to the neonatal infants. In this modern scientific community, the majority of the parents are occupied with their work and have fewer prospects to care for their infants. Thinking about the present circumstance, the usual neonatal incubator finds difficult to give dedicated care to infants. Along these lines, an alternative approach in developing an advanced neonatal incubator ought to be inquired about which could self-adaptively take decision on the condition based on the sensors input acquired from the infant. As the innovation of the Internet and system is concentrated by an ever-increasing number of researchers, the IoT system has been created [28]. Similarly, as with some other electro-clinical hardware, a neonatal incubator needs to be calibrated intermittently in light of the fact that its malfunction may pose a threat to the infant's wellbeing or even cause death. The standard IEC 60601-2-19 builds up working specifications for neonatal incubators so that a technical domain can be offered for infants. These technical specifications are confirmed by executing a few tests, comprising the use of information about the temperature and relative humidity, for examining the conduct of the previously mentioned parameters with explicit focus inside the incubator. Along these lines, the advanced sensor network with IoT-based system incorporating machine learning techniques can decrease the time of the assessment and calibration of the incubator; thereby, eventually enhancing the quality of life of newborn babies.

5.7 CONCLUSION

This chapter aims to highlight the importance of temperature and humidity regulation in a neonatal incubator. It also explains the overall system design of a currently available neonatal incubator. The review focuses on the portable neonatal incubator and IoT-based neonatal incubators. This review has discussed the methods and techniques used in measuring the parameters and other developments in a neonatal incubator. Finally, an advanced neonatal incubator has been proposed with sophisticated sensor technologies and machine intelligence. There are still improvements taking place toward developing a smart and portable neonatal incubator. These types of incubators help in data transmission to long distances.

REFERENCES

1. Rajalakshmi, A., K.A. Sunitha, and R. Venkataraman. 2019. A survey on neonatal incubator monitoring system. *Journal of Physics: Conference Series* 1362(012128): 1–8.
2. Sinclair, J.C. 2002. Servo-control for maintaining abdominal skin temperature at 36C in low birth weight infants. *Cochrane Database of Systematic Reviews* CD001074(1): 1–2.
3. Shaib, M., M. Rashid, L. Hamawy, M. Arnout, I. El Majzoub, and A.J. Zaylaa. 2017. Advanced portable preterm baby incubator. In *Proceedings of the Fourth International Conference on Advances in Biomedical Engineering*, Hadat-Beirut, Lebanon, 1–4.
4. Zaylaa, A.J., M. Rashid, M.Shaib, and I. El Majzoub. 2018. A handy preterm infant incubator for providing intensive care: simulation, 3D printed prototype, and evaluation. *Journal of Healthcare Engineering* 2018(8937985): 1–4.
5. Nisha, Z., and A. Elahi. 2014. Low-cost neonatal incubator with smart control system. In *Proceedings of the 8th International Conference on Software, Knowledge, Information Management & Applications*, Dhaka, Bangladesh.
6. Tisa, T.A., Z.A. Nisha, and M.A. Kiber. 2013. Design of an enhanced temperature control system for neonatal incubator. *Bangladesh Journal of Medical Physics* 5(1): 53–61.
7. Abdiche, M., G. Farges, S. Delanaud, V. Bach, P. Villon, and J.P. Libert. 1998. Humidity control tool for neonatal incubator. *Medical & Biological Engineering & Computing* 36(2): 241–45.

8. Costa, J.L., C.S. Freire, B.A. Silva, M.P. Cursino, R. Oliveira, A.M. Pereira, and F.L. Silva. 2009. Humidity control system in newborn incubator. In *XIX IMEKO World Congress, Fundamental and Applied Metrology*, 1760–64.

9. Silverman, W.A., and W.A. Blanc. 1957. The effect of humidity on survival of newly born premature infants. *Pediatrics* 20(3): 477–87.

10. Harpin, V.A., and N. Rutter. 1985. Humidification of incubators. *Archives of Disease in Childhood* 60(3): 219–24.

11. Lepcha, S., S.S. Jain, K. Sonal, C. Puneeth, N. Singhal, A.K. Saw, A.S. Batra, S. Shukla, and A. Unal. 2020. Smart portable neonatal intensive care for rural regions. In *ICT Analysis and Applications*. Lecture Notes in Networks and Systems, ed. S. Fong, N. Dey, and A. Joshi, 93, 375–85, Singapore: Springer.

12. Vignesh, T.C., J. Jai Kumar, R. Hari Krishna, and M. Krishna Raj. 2019. Neonatal incubator monitoring system using IoT. *International Journal of Advance Research, Ideas and Innovations in Technology* 5(2): 606–7.

13. Mittal, H., L. Mathew, and A. Gupta. 2015. Design and development of an infant incubator for controlling multiple parameters. *International Journal of Emerging Trends in Electrical and Electronics* 11(5): 65–72.

14. Shakunthala, M., R. Jasmin Banu, L. Deepika, and R. Indu. 2018. Neonatal healthcare monitoring incubator using IoT. *International Journal of Electrical, Electronics and Data Communication* 6(6): 59–64.

15. Sengeni, D. 2019. Implementation of IoT based smart incubator unit in LabVIEW environment. *International Journal of Advanced Research in Management, Architecture, Technology and Engineering* 5(11): 1–7.

16. Sowmiya, S., V. Smrithi, and G. Irin Loretta. 2018. Monitoring of incubator using IoT. *International Research Journal of Engineering and Technology* 5(4): 635–38.

17. Sivamani, D., R. Sagayaraj, R. Jai Ganesh, and A. Nazar Ali. 2018. Smart incubator using internet of things. *International Journal for Modern Trends in Science and Technology* 4(9): 23–27.

18. Romansky, R. 2017. A survey of digital world opportunities and challenges for user's privacy. *International Journal on Information Technologies and Security* 9(4): 97–112.

19. Jabbar, W.A., H.K. Shang, S.N. Hamid, A.A. Almohammedi, R.M. Ramli, and M.A. Ali. 2019. IoT-BBMS: internet of things-based baby monitoring system for smart cradle. *IEEE Access* 7: 93791–93805.

20. Malik, A., S. Gautam, S. Abidin, and B. Bhushan. 2019. Blockchain technology-future of IoT: including structure, limitations and various possible attacks. In *Proceedings of the 2nd International Conference on Intelligent Computing, Instrumentation and Control Technologies*, Kannur, Kerala, India, 1100–4.

21. Neshenko, N., E. Bou-Harb, J. Crichigno, G. Kaddoum, and N. Ghani. Demystifying IoT security: an exhaustive survey on IoT vulnerabilities and a first empirical look on internet-scale IoT exploitations. *IEEE Communication Surveys & Tutorials* 21(3): 2702–33.

22. Hassija, V., V. Chamola, V. Saxena, D. Jain, P. Goyal, and B. Sikdar. 2019. Survey on IoT security: application areas, security threats, and solution architectures. *IEEE Access* 7: 82721–43.

23. Shin, D., K. Yun, J. Kim, P.V. Astillo, J.-N. Kim, and I. You. 2019. A security protocol for route optimization in DMM-based smart home IoT networks. *IEEE Access* 7: 142531–50.

24. Sharma, T., S. Satija, and B. Bhushan. 2019. Unifying blockchain and IoT: security requirements, challenges, applications and future trends. In *Proceedings of the International Conference on Computing, Communication, and Intelligent Systems*, Greater Noida, India, 341–46.

25. Wu, H., D. Sun, L. Peng, Y. Yao, J. Wu, Q.Z. Sheng, and Y. Yan. 2020. Dynamic edge access system in IoT environment. *IEEE Internet of Things Journal* 7(4): 2509–20.

26. Mbarek, B., M. Ge, and T. Pitner. 2020. An efficient mutual authentication scheme for Internet of Things. *Internet of Things* 9: 100160.

27. Shanmathi, N., and M. Jagannath. 2018. Computerised decision support system for remote health monitoring: a systematic review. IRBM 39(5): 359–67.

28. Arora, A., A. Kaur, B. Bhushan, and H. Saini. 2019. Security concerns and future trends of internet of things. In *Proceedings of the 2nd International Conference on Intelligent Computing, Instrumentation and Control Technologies*, Kannur, Kerala, India, 891–96.

6 Malware Threat Analysis of IoT Devices Using Deep Learning Neural Network Methodologies

Moksh Grover and Nikhil Sharma
HMR Institute of Technology & Management

Bharat Bhushan
School of Engineering and Technology, Sharda University

Ila Kaushik
Krishna Institute of Engineering & Technology

Aditya Khamparia
Lovely Professional University

CONTENTS

6.1   INTRODUCTION

The interrelationship between materialistic mobile objects installed with sensors, electronic chips, and other hardware forms called "Things" using the internet is known as the Internet of Things (IoT) [1]. Every machine can be controlled and audited remotely as they interact with other linked nodes and are uniquely identified globally by the use of RFID or Radiofrequency-Identifier Tags [2]. We are granted universal connectivity to a boundless range of cloud computing services, service industries [3], smart physical objects, and applications. E-banking, e-shopping, education system, management industry, healthcare, smart cities, entertainment, and protection of human beings are the various uses of technologies powered by the IoT [4]. IBM previously stated that by the year 2020 the sum of devices linked over the Internet is assumed to rise to 50,000 million [5]. It will boost "Big Data" that can be distributed through cloud services in addition to an increment in the amount of connections between smart objects, which, in turn, will increase the tally of communication net [6].

The constant availability of network can lead to an open attack on IoT devices. Pirated software and malware infection can easily target industrial IoT cloud for harmful usage along with compromising the security [7]. Development of software by disguising the illegal reuse of someone else's work such as source codes and presenting it as their original product is known as "software piracy" [8]. By utilizing reverse engineering practices, the programmer may duplicate the logic of the original software and then in another type of source code produce the same logic [9]. It gives passage to unrestricted downloads to open source codes, pirated software, as well as popularize and promote the pirated variations of software, which is a serious threat for cyberspace security [10]. It upsurges rapidly every year and causes huge economic losses to all software corporations [11]. Every year there are business damages of up to 52.2 billion dollars as a result of public software piracy rate of approximately 39%, as stated by the 2016 report of Business Software Alliance (BSA). It has been shown by various researches that every software is made up of source code that is plagiarized in the context of logic within a 5%–20% range [12]. To identify the stolen source codes used to develop pirated software, various resourceful software plagiarism approaches are required. Source code similarity identification, software birthmark investigation, clone detection, and software bug analyses are several proposed plagiarism detection systems which are mainly text-based and employ structure analysis [13]. The structure-based analysis technique inspects the primary structure of syntax trees, source codes, function call graph of subroutines, and graph behavior. Consequently, it is restricted to a specific programming language structure. Therefore, it is hard to catch a programmer if the logic of software is reused to create a program in a different programming language due to different structure behavior. To protect and secure various smart devices, we may use the Industrial Internet of Things (IIoT) by scheming various intelligent malware detection and software plagiarism techniques.

Because of the expanding number of IoT networks, malicious attacks are more unexpected at present. Malware infections are generally outlined to corrupt the confidentiality of smartphones, IoT nodes, and computer systems connected over the Internet [14]. To detect and identify windows-based malware, several scanning techniques are purposed by testing specific signatures. Dynamic and static-based analysis are the two types of malware identification analysis methods [15]. There is no need to execute source codes in real time during static malware analysis. To capture format information of malware executable binaries, we may use static malware analysis. The signature-based methods of malware recognition such as opcode frequency, CFG, string signature, and N-grams are static-based. To divulge the executables, disassembling tools, such as OllyDbg and IDA Pro [16], are applied before implementing static-based algorithms. To excerpt the hidden patterns from the binary executables, we use these disassembling tools, which are, in turn, used to fetch encoded string from these binary executables. Types of static-based analysis include the byte sequence technique and the function call graph [17].

In a real-time virtual environment, dynamic analysis learns the patterns of malware files while executing the codes [18]. Function calls, information flow, virtual analysis of codes, instruction tracing, and function parameter analysis are various practices that can be used to observe malicious behavior [19]. Because the dynamic analysis method monitors the dynamic behavior of source codes, it is more time consuming [20]. To investigate the dynamic behavior of malicious codes, there are several automated tools available online such as TT Analyzer, Anubis, and CW Sandbox.

The rest of the chapter is organized as follows. Section 6.2 contains a literature review which discusses previous studies. Section 6.3 shows the challenges faced while handling malware and software piracy threat detection. Section 6.4 explains the engineering model designed around threat handling methodologies. Section 6.5 describes the implementation process. Section 6.6 illustrates the results obtained from our implementation followed by the conclusion of the paper in Section 6.7.

6.2 LITERATURE REVIEW

To reduce time and cost and achieve significant identification performance, further studies have been conducted for the productive detection of malware and software plagiarism threats in the IoT environment. An impression of various aspects of software plagiarism detection can be observed in distinct studies where most of the preceding production are completed on a single programming language, that is, the present literature is valuable if a coder can generate source code in a programming language by altering its logic to another data structure.

Yasaswi et al. [21] proposed to extract similarity features based on compiler infrastructure from intermediate code generation. Furthermore, to measure the similarity between the source codes, unsupervised learning was used, whereas plagiarism was detected by similar functionalities depicted by contrasting source codes. In Ref. [22], in java source codes, software benchmark was used to compare the source codes to compute the similarities for threat detection. By running source codes, it captures its structural characteristics by selecting the control flow information. In Ref. [23], plagiarism in student's assignments was identified using the latent semantic analysis where it was combined with PlaGate. This combination was used to audit the linguistic parallels between various documents. Based on the parse tree, a syntax tree was drawn from any given source code. On the basis of their syntax tree, different source codes could be compared. Cosma et al. [24] developed a Source Forager search engine that fetched various properties from the code example, such as functionalities between C++ and C codes as a feedback to the user questions, and processed them in the shape of "k" number of functionalities from the corpus. The developed software could detect software resemblance and the logic was the conceptual structure of the program. Kashyap et al. [25] extracted similar texts using the parse tree kernel method between various java source codes. In core functionalities, there were irregular variations of nodes due to which this technique did not produce a better outcome. Therefore, to extract the resemblance between various source codes, the fingerprinting method was designed. In Ref. [26], to compute the behavior of dissimilarity between various source codes, a logic-based approach was employed. To obtain semantics for dissimilarities from execution paths, symbolic execution and precondition reasoning were used, and if there were no dissimilarities, then it originated in the plagiarism problem. A detailed summary of the types of analysis for malware detection is explored in the subsections below.

6.2.1 STATIC ANALYSIS

These types of analysis techniques usually consist of feature extraction from binary files using static means, that is, using binary data extraction tools. Zhang et al. [27] proposed an approach applying machine learning using SVM classifier based on N-Opcode sequences for malware detection with an accuracy of 98%. For malware identification process, critical instruction sequence techniques and cosine similarity were used, and opcodes were the machine code mnemonics.

It was shown that some common core signatures are shared by every interpretation of similar malware family and can use API call sequences to capture them. In Ref. [28], based on machine learning, the author could generate a classifier using classify worms from binaries of benign files and sequence of variable length instructions with an accuracy of 96% while working on a dataset consisting of 1330 benign files and 1444 worm files. Siddiqui et al. [29] proposed an obfuscation scheme to test the constraints faced while using the static analysis approach. The static analysis independently is not satisfactory enough for the effective analysis of malware samples based on experimental results.

6.2.2 DYNAMIC ANALYSIS

If the malware is obfuscated or packed, we can easily avoid the static analysis approach. Therefore, during analysis, some potent behavioral features are mandatory, and such types of malware detection are known as the dynamic analysis method. To extract features within a virtual machine, such a concept is commonly established according to the execution of binary samples in a controlled environment. In Ref. [30], based on their execution behavior, large malware samples were grouped or divided into classes of clusters. This cluster-based system augmented the Anubis system for tainted tracking and additional network analysis. For selected malware samples, this system produces an automated truce report. Their augmented system was more abstractly able to portray various activities of the program. To identify the malicious behavior which was not detected beforehand by a benign application within the program, Christodorescu et al. [31] proposed a manual process for malware analysis. This approach can provide significant insights as it only provides us with a limited amount of information in regard to malicious behavior to understand it.

6.2.3 HYBRID ANALYSIS

Hybrid techniques were brought in to overcome the matter of computation time as well as the limitations of static approaches to improve malware detection systems. Results show that rather than executing static and dynamic approaches independently, hybrid approaches perform much better. Santos et al. [32] extracted features from malware samples using both static and dynamic analysis to train and generate a malware classifier named OPEM. To train their classification model as dynamic features, they used system calls, execution traces of executable files, and exceptions; and for static features, they used frequency of occurrence of operation codes.

6.2.4 VISUALIZATION ANALYSIS

To boost the results given by a classifier, numerous studies have been directed toward the visualization of malware features accounting for the reduction in resource overhead, size, and time consumed. For example, Kumar et al. [33] devised a deep learning model with 98% accuracy for 9339 samples for malware detection. Kalash et al. [34] developed a convolutional neural network (CNN) and image-based malware classification method, in which during testing they separated only 10% of samples in a family and achieved an accuracy of 98.52%. In Ref. [35], with an accuracy of 94.5% accuracy, Cui et.al. proposed a CNN-based malware classification method.

6.3 CHALLENGES

Various challenges faced while handling malwares and mitigating software piracy are explored in the subsections below.

6.3.1 MALWARE THREAT DETECTION

High computational cost is required in regard to texture feature mining practiced by applying malware visualizations even though code obfuscation concerns may be solved using conventional methods [36]. Though these types of feature extraction practices do not function adequately with a considerable amount of malware data analysis, we have proposed malware detection methods that tackle the following queries: as malware extraction is a costly process, how to extract malware features and maintain low computational cost? How to improve our accuracy while processing big malware datasets? How to distinguish between malware with reduced overhead and goodware? Over time, detection becomes more challenging as new malware is constantly generated, updated, and manipulated [37].

6.3.2 SOFTWARE PIRACY THREAT DETECTION

Authorship rights of software and intellectual digital property are hard to sustain because they are not physical due to global accessibility on the Internet [38]. Currently, one-third of the installed software applications are pirated globally. The programmer may crack and redesign the logic of original software into another type of programming language's source code. One can get paid for the ability to provide solutions to the real world but cannot get paid for his/her own ideas [39]. MoHCA-JAVA is an example of various tools available to the code-crackers that can translate one type of source code into the source code of another type of programming language as each programming language has different semantic structures and syntax, which is the reason that it is really difficult to detect the code-cracker's malicious activities in cross-domain source codes [40].

6.4 ENGINEERED MODEL

Figure 6.1 shows an engineered model for protection measures and threats of cybersecurity in the IIoT. A high time cost is required to process a colossal amount of data. Therefore, for the following engineered model in the cloud storage, we deploy four databases – new detected malware attack's signatures are stored in

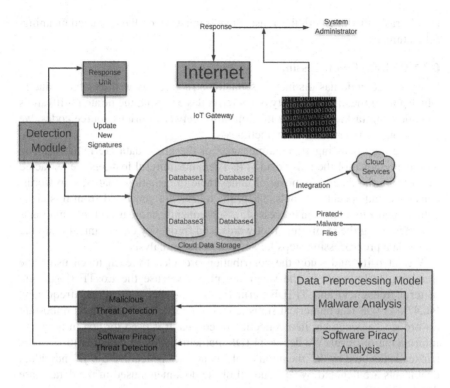

FIGURE 6.1 IoT – Cybersecurity model.

the first database; the second one consists of the raw network traffic; the third one stores a list of previous datasets; and in the fourth database the cracker or programmer stores pirated versions of the software provided by IoT devices. The raw data files stored in the second database are passed through the preprocessing data module which is further passed on to the detection module [41]. By learning from data of various signatures provided by databases one, three, and four, the detection modules capture malware and pirated software attacks. The administrator is warned by the system to take relevant action if in the given network any malicious activity is detected [42].

6.4.1 SOFTWARE PIRACY THREAT DETECTION METHODOLOGY

A deep learning methodology is designed to detect pirated versions of the software among a number of source codes. The cracker creates a pirated copy of the software using the logics used in the original software, which is known as the plagiarized version of the software [43]. Initially, in the first step, to extract meaningful features and to reduce the data's dimensions, the source codes are broken down into a number of tokens. In the second step, to identify the plagiarism between various source codes, a deep learning algorithm is applied, which uses Keras API along with the TensorFlow framework on the extracted meaningful features.

6.4.1.1   Data Preprocessing

Every source code has distinctive semantic structures as well as syntax due to which amidst the diverse variety of source codes exposing the pirated software is a challenging task. To identify the similarity between various source codes, we use various software plagiarism methods.

In the preprocessing steps, dimensions of the given data are reduced along with the removal of the noisy data [44]. Then, meaningful features are extracted by cleaning the data by removal of undesirable information, that is, stop words, constants, and special symbols. Then, the tokens are generated from this clean data using the tokenization process [45]. Subsequently, more useful information is extracted using the stemming, root words, and frequency constraints. Figure 6.2 shows data preprocessing steps for software piracy analysis.

We intensify and study the contribution provided by each token using the weighting techniques. In the weighting phase, we use the LogTF (logarithm of term frequency) and TFIDF (term frequency–inverse document frequency) [46,47]. TFIDF is a numerical statistic that is intended to reflect how important a word is to a document in a collection or corpus. It is used for text mining and information retrieval. We define TFIDF mathematically in equation (6.1) where a token is denoted by "t," the number of frequencies is denoted by "f," individual documents are denoted by "d," and all of the documents used in the dataset are denoted by "D."

$$tfidf\left(t, d, D\right) = tf\left(t, d\right) \times idf\left(t, D\right) \tag{6.1}$$

6.4.1.2   TensorFlow and Keras API – Deep Learning

In a complex environment for performing high-level computations, a machine learning system called "TensorFlow" can be used. Various types of deep and machine learning algorithms [48] can be implemented by calling various APIs of TensorFlow in a program. For training data, complex computations, and supervising the state run of each function, a number of different layers of TensorFlow

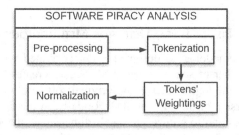

FIGURE 6.2   Software piracy analysis – data preprocessing.

can be configured [49]. To identify various source codes that are identical and are created using various types of programming languages, an in-depth approach is designed using the TensorFlow framework [50]. Then, to identify the pirated software the similar extracted codes are used. For the input values to the deep learning model, weighting values are used [51]. Then, the fully connected layer or the dense layer is configured for input as well as output data. A total of 100, 50, and 30 neurons are used to configure three dense layers, where the first layer is used as an input layer that receives the data with an input shape parameter. The current layer passes on the information to the next layer, and thus can be said as densely connected. While the fourth and final layer is used as an output layer which produces an output variable that is used to point out and distinguish between an original and a plagiarized code. Figure 6.3 shows the software piracy analysis process in deep learning.

Using the dropout layer we solve the overfitting problem along with enhancing the deep learning approach in the context of the optimizer, loss function, activation function, and the learning error rate. In the deep learning model for compiling and optimizing, the "Adam optimizer" is used, and to get the patterns of the data received through input variables, "ReLu" activation method is used, which is shown mathematically in the equation (6.2), where the input to equivalent neurons is denoted by "x."

$$f(x) = x^+ = \max(0, x) \tag{6.2}$$

The sigmoid function defined mathematically in equation (6.3) is a logistical method which is used to grip the multiclass problem, where the sigmoid function is denoted by "S,"

$$s(x) = \frac{1}{1 + e^{-x}} \tag{6.3}$$

For each limitation, the Adam optimizer calculates discrete adaptive learning rates. In equations (6.4) and (6.5), the decaying means of pas squared gradients

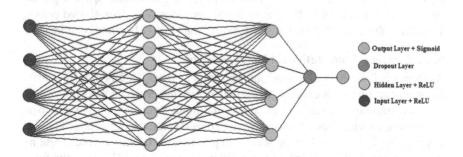

FIGURE 6.3   Software piracy analysis – deep learning.

are shown, where "m_t" denotes the predictable means of first instant gradient and "v_t" denotes the predictable means of second instant gradient.

$$m_t = \beta_1 m_{t-1} + (1 - \beta_1) g_t \qquad (6.4)$$

$$v_t = \beta_2 v_{t-1} + (1 - \beta_2) g^2 \qquad (6.5)$$

6.4.2 MALWARE THREAT DETECTION METHODOLOGY

The steps for malware threat detection are discussed below.

6.4.2.1 Data Preprocessing

To transform the malware detection problem from the raw binary files, color images are generated using the image classification problem because when compared to grayscale images, the color images can retrieve better features, as grayscale images only consist of 256 colors. Any reverse engineering tools for instance disassembler and decompiler are not required for this method. For classifying malware families, the improved features of malware images can perform exceptionally well.

Machine learning algorithms based on malware detection techniques offered better results using grayscale images [52]. To classify malware types, feature extraction techniques were used after grayscale images were generated from the colored images. To decrease the feature set, we use feature reduction methods, which, in turn, helps to boost classification performance. Machine learning generates exponential values when used with colored images [53]. Therefore, for malware threat detection, machine learning algorithms are not a better choice. Figure 6.4 shows the malware analysis process in data preprocessing.

As the deep learning methods use filters to automatically reduce noise, these types of methods outperform when used with big malware datasets. Thus, deep learning techniques generate better results when used with colored images. First, from raw binary files hexadecimal streams (0–15) are produced. Then in the second stage, these hexadecimal streams are divided into a chunk of eight-bit vector. Each of these eight-bit segments is measured as an unsigned integer value ranging from 0–255. Two-dimensional matrix spaces are generated from each of these eight-bit vectors in the third stage. Fourth, from two-dimensional space, eight-bit integers are generated which are plotted with red, blue, and green-shaded colors. These four phases combine to transform the malware binary files into color images.

6.4.2.2 Deep Convolutional Neural Network

As shown in Figure 6.5 to conduct malware data analysis, DCNN or deep convolutional neural network contains five modules. For the designed model of neural network, training images are passed into the input layer. Initially, to give better signal characteristics and decrease the noise the convolutional layer is used. The pooling layer is used in the second step to reduce the data overhead retaining useful information. Third, to convert the 2D array into a 1D array, the fully

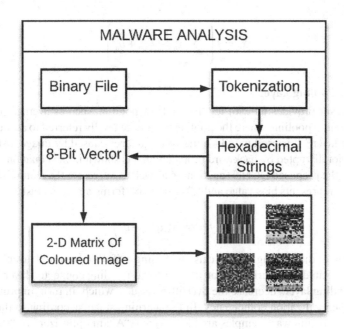

FIGURE 6.4 Malware analysis – data preprocessing.

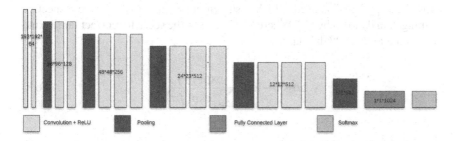

FIGURE 6.5 Malware analysis – deep CNN.

connected layer is used and is input into the specific classifier. Using the classifier, the malware families are identified from the respective images in the fourth step.

6.4.2.3 Convolution Layer

Image of parameters is reduced using the convolutional layer, which, in turn, helps to extract meaningful features. It offers the generalization approach to primary architecture and decreases the overfitting problem. As shown in equation (6.6), the input of the convolutional layer consists of a handful number of maps; where "b_n^{l}" is the "mth" feature map's bias consistent; "M_n" denotes the clusters of given maps; "k_{mn}^{l}" is utilized for connecting the "mth" input feature map to the "nth" feature map, which is used to define the convolution kernel. "b_n^{l}" is also the activation function.

$$x_n^l = f\left(\sum_{m \in M_n} x_n^{l-1} * k_{mn}^l + b_n^l\right) \tag{6.6}$$

6.4.2.4 Pooling Layer

Pooling layer provides us with few methods of pooling, such as average pooling and maximum pooling, where the pooling layer is generally referred to as the "subsampling layer." It is used to reduce the consequences caused by image distortion and it is not disrupted by backward propagation. As shown in the equation (6.7), it increases the proposed DCNN functioning as well as decreases the feature's factor, where "b" represents bias value and "." or downperforms a pooling task.

$$x_n^l = f\left(\text{down}\left(x_n^{l-1}\right) + b_n^l\right) \tag{6.7}$$

The output of the pooling layer is classified using a "fully connected layer." Every neuron is related to the prior neuron using a corresponding connected layer. These corresponding layers decrease the overfitting issues, which, in turn, improves the model's generalization competency. In the learning stage, according to the family names, the malware samples are categorized. "Adam optimizer" or "stochastic descent gradient" is used which minimizes the acquired losses as the model learns the parameters of the training data. To train this DCNN model, "Softmax-Cross-Entropy-Loss" is used, which is shown in equation (6.8), for the respective training data denoted by "k." Where "fzt" denotes the score for correct family and "fzt" is the rank for "kth" class.

$$\text{Loss} = -\log\left(\frac{\exp\left(fzt\right)}{\sum_k \exp\left(fzt\right)}\right) \tag{6.8}$$

6.5 IMPLEMENTATION

In this study, based on our knowledge of deep learning, we have proposed the following two neural network methodologies for malware threat detection.

6.5.1 CONVOLUTIONAL NEURAL NETWORK

Here, we propose a CNN-based model in which we work on two datasets of 32×32 (1024 pixels)-sized images of malware and goodware denoted by specific hash values, as shown in Figure 6.6. On this dataset of images, we apply convolutional layers with activation functions such as "ReLu" and "Sigmoid" along with layers of MaxPooling. The purpose of these convolutional layers is to create a feature map by finding features in images using feature detectors. The purpose of the pooling layers is to preserve features from any kind of distortions. Then, we apply the flatten layer that reduced all of the input dimensions into one dimension.

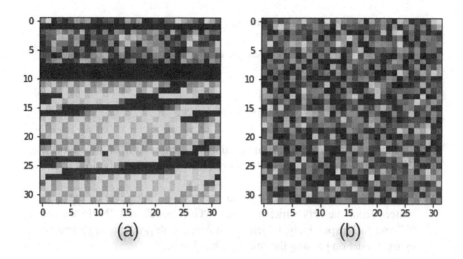

FIGURE 6.6 (a) Malware image – hash value – e44fea4913fc9fd91b8b07c4670aeac4. (b) Goodware image – hash value – 7005e2d92aebcd7164b2b270a8ba22fb.

This flattened matrix is passed through a fully connected layer for classification between malware and goodware. In total, 1204 pixel positions and values of the images of malware and goodware are provided as input by extracting them from the images. Moreover, the output obtained after prediction is compared with the class indices of malware and goodware generated from the training set.

CNN Model Algorithm

1. We set "Keras Sequential Model" as the classifier.
2. The input layer or first convolutional layer takes in the following as its arguments: (a) input shape, (b) output dimensions, and (c) activation function.
3. This convolutional layer is accompanied by a "MaxPooling" layer with pool size=(2,2) as an argument. In this layer, we get the data with the maximum features.
4. The third layer consists of another convolutional layer followed by a MaxPooling layer as a fourth layer again with pool size=(2,2) as an argument.
   - In the above convolutional layers, that is, input layer and third layer the activation function applied is "Rectified Linear Unit" or "ReLu." The property of ReLu is that all of the neurons are not activated at the same time, that is, $R(X) = \max(0, X)$, If, $X != 0$
   - The neuron is not activated, that is, $R(X)=0$, If, $X < 0$.
5. The data obtained from the fourth layer is flattened and passed on to the next layer.

6. The final layer is the fully connected layer, that is, output layer with output dimensions reduced to "2."
   - The activation function used by output layer is "Sigmoid" function which is denoted by $f(x)$,

$$f(x) = \sigma(x) = \frac{1}{1+e^{-x}}$$

7. Further, we use Adam optimizer during compilation of this model. When the dataset is trained, Adam optimizer optimizes the loss in the machine being trained.
8. The model is then fitted with training and test sets. We generate the class indices from the sets which represent "Goodware" and "Malware" as "0" and "1," respectively, which is compared with predictions generated by the model on passing the image to be classified.

6.5.2 ARTIFICIAL NEURAL NETWORK (ANN)

Here we propose an ANN-based model in which we work on a categorized dataset of malware and goodware. As shown in Table 6.1, the dataset consists of 51,960 rows of hash values, as well as pixel position values along with classification value denoting malware as "1" and goodware as "0." On this dataset, we apply an input layer, with few hidden layers, and an output layer with activation functions such as "ReLu" and "Sigmoid."

This dataset is split up into training and test set using SkLearn's Train–Test–Split. The train values are used to train the model, whereas the test values are used to test and calculate the accuracy of our model. The pixel position values are passed as input values to the input layer which has input dimensions of 1024, and the classification value of malware and goodware is set as the target value. The output values obtained from the output layer are stored as the predictions and are compared with the actual values of the test target values and accuracy is calculated.

TABLE 6.1

ANN Malware–Goodware Dataset

S.no	Hash	pixel_0	pixel_1	...	pixel_1023	Malware
1.	b324140e1fb35dc6b694879ba1f2be45	15	15		122	1
2.	1d32b1326a524b163eb74af645cd34d5	234	196		193	1
3.	e44fea4913fc9fd91b8b07c4670aeac4	196	255		233	1
4.	ba04584917d498daf6b054f1476e93e1	199	15		204	0
5.	de6fb739ae97e914affa2b857aa04e48	2	155		0	0

ANN Model Algorithm

1. We set "Keras Sequential Model" as the classifier.
2. The input layer, that is, the first layer takes in the following as its arguments: (a) input shape, (b) output dimensions, and (c) activation function.
3. This is followed by four hidden layers that take activation function and output dimensions to the next layer as its arguments.
   - In the above layers, the activation function applied is "Rectified Linear Unit" or "ReLu." The property of ReLu is that all of the neurons are not activated at the same time, that is, $R(X) = \max(0, X)$, If, $X != 0$
   - The neuron is not activated, that is, $R(X) = 0$, If, $X < 0$.
4. The final layer is the output layer that receives the data from previous layer and provides the output with output dimension set to "1."
   - The activation function used by the output layer is "Sigmoid" function which is denoted by $f(x)$,

$$f(x) = \sigma(x) = \frac{1}{1 + e^{-x}}$$

5. Further we use Adam optimizer during compilation of this model. When the dataset is trained, Adam optimizer optimizes the loss in the machine being trained.
6. The model is then fitted with training and test sets and predictions are generated.

6.6 RESULT

Upon performing the two proposed methodologies in Section 6.5, we obtained the following results.

6.6.1 CNN

Training of the dataset of images was done using CNN in which multiple layers were added for better predictions. Our CNN model makes the predictions and classifies the received image as either "Malware" or "Goodware." The testing image is received and is first converted to an array, and then its dimensions are expanded before classification is done. Subsequently, the output received is used to decide whether the image detected is either a "Malware" or a "Goodware" with a prediction accuracy of 94.75%. The efficiency of our model can be viewed in Figure 6.7, which is the relation between the epochs and the accuracy of the model during training.

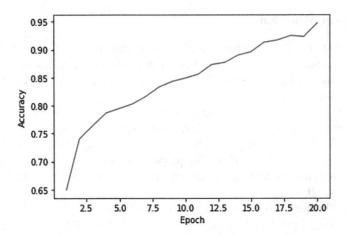

FIGURE 6.7   Epochs versus training accuracy of the CNN model.

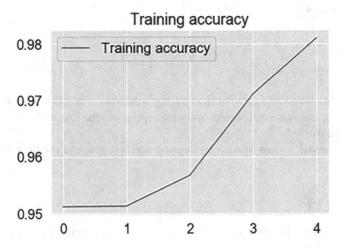

FIGURE 6.8   Epochs versus training accuracy of the ANN model.

6.6.2   ANN

Training for the dataset of pixel positions was done using ANNs. In this, we have used multiple hidden layers for better prediction and accuracy. The data is split up in test and training data. After the training is done the predictions are stored and compared with the actual data.

The efficiency of our model can be measured from the following images, where Figure 6.8 depicts the relation between epochs and training accuracy and Figure 6.9 depicts the relation between epochs and training loss.

Our model makes the prediction and classifies the data as that of "Malware" or "Goodware" with an accuracy score of 94.62%, which is also depicted as the confusion matrix in the image below, where TP signifies "True Positive," that

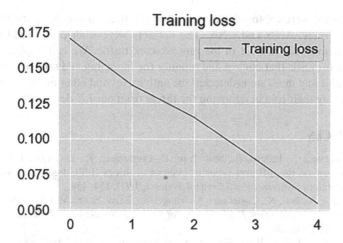

FIGURE 6.9 Epochs versus training loss of the ANN model.

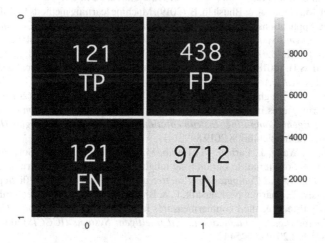

FIGURE 6.10 Confusion matrix of the ANN model.

is, you predicted positive and it was true; TN signifies "True Negative," that is, you predicted negative and it was true; FP signifies "False Positive," that is, you predicted positive and it was false; FN signifies "False Negative," that is, you predicted negative and it was false. According to confusion matrix, shown in Figure 6.10, TP= 121, FP=438, FN= 121, TN=9712; therefore, overall, we can say that 9833 predictions were right out of 10,392 predictions.

6.7 CONCLUSION

In this chapter, the issue of open malware attacks and software piracy threats on IoT devices has been investigated. Various types of datasets of different malware and goodware were captured, classified, and predicted using two deep learning methods, that is, ANN and CNN. The datasets were studied and used to recognize the malware present in the raw network traffic. We can implement our models as firewalls that control and monitor the incoming and outgoing network traffic, and if any malware is detected the antiviruses and other malware removal tools can play their role in removing the threat from our IoT devices.

REFERENCES

1. Srinivasan, C., Rajesh, B., Saikalyan, P., Premsagar, K., & Yadav, E.S. (2019). A review on the different types of internet of things (IoT). *Journal of Advanced Research in Dynamical and Control Systems, 11*(1), 154–158.
2. Zanella, A., Bui, N., Castellani, A., Vangelista, L., & Zorzi, M. (2014). Internet of Things for smart cities. *IEEE Internet of Things Journal, 1*(1), 22–32. doi: 10.1109/jiot.2014.2306328
3. Joyia, G.J., Liaqat, R.M., Farooq, A., & Rehman, S. (2017). Internet of Medical Things (IOMT): Applications, benefits and future challenges in healthcare domain. *Journal of Communications.* doi: 10.12720/jcm.12.4.240-247
4. Jindal, M., Gupta, J., & Bhushan, B. (2019). Machine learning methods for IoT and their Future Applications. In *2019 International Conference on Computing, Communication, and Intelligent Systems (ICCCIS).* doi: 10.1109/icccis48478.2019.8974551
5. Jabbar, S., Malik, K.R., Ahmad, M., Aldabbas, O., Asif, M., Khalid, S., Han, K., & Ahmed, S.H. (2018). A methodology of real-time data fusion for localized big data analytics. *IEEE Access, 6*, 24510–24520.
6. Sharma, A., Singh, A., Sharma, N., Kaushik, I., & Bhushan, B. (2019). Security countermeasures in web based application. In *2019 2nd International Conference on Intelligent Computing, Instrumentation and Control Technologies (ICICICT).* doi: 10.1109/icicict46008.2019.8993141
7. Ullah, F., Wang, J., Farhan, M., Habib, M., & Khalid, S. (2018). Software plagiarism detection in multiprogramming languages using machine learning approach. *Concurrency and Computation: Practice and Experience.* doi: 10.1002/cpe.5000
8. Varshney, T., Sharma, N., Kaushik, I., & Bhushan, B. (2019). Architectural model of security threats & their countermeasures in IoT. In *2019 International Conference on Computing, Communication, and Intelligent Systems (ICCCIS).* doi: 10.1109/icccis48478.2019.8974544
9. Chae, D.-K., Ha, J., Kim, S.-W., Kang, B., & Im, E.G. (2013). Software plagiarism detection. In *Proceedings of the 22nd ACM International Conference on Conference on Information & Knowledge Management – CIKM 13.* doi: 10.1145/2505515.2507848
10. Akbulut, Y., & Dönmez, O. (2018). Predictors of digital piracy among Turkish undergraduate students. *Telematics and Informatics, 35*(5), 1324–1334. doi: 10.1016/j.tele.2018.03.004
11. Goyal, S., Sharma, N., Kaushik, I., Bhushan, B., & Kumar, A. (2020). Precedence & issues of IoT based on edge computing. In *2020 IEEE 9th International Conference on Communication Systems and Network Technologies (CSNT).* doi: 10.1109/csnt48778.2020.9115789
12. Jaidka, H., Sharma, N., & Singh, R. (2020). Evolution of IoT to IIoT: Applications & challenges. *SSRN Electronic Journal.* doi: 10.2139/ssrn.3603739
13. Mythili, S., & Sarala, S. (2015). A measurement of similarity to identify identical code clones. *The International Arab Journal of Information Technology, 12*, 735–740.

14. Ragkhitwetsagul, C. (2016). Measuring code similarity in large-scaled code corpora. In *2016 IEEE International Conference on Software Maintenance and Evolution (ICSME)*. doi: 10.1109/icsme.2016.18

15. Imran, S., Khan, M.U., Idrees, M., Muneer, I., & Iqbal, M.M. (2018). An enhanced framework for extrinsic plagiarism avoidance for research article. *Technical Journal, 23*(1), 84–92.

16. Vukasinovic, N., & Duhovnik, J. (2019). Introduction to reverse engineering. In *Advanced CAD Modeling. Springer Tracts in Mechanical Engineering.* Cham: Springer. doi: 10.1007/978-3-030-02399-7_7

17. Gandotra, E., Bansal, D., & Sofat, S. (2014). Malware analysis and classification: A survey. *Journal of Information Security, 5*(2), 56–64. doi: 10.4236/jis.2014.52006

18. Shabtai, A., Moskovitch, R., Elovici, Y., & Glezer, C. (2009). Detection of malicious code by applying machine learning classifiers on static features: A state-of-the-art survey. *Information Security Technical Report, 14*(1), 16–29. doi: 10.1016/j. istr.2009.03.003

19. Egele, M., Scholte, T., Kirda, E., & Kruegel, C. (2012). A survey on automated dynamic malware-analysis techniques and tools. *ACM Computing Surveys, 44*(2), 1–42. doi: 10.1145/2089125.2089126

20. Ghafir, I., Saleem, J., Hammoudeh, M., Faour, H., Prenosil, V., Jaf, S., … Baker, T. (2018). Security threats to critical infrastructure: The human factor. *The Journal of Supercomputing, 74*(10), 4986–5002. doi: 10.1007/s11227-018-2337-2

21. Yasaswi, J., Kailash, S., Chilupuri, A., Purini, S., & Jawahar, C. V. (2017). Unsupervised learning based approach for plagiarism detection in programming assignments. In *Proceedings of the 10th Innovations in Software Engineering Conference on ISEC 17*. doi: 10.1145/3021460.3021473

22. Lim, H.-I., Park, H., Choi, S., & Han, T. (2009). A method for detecting the theft of Java programs through analysis of the control flow information. *Information and Software Technology, 51*(9), 1338–1350. doi: 10.1016/j.infsof.2009.04.011

23. Cosma, G., & Joy, M. (2012). An approach to source-code plagiarism detection and investigation using latent semantic analysis. *IEEE Transactions on Computers, 61*(3), 379–394.

24. Kashyap, V., Brown, D.B., Liblit, B., Melski, D., & Reps, T.W. (2017). Source forager: A search engine for similar source code. ArXiv, abs/1706.02769.

25. Son, J.W., Noh, T., Song, H., & Park, S. (2013). An application for plagiarized source code detection based on a parse tree kernel. *Engineering Applications of Artificial Intelligence, 26*, 1911–1918.

26. Zhang, F., Wu, D., Liu, P., & Zhu, S. (2014). Program logic based software plagiarism detection. In *2014 IEEE 25th International Symposium on Software Reliability Engineering*, 66–77. doi: 10.1109/issre.2014.18

27. Kang, B., Yerima, S.Y., McLaughlin, K., & Sezer, S. (2016). N-opcode analysis for android malware classification and categorization. In *2016 International Conference on Cyber Security and Protection of Digital Services (Cyber Security)*, 1–7. doi: 10.1109/cybersecpods.2016.7502343

28. Siddiqui, M., M.C. Wang, & J. Lee (2009). Detecting internet worms using data mining techniques. *Journal of Systemics, Cybernetics and Informatics, 6*(6), 48–53.

29. Moser, A., C. Kruegel, & E. Kirda (2007). Limits of static analysis for malware detection. In *Twenty-Third Annual Computer Security Applications Conference (ACSAC 2007)*. IEEE. doi: 10.1109/acsac.2007.21

30. Bayer, U., Comparetti, P.M., Hlauschek, C., Krügel, C., & Kirda, E. (2009). *Scalable, Behaviour-Based Malware Clustering.* New York: NDSS.

31. Christodorescu, M., S. Jha, & C. Kruegel (2007). Mining specifications of malicious behavior. In *Proceedings of the 1st Conference on India Software Engineering Conference – ISEC '08*. ACM. doi: 10.1145/1342211.1342215

32. Santos, I., J. Nieves, & P.G. Bringas (2011). Semi-supervised learning for unknown malware detection. In *Advances in Intelligent and Soft Computing International Symposium on Distributed Computing and Artificial Intelligence*, 415–422. Springer. doi:10.1007/978-3-642-19934-9_53

33. Kumar, R., Zhang, X., Khan, R.U., Ahad, I., & Kumar, J. (2018). Malicious code detection based on image processing using deep learning. In *Proceedings of the 2018 International Conference on Computing and Artificial Intelligence – ICCAI 2018*. doi: 10.1145/3194452.3194459

34. Kalash, M., Rochan, M., Mohammed, N., Bruce, N.D., Wang, Y., & Iqbal, F. (2018). Malware classification with deep convolutional neural networks. In *2018 9th IFIP International Conference on New Technologies, Mobility and Security (NTMS)*, 1–5. doi: 10.1109/ntms.2018.8328749

35. Cui, Z., Xue, F., Cai, X., Cao, Y., Wang, G., & Chen, J. (2018). Detection of malicious code variants based on deep learning. *IEEE Transactions on Industrial Informatics, 14*, 3187–3196.

36. Goel, A. K., Rose, A., Gaur, J., & Bhushan, B. (2019). Attacks, countermeasures and security paradigms in IoT. In *2019 2nd International Conference on Intelligent Computing, Instrumentation and Control Technologies (ICICICT)*. doi: 10.1109/icicict46008.2019.8993338

37. Tiwari, R., Sharma, N., Kaushik, I., Tiwari, A., & Bhushan, B. (2019). Evolution of IoT & data analytics using deep learning. *2019 International Conference on Computing, Communication, and Intelligent Systems (ICCCIS)*. doi: 10.1109/icccis48478.2019.8974481

38. Moore, A. (2017). *Intellectual Property and Information Control: Philosophic Foundations and Contemporary Issues*. London: Routledge.

39. Manchanda, C., Rathi, R., & Sharma, N. (2019). Traffic density investigation & road accident analysis in india using deep learning. In *2019 International Conference on Computing, Communication, and Intelligent Systems (ICCCIS)*. doi: 10.1109/icccis48478.2019.8974528

40. Malabarba, S., Devanbu, P., & Stearns, A. (1999). MoHCA-Java. *Proceedings of the 21st International Conference on Software Engineering – ICSE '99*. doi:10.1145/302405.302918.

41. Harjani, M., Grover, M., Sharma, N., & Kaushik, I. (2019). Analysis of various machine learning algorithm for cardiac pulse prediction. In *2019 International Conference on Computing, Communication, and Intelligent Systems (ICCCIS)*. doi: 10.1109/icccis48478.2019.8974519

42. Agostinelli, F., Hoffman, M.D., Sadowski, P.J., & Baldi, P. (2014). Learning activation functions to improve deep neural networks. *CoRR, abs/1412.6830.*

43. Singh, A., Sharma, A., Sharma, N., Kaushik, I., & Bhushan, B. (2019). Taxonomy of attacks on web based applications. In *2019 2nd International Conference on Intelligent Computing, Instrumentation and Control Technologies (ICICICT)*. doi: 10.1109/icicict46008.2019.8993264

44. Elfwing, S., E. Uchibe, & K. Doya (2018). Sigmoid-weighted linear units for neural network function approximation in reinforcement learning. *Neural Networks, 107*, 3–11.

45. Grover, M., Verma, B., Sharma, N., & Kaushik, I. (2019). Traffic control using V-2-V based method using reinforcement learning. In *2019 International Conference on Computing, Communication, and Intelligent Systems (ICCCIS)*. doi: 10.1109/icccis48478.2019.8974540

46. Abadi, M., Barham, P., Chen, J., Chen, Z., Davis, A., Dean, J., Devin, M., Ghemawat, S., Irving, G., Isard, M., Kudlur, M., Levenberg, J., Monga, R., Moore, S., Murray, D.G., Steiner, B., Tucker, P.A., Vasudevan, V., Warden, P., Wicke, M., Yu, Y., & Zhang, X. (2016). TensorFlow: A system for large-scale machine learning. In *OSDI*. https://www.usenix.org/system/files/conference/osdi16/osdi16-abadi.pdf

47. Baylor, D., Breck, E., Cheng, H., Fiedel, N., Foo, C.Y., Haque, Z., Haykal, S., Ispir, M., Jain, V., Koc, L., Koo, C.Y., Lew, L., Mewald, C., Modi, A.N., Polyzotis, N., Ramesh, S., Roy, S., Whang, S.E., Wicke, M., Wilkiewicz, J., Zhang, X., & Zinkevich, M. (2017). TFX: A TensorFlow-based production-scale machine learning platform. In *Proceedings of the 23rd ACM SIGKDD International Conference on Knowledge Discovery and Data Mining*. doi:10.1145/3097983.3098021.

48. Zhang, Z. (2018). Improved Adam optimizer for deep neural networks. In *2018 IEEE/ACM 26th International Symposium on Quality of Service (IWQoS)*. IEEE. doi:10.1109/iwqos.2018.8624183

49. Sharma, N., Kaushik, I., Rathi, R., & Kumar, S. (2020). Evaluation of accidental death records using hybrid genetic algorithm. *SSRN Electronic Journal*. doi: 10.2139/ssrn.3563084

50. Paik, J.H. (2013). A novel TF-IDF weighting scheme for effective ranking. In *Proceedings of the 36th International ACM SIGIR Conference on Research and Development in Information Retrieval – SIGIR '13*. ACM. doi:10.1145/2484028.2484070

51. Rustagi, A., Manchanda, C., & Sharma, N. (2020). IoE: A boon & threat to the mankind. In 2020 *IEEE 9th International Conference on Communication Systems and Network Technologies (CSNT)*. doi: 10.1109/csnt48778.2020.9115748

52. Haddi, E., X. Liu, & Y. Shi (2013). The role of text pre-processing in sentiment analysis. *Procedia Computer Science, 17*, 26–32.

53. Bouvrie, J. (2006). Notes on convolutional neural networks. http://cogprints.org/5869/1/cnn_tutorial.pdf

7 Data Encryption for IoT Applications Based on Two-Parameter Fuss–Catalan Numbers

M. Saračević
University of Novi Pazar

A. Selimi
International Vision University

CONTENTS

7.1 INTRODUCTION

Modern cryptography is based primarily on the application of mathematical systems that belong to the number theory. The efficient generation of a parameter that generates the key is the basic precondition for developing public-key cryptosystems. Cryptography as the science of designing cryptosystems would be too limited if the algorithm would be initialized with only one parameter. Due to this, different cryptosystems use different initialization parameters to implement all the functionality of the selected cipher. Cryptography is a rapidly evolving science, and new ways of encrypting and hiding information are generated on a daily basis. Encryption protects data from unauthorized access or a potential attacker using an algorithm and a key to transform a plaintext into an encrypted text. For this coding system to provide perfect secrecy, it is necessary to fully meet certain conditions. We start from the assumption that the algorithm is secure only if the encryption procedure uses a secret key that has an extremely large space key-space. In this case, our first task is to generate a random binary sequence that satisfies Catalan key properties. It is important to note that by increasing the Catalan basis of a generated key, we noticed the drastic increase of the keyspace, and thus reducing the possibility of breaking the cipher for an adversary. The secrecy in computer-based cryptosystems is based on the assumption that the opponent does not have enough time to compute. Theoretically, practical cryptosystems can be broken fully or partially, but most commonly such attacks are unattainable due to the time needed to perform such attacks.

In this chapter, we present an approach that uses more initializing parameters, more precisely a combination of Fuss–Catalan objects (two-parameter) and lattice path combinatorics. An important segment in Internet of Things (IoT) applications is data confidentiality. When developing an IoT platform, it is necessary to consider the required functional elements, and one of the most important elements of any IoT system is the security component. The solution to the security problems in the transport of data is seen in the rarity of the encrypted connections. One of the obstacles to the dynamic development of IoT applications is the constant threat to device security. The main risk is the danger that unauthorized parties will take control of the IoT device, or that personal and important information will be stolen.

Blockchain is a distributed and decentralized data warehouse that records all data exchanges between different parties in a secure and encrypted form. This technology provides a faster and more secure system for transferring and connecting pages. Blockchain storage and encryption applications are particularly popular. Blockchain technology has tremendous capabilities to enhance encryption and authentication. Moreover, blockchain has the potential to enhance other aspects such as data integrity and enhancement of digital identities, as well as enabling the best possible security of IoT devices, to prevent various types of attacks, including DDoS attacks. In a word, blockchain improves data confidentiality and integrity and fills in the deficiencies in poor security and lack of reliability. It is resistant to data modification and gives tremendous benefit for IoT security. Although blockchain was created without specific access controls, owing to its public distribution, some blockchain implementations now specifically address the issues of confidentiality and data access control. This is a critical challenge in an age where data can easily be spied on or modified, but fully encrypting blockchain data ensures that this data will not be accessible to unauthorized parties.

The main contribution of this chapter is a specific novel method for data encryption of text and images based on the Fuss–Catalan object (FC key) and one combinatorial problem, more precisely, the lattice path problem. In our previous research, we only used Catalan numbers as cryptographic keys. In this method, we have included two-parameter FC objects as cryptographic keys for encryption of files. Using the FC key or object in encryption, we get a much larger keyspace than the Catalan objects, which has exceptional importance in cryptosystem security and cryptanalysis. In the experimental part of this chapter, we present a comparative analysis of the proposed method and some of the previously published methods, where better results were obtained from the aspect of the speed of encryption (CPU) and cryptanalysis of the keyspace (total search method).

In Amounas et al. (2013), novel cryptographic algorithms were proposed based on sequences of Catalan numbers. In particular, integer sequences play a very important role in cryptography. Manimaran et al. (2016) proposed a method of encryption of plaintext using algebraic chessboard mapping (or chess notation). In Sivakumar and Venkatesan (2016), a novel encryption method for the image was proposed based on the concrete implementation of pixel scan with random numbers and Knight's travel problem. Thamizhvanan et al. (2017) proposed a method of encryption of text or image using a chess cipher algorithm. Singha et al. (2015) introduced an approach in the visual cryptography by Knight's tour problem. Prajit (2018) showed the chessboard cryptographic algorithm as an algorithm to encrypt N-bit binary data using a key of size N (the process of movement on the chessboard). An important feature is that the encryption and decryption functions are not inverse as in typical cryptographic algorithms, but are already the same. Elkies (2005) presented new directions in enumerative chess problems.

The rest of this chapter is organized into six sections. The second section discusses the basic properties of Fuss–Catalan numbers and objects and their relationship with the lattice path combinatorics. The third section discusses the importance of the application of blockchain technologies in cryptography and in the development of IoT applications. In addition, we considered some applications combinatorics in data encryption and some possibilities of the use of Fuss–Catalan numbers (objects). In the fourth section, examples for encryption of text and images based on the FC key and lattice path problem are presented. In the fifth section, we present a comparative analysis of the proposed method and some of the previously published methods. A comparison is given for data encryption methods using Catalan keys, such as encryption based on Ballot combinatorial problem and stack permutations method based on Catalan objects and lattice path and encryption based on Fuss–Catalan and lattice path. Furthermore, this section lists some scenarios and proposals for applications of Fuss–Catalan objects in the development of blockchain and IoT technologies. Concluding remarks and proposals for further research are given in the last section.

7.2 RELATED WORKS

When it comes to the security of IoT applications, issues of data confidentiality and access control occupy a special place. Special attention is paid to encrypt data in IoT applications. Hussain et al. (2017) and Sung et al. (2018) presented the application of DES, AES, and RSA algorithms in data encryption for IoT applications. Unde and Deepthi (2020), Sudhakaran and Malathy (2019), Chaudhry (2018), Chandu et al. (2017), Boutros et al. (2017), and Perez et al. (2017) showed different versions of data encryption for IoT applications.

The importance of the application of blockchain technologies in cryptography and in the development of IoT applications is presented in Kshetri (2017), Zhu and Fan (2019), Cachin (2017), and Li et al. (2019). Fernandez-Carames and Fraga-Lamas (2018) reported a thorough review of how to adapt blockchain to the specific needs of IoT to develop blockchain-based IoT applications. Moreover, the most relevant blockchain IoT applications are described with the objective of emphasizing how blockchain can impact traditional cloud-centered IoT applications. Banerjee et al. (2018) presented a blockchain future for IoT security and give the potentially sensitive nature of IoT datasets. In addition, this paper discussed the opportunities of blockchain technology in facilitating the secure sharing of data in IoT datasets and securing IoT systems. Khan and Salah (2018) surveyed major security issues for IoT. They also discussed how blockchain can be a key enabler for solving many IoT security problems. The paper also identified open research problems and challenges for IoT security.

Huh et al. (2017) proposed using blockchain to build the IoT system. The authors managed keys using RSA public-key cryptosystems and used a few IoT devices instead of a full IoT system, which consists of thousands of IoT devices. Panarello et al. (2018) presented a comprehensive survey on blockchain and IoT integration. The objective of this paper was to analyze the current research trends and the main challenges in the integration of IoT and blockchain. Raikwar et al. (2019) thoroughly reviewed all cryptographic concepts which are already used in blockchain. In addition, authors gave a list of cryptographic concepts that have big potential to improve the current blockchain solutions and include possible instantiations of these cryptographic concepts in blockchain technologies. Zhai et al. (2019) outlined the infrastructure of blockchain, including the data layer, network layer, consensus layer, contract layer, and application layer. The principles of encryption technology are introduced briefly, such as asymmetric cryptosystem and digital signature. Blockchain-enabled multimedia in industrial IoT is proposed by Wan et al. (2020). In Higgins (2008) and Horak et al. (2015), stated importance of number theory and combinatorics in cryptography is discussed. The purpose of our papers (Saračević et al. 2017, 2018, 2019) was to investigate the properties of Catalan objects and their application in encryption and steganography. For more details on Fuss–Catalan numbers, their properties, and possible applications, see Di Francesco (1998), Landau (2001), Aval (2008), Collins et al. (2010), Bacher and Krattenthaler (2011), Chou et al. (2018), Ballot (2018), Hussein (2019), Qi and Cerone (2018), Qi et al. (2019).

For the Catalan encryption method, we provided time and storage complexity of Catalan keys, from which we can conclude that such a process is extremely burdensome for the computer resources, such as CPU time and working memory. NIST statistical tests were performed for the generation of Catalan keys through several aspects, where good results were obtained (Saračević and Adamovic, 2018). In the additional testing of the quality of the Catalan keys generated, we can conclude that the result of advanced analysis (such as approximate entropy, random digression, and nonoverlapping or overlapping template matching) satisfies all NIST requirements.

7.3   PRELIMINARIES FOR THE PROPOSED METHOD

The Catalan numbers are defined as (Koshy 2009):

$$C_n = \frac{(2n)!}{(n+1)!\,n!}$$ (7.1)

Generalization of Catalan numbers is called the *Fuss–Catalan numbers* (Koshy 2009):

$$A_n(p,r) = \frac{r}{(np+r)}\binom{np+r}{n}$$ (7.2)

In our previous papers, we analyzed some applications of the Catalan objects and the corresponding combinatorial approaches. Saračević and Adamovic (2018) analyzed the properties of the Catalan objects and their relation to the combinatorial problem called lattice path in cryptography. Saračević and Koricanin (2017) examined the application of concrete combinatorial problems in encryption and decryption of text or image. For the proposed methods, we provided time and storage complexity of Catalan keys, from which we can conclude that such a process is extremely burdensome for the computer resources, such as CPU time and working memory. Moreover, NIST statistical tests were performed for the generation of Fuss–Catalan keys through several aspects (such as approximate entropy, random digression, and nonoverlapping or overlapping template matching), where good results were obtained. Two-parameter FC object does not need to have representation in the square network (see Figure 7.1).

In this chapter, we present an encryption method based on the Fuss–Catalan (FC) key where the rule of bit-balance does not apply. The FC keys space is much larger than the Catalan space, as shown in Table 7.1.

As can be seen from Table 7.1, the case $p = 2$ and $r = 1$ corresponds to the Catalan number. Increasing the parameter r and p produces the drastic increase of the keyspace, that is, the number of possible paths in the lattice path space.

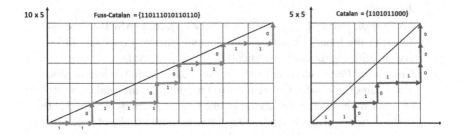

FIGURE 7.1   Fuss–Catalan object and lattice path (a); Catalan object and lattice path (b).

TABLE 7.1

Number of the Path in Lattice Space Based on a Formula (7.2)

p, r	Fuss–Catalan Numbers	OEIS: The On-Line Encyclopedia of Integer Sequences (https://oeis.org/)
(2,1)	1,1,2,5,14,42,132,429,1430,4862	*Catalan numbers: C(n)*
(2,3)	1,3,9,28,90,297,1001,3432,11934,41990	$a(n) = 3 \times (2 \times n)!/((n+2)! \times (n-1)!)$
(2,4)	1,4,14,48,165,572,2002,7072,25194,90440	*Fourth convolution of Catalan numbers*
(3,1)	1,1,3,12,55,273,1428,7752,43263,246675,1430715	$a(n) = binomial\ (3n, n)/(2n+1)$
(3,2)	1,2,7,30,143,728,3876,21318,120175,690690	$a(n) = binomial(3 \times n+1, n)/(n+1)$
(3,4)	1,4,18,88,455,2448,13566,76912,444015,2601300	*Self-convolution 4th power of (3,1)*
(4,1)	1,1,4,22,140,969,7084,53820,420732,3362260, ...	$a(n) = binomial\ (4n, n)/(3n+1)$
(4,2)	1,2,9,52,340,2394,17710,135720,1068012,8579560	$a(n) = binomial(4 \times n+1, n) \times 2/(3 \times n+2)$
(4,3)	1,3,15,91,612,4389,32890,254475,2017356,16301164	$a(n) = 3 \times binomial(4 \times n-1, n-1)/(4 \times n-1)$

7.4 ENCRYPTION METHOD BASED ON FUSS–CATALAN OBJECTS AND LATTICE PATH

In this section, we present one method that uses the Fuss–Catalan objects as a cryptographic key (hereinafter FC object or key) for encryption of text or image based on Lattice path combinatorics.

Our method consists of five phases and is implemented in *Java Net Beans* GUI environment. The proposed application has two segments: a text encryption module and an image encryption module. The basic steps in the proposed method for encrypting text are:

1. *Definition Phase*: Input parameters are r and p for FC key generation and forming block. The other parameters are plaintext, block bit organization (horizontal or vertical), and block bit selection (above or below diagonal).
2. *Conversion Phase*: Convert text to binary sequence (ASCII to the bin) and split the binary sequence into blocks $n \times m$. At this stage, the bit organization (horizontal or vertical) is also determined.
3. *Generation Phase*: Generating the FC key based on dimensions of block and parameters r and p. On the basis of formula (7.2), the total number of FC keys are generated. In this set are the keys that represent a valid path in the lattice path space. At this stage, one random FC object is selected in the generated set. We will use that object as a cryptographic key to select and invert bits.

4. *Selection Phase*: Determines the method of selection based on the FC key on lattice path space (below or above the main diagonal). Bit "1" in FC key represents moving down and "0" represents moving right.
5. *Inversion Phase*: Selected bits are inverted ($1 \rightarrow 0, 0 \rightarrow 1$). It is possible to invert bits only in the selected path, as well as to invert bits below or above the path.

Finally, we get the ciphertext (see Figure 7.2).

Example 7.1: We'll show you how to encrypt a short text based on a random FC (object) key and lattice path combinatorics.

- Definition of input parameters:
  1. $r=3$ and $p=2$.
  2. Plaintext=*"ALPHANUMERICALLY."*
  3. Block dimension: *16×8*

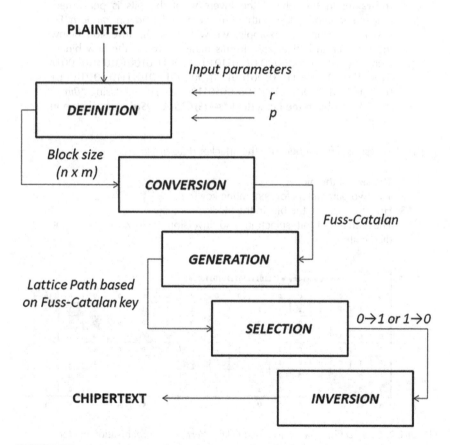

FIGURE 7.2   A general scenario for the proposed method.

4. Bit organization method in block=*"Vertical."*
5. Selection variant in block=*"below diagonal."*

- **Conversion:** The plaintext is converted to binary notation. We get a binary sequence P(bin)=*"01000001 01001100 01010000 01001000 01000001 01001110 01010101 01001101 01000101 01010010 01001001 01000011 01000001 01001100 01001100 01011001."* At this stage, the bit organization (vertical) is also determined (see Figure 7.3).

- **Generation:** This binary string can in this case be divided into *16×8* binary block. In this phase, a random key is generated based on *generated binary block*. In this case, a randomly selected key from a set of Fuss–Catalan objects, based on a formula (7.2), is cryptographic key K = *{1111011101101011101010}*.

- **Selection:** Based on the binary notation of the already generated key, a movement path through the Lattice path is formed. Determines the method of selection based on random FC key (below the main diagonal, as shown in Figure 7.3). In this case, bit *"1"* in Fuss–Catalan key represents moving down and *"0"* represents moving right.

- **Inversion:** In this phase, the inversion of the bits is performed, which encompass the path of movement based on random FC key. In this concrete example, we will show the inversion below the diagonal in lattice path. In this manner, we get the new binary string: *C(bin)=*"01000000 01001101 01010011 01001010 01000011 01001100 01010011 01001001 01000001 01011110 01010001 01010011 01110001 01101100 00101100 10011001." Using *"Bin to ASCII"* we obtain the ciphertext *"@MSJCLSIA^QSql,?"* (as shown in Figure 7.4).

Cryptanalysis is difficult because the attacker does not know:

1. The size of the block.
2. The two parameters for generating key (r, p).
3. How to organize the bits in the block.
4. The variant of bit selection and inversion (below or above the diagonal).

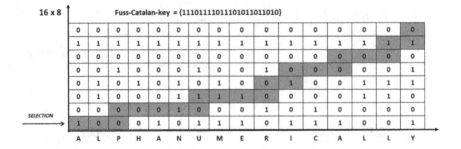

FIGURE 7.3 Selection phase based on FC key: *Vertical* bit organization in block.

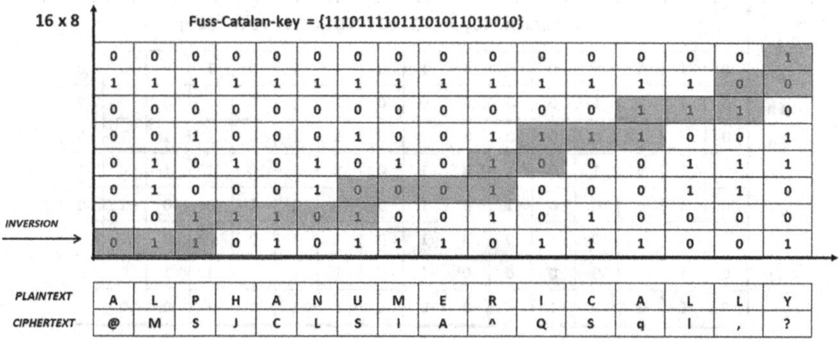

16 x 8   Fuss-Catalan-key = {11101110111101011011010}

0	0	0	0	0	0	0	0	0	0	0	0	0	0	0	1
1	1	1	1	1	1	1	1	1	1	1	1	1	1	0	0
0	0	0	0	0	0	0	0	0	0	0	0	1	1	1	0
0	0	1	0	0	0	1	0	0	1	1	1	1	0	0	1
0	1	0	1	0	1	0	1	0	1	0	0	0	1	1	1
0	1	0	0	0	1	0	0	0	1	0	0	0	1	1	0
0	0	1	1	1	0	1	0	0	1	0	1	0	0	0	0
0	1	1	0	1	0	1	1	1	0	1	1	1	0	0	1

INVERSION →

PLAINTEXT	A	L	P	H	A	N	U	M	E	R	I	C	A	L	L	Y
CIPHERTEXT	@	M	S	J	C	L	S	I	A	^	Q	S	q	I	,	?

FIGURE 7.4 Inversion of selected bits.

We will analyze the second scenario where the organization of the bits in the block is horizontal.

Example 7.2: If we select the Fuss–Catalan key K_{bin} = {1111011101101101101010} and let the same plaintext P = "alphanumerically." In this case, we will choose other parameters (bit organization is *horizontal* in block *16×8*), as shown in Figure 7.5.

We invert selected bits to obtain a binary string C(bin)="*011000010110110 101 11000001101011011000010110100001110101010100010110010110010010011 01010111000110111111101101100100110001111001.*" Using "Bin to ASCII," we get different ciphertext "**ampkahuQe?jã?l?y**" (as shown in Figure 7.6).

As noted above, the third side does not know the number of paths of inverted bits. The application also supports an enhanced version of encryption, which is an additional inversion of one more path in lattice path from the diagonal.

For the realization and testing purposes of the proposed encryption method, we have implemented our solution in the *Java* programming language. The proposed solution consists of three main classes (Appendix includes the important segments of these classes):

- Class for cryptographic key generation (*FussCatalanKey*)
- Class for encryption of text or image (*EncFussCatalanPath*)
- Class for decryption of text or image (*DecFussCatalanPath*)

The developed software solution has the following functionalities:

1. Generation of the random FC key based on parameters r, p, and dimensions of lattice path space;
2. Loading of text (module 1) or image (module 2) in the process of encryption;

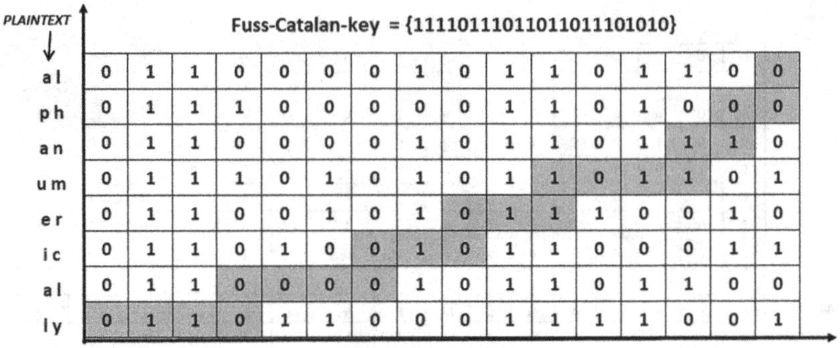

PLAINTEXT — Fuss-Catalan-key = {11110111011011011101010}

a l	0	1	1	0	0	0	0	1	0	1	1	0	1	1	0	0
p h	0	1	1	1	0	0	0	0	0	1	1	0	1	0	0	0
a n	0	1	1	0	0	0	0	1	0	1	1	0	1	1	1	0
u m	0	1	1	1	0	1	0	1	0	1	1	0	1	1	0	1
e r	0	1	1	0	0	1	0	1	0	1	1	1	0	0	1	0
i c	0	1	1	0	1	0	0	1	0	1	1	0	0	0	1	1
a l	0	1	1	0	0	0	0	1	0	1	1	0	1	1	0	0
l y	0	1	1	0	1	1	0	0	0	1	1	1	1	0	0	1

FIGURE 7.5   Selection: *Horizontal* bit organization in block.

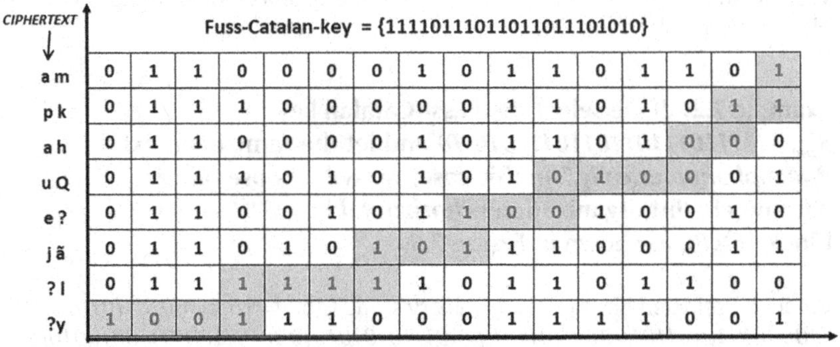

CIPHERTEXT — Fuss-Catalan-key = {11110111011011011101010}

a m	0	1	1	0	0	0	0	1	0	1	1	0	1	1	0	1
p k	0	1	1	1	0	0	0	0	0	1	1	0	1	0	1	1
a h	0	1	1	0	0	0	0	1	0	1	1	0	1	0	0	0
u Q	0	1	1	1	0	1	0	1	0	1	0	1	0	0	0	1
e ?	0	1	1	0	0	1	0	1	1	0	0	1	0	0	1	0
j ã	0	1	1	0	1	0	1	0	1	1	1	0	0	0	1	1
? l	0	1	1	1	1	1	1	1	0	1	1	0	1	1	0	0
? y	1	0	0	1	1	1	0	0	0	1	1	1	1	0	0	1

FIGURE 7.6   Inversion of selected bits.

3. Selection of encryption mode and entering additional parameters: the size of the block, bit organization method in the block, selection variant in the block.

Based on the key and the selected encryption mode, the selection and inversion of the bits in the cipher are done. In addition, the software solution provides a reversible process, that is, it allows loading the cipher and the Fuss–Catalan keys in decryption method selection. It is important to note that the text encryption principle shown can also be applied to image encryption. The only difference is in the second phase (*Conversion*), one step more is added, which converts the image to Base64 and then to the binary sequence.

7.5 EXPERIMENTAL RESULTS

In this section, we present a comparative analysis of the proposed method and some of the previously published methods. A comparison is given for data encryption methods using Catalan keys, such as:

1. encryption based on Catalan objects and lattice path (Saračević and Adamovic 2018)
2. encryption based on ballot combinatorial problem and stack permutations (Saračević and Koricanin 2017)

The two methods are compared with the proposed method based on Fuss–Catalan and lattice path. In the first two methods, the encryption is based on the bit-balance properties of Catalan objects, and in this proposed encryption method, it is based on Fuss–Catalan objects (where the bit-balance rule, that is, the equal counter of bits 1 and 0, does not apply).

Testing was done using the following three aspects:

1. Comparison of data encryption methods using Catalan key (CPU speed).
2. Comparison of the size of the keyspace (memory space).
3. Comparison of CPU decryption rate (in seconds, average 1000 per second).

Table 7.2 presents the results of comparison (text encryption speed for data encryption methods implemented in the *Java* programming language).

As can be observed, the efficiency in encrypting the data is higher and not complicated because we apply more parameters than the previous methods. In our application, the efficiency is much better compared to the previous methods, as shown in Figure 7.7.

Table 7.3 shows a comparison of the five models of the cryptographic key generation for a small n (from 1 to 10). The first model is the classic (one parameter) Catalan objects we used in our papers (Saračević and Adamovic 2018; Saračević and Koricanin 2017). The other four models are a combination of two-parameter Fuss–Catalan keys with p and r (Saračević et al. 2020).

TABLE 7.2

Comparison of Data Encryption Methods Using Catalan Key and Fuss–Catalan Key (Results in Seconds)

Test	Length Text (Numbers of Characters, Including Spaces)	Catalan Lattice Path	Catalan Stack Permutations	Catalan Ballot	Fuss–Catalan Lattice Path
1	1000	0.0012	0.0015	0.0013	0.0012
2	5000	0.0034	0.0041	0.0042	0.0033
3	30,000	0.0241	0.0262	0.0254	0.0240
4	50,000	0.0403	0.0411	0.0405	0.0398
5	100,000	0.0854	0.0871	0.0857	0.0835
6	500,000	0.4074	0.4173	0.4113	0.3992
7	1,000,000	0.9891	0.9952	0.9912	0.9691
8	5,000,000	4.8943	5.1057	5.0094	4.7963

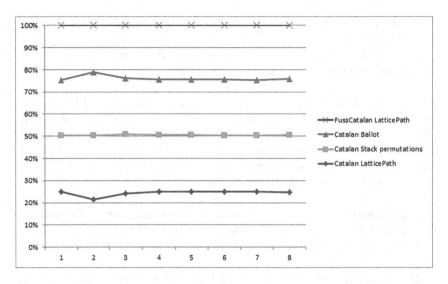

FIGURE 7.7   Comparative analyses: encryption speed.

Figure 7.8 shows the relationship in complexity, respectively keyspace based on two parameters.

Table 7.4 shows the comparative analysis in the decryption process. In the decryption process of this test, in one second, an average of about 1000 keys can be tested.

We conclude that by increasing the two-parameter FC key, we significantly affect the security of the encrypted data.

7.6   CONCLUSION

According to many scientific papers and research studies, quantum cryptography will play a very important role in the future of secure communications. Many studies have analyzed the use of number theory in the implementation of techniques that deal with secure data exchange. By merging quantum computing and IoT, new interoperability protocols, complex autonomous systems, cybersecurity, privacy, and surveillance measures are expected to evolve. In addition, it is important to mention the multiplicity of data encryption supplements based on Catalan objects, such as the application of two-parameter FC objects in data encryption.

We can conclude that new digital technologies including IoT and blockchain will transform the world we live in. There is no doubt that technologies such as IoT will provide insight into user behaviors in all areas, enabling a greater level of personalized communication. When it comes to data protection and control, blockchain technology in conjunction with IoT will bring many opportunities. They together have significant applications in automating business processes, crypto-currencies, smart contracts, financial services, supply chains, and many

TABLE 7.3
Testing in the Generation of All Keys and Memory Space

n	Catalan Key		Fuss–Catalan Key (2,3)		Fuss–Catalan Key (3,3)		Fuss–Catalan Key (3,4)		Fuss–Catalan Key (4,3)	
	Number of Keys	Memory Space (Kb)	Number of Keys	Memory Space (Kb)	Number of Keys	Memory Space (Kb)	Number of Keys	Memory Space (Kb)	Number of Keys	Memory Space (Kb)
1	1	0.000	1	0.000	1	0.000	1	0.000	1	0.000
2	2	0.002	3	0.002	3	0.001	4	0.002	3	0.0011
3	5	0.002	9	0.004	12	0.006	18	0.009	15	0.0073
4	14	0.009	28	0.017	55	0.034	88	0.054	91	0.0561
5	42	0.031	90	0.066	273	0.200	455	0.333	612	0.4481
6	132	0.113	297	0.254	1428	1.220	2448	2.092	4389	3.7502
7	429	0.419	1001	0.978	7752	7.570	23,566	23.014	328,90	32.1191
8	1430	1.571	3432	3.771	43,263	47.530	76,912	84.498	254,475	279.5752
9	4862	5.935	11,934	14.568	246,675	301.11	444,015	542.01	2,017,356	2462.5513
10	16,796	22.553	41,990	56.383	1,430,715	1921.1	2,601,300	3492.9	16,301,164	21,888.523

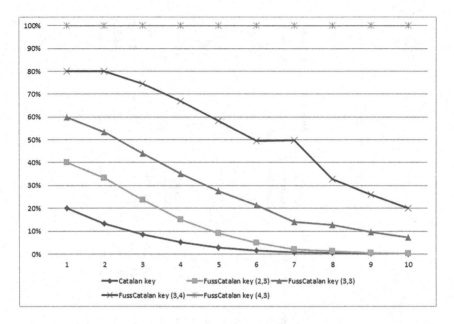

FIGURE 7.8 Comparative analyses: memory space.

TABLE 7.4

Comparison of CPU Decryption Rate (in Seconds, Average 1000 per Second)

T	Catalan Key	Fuss–Catalan Key (2,3)	Fuss–Catalan Key (3,3)	Fuss–Catalan Key (3,4)	Fuss–Catalan Key (4,3)
1	0.00	0.00	0.00	0.00	0.00
2	0.00	0.00	0.00	0.00	0.00
3	0.01	0.01	0.01	0.02	0.02
4	0.01	0.03	0.06	0.09	0.09
5	0.04	0.09	0.27	0.46	0.61
6	0.13	0.30	1.43	2.45	4.39
7	0.43	1.00	7.75	23.57	32.89
8	1.43	3.43	43.26	76.91	254.48
9	4.86	11.93	246.68	444.02	2017.36
10	16.80	41.99	1430.72	2601.30	16,301.16

others. Blockchain will allow the sharing of a single database between multiple parties, without the need for a central authority and where access control is determined at the software level. Blockchain is resistant to attempting to modify data once entered and provides high efficiency.

In this chapter, we proposed a novel data encryption method for IoT applications based on the FC key and lattice path combinatorics. In the presented method, we have included two-parameter FC objects as cryptographic keys for encryption of files. Using the FC object as a key in encryption, we get a lot. In the experimental part of this chapter, we presented a comparative analysis of the proposed method and some of the previously published methods. A comparison is given for data encryption methods using Catalan keys (using lattice path, ballot problem, and stack permutations) and encryption based on Fuss–Catalan objects and lattice path combinatorics. Testing was done through the CPU speed comparison of data encryption methods based on Catalan key, comparison of the size of the keyspace (memory space), and comparison of CPU decryption rate.

APPENDIX

Java implementation details:

A. The method in class *FussCatalanKey* which is used to generate binary records of all keys for the given basis *n*.

```
public static intgenKey(File file) throws IOException {
short[] A = new short[128];
short n, i;
long m, s, count, cstart, cend;
FileWriter writer = new FileWriter(file);
        s = cstart;
while (s <cend) {
             m = s;
             n = 0;
while (m > 1) {
if ((m & 1) > 0) {
        A[n] = 1;
    } else {
        A[n] = 0;
    }
n++;
    m = m >> 1;
             }
             A[n] = 1;
n++;
count = 0;
i = (short) (n - 1);
while ((i >= 0) && (count >= 0)) {
if (A[i] == 1) {
count++;
    } else {
count--;
    }
i--;
    }
```

B. Part of the class*EncFussCatalanPath* which as input parameters expects the file (text or image) and already generates a 128-bit Catalan key.

```
File encFussCatalanPath(File file, String key) {
short[] K = new short[128];
byte[] text = new byte[segment];
byte[] path = new byte[segment];
byte[] ciphertext = new byte[segment];
short n, i,  toGoal, free, busy;
long j, s;
BigInteger start;
long count = 0;
        File fileOut = null;
try {
fileOut = new File(file.getParent() + "\\CIPHERTEXT.txt");
FileOutputStream out;
try (FileInputStream in = new FileInputStream(file)) {
out = new FileOutputStream(fileOut);
start = new BigInteger(key);
   n = 0;
```

C. A part relating to text encryption, that is, a bit substitution procedure based on the principle of movement on lattice path and based on the already generated cryptographic key.

```
// read binary notation of Fuss-Catalan key
while (BigInteger.valueOf(1).compareTo(start) < 0) {
if (start.and(BigInteger.valueOf(1)).compareTo(BigInteger.
valueOf(0))> 0) {
   K[n] = 1;
   } else {
       K[n] = 0;
   }
n++;
start = start.shiftRight(1);
   }
//read message
   K[n] = 1;
count = file.length()/segment;
for (j = 0; j < count; j++) {
in.read(text, 0, segment);
toGoal = segment - 1;
free = busy = 0;
// read the binary notation of the key (n-arrangement of
pairs 0 and 1)
for (i=n; i>=0; i--) {
// move right (forward if bit 1)
```

```
if (K[i] == 1)
            {
path[free] = text[toGoal];
free++;
toGoal--;
            } // scroll up (if bit 0)
else {
ciphertext[busy] = path[free-1];
busy++;
free--;
path[free] = 0;
            }
        }
out.write(ciphertext);
    }
count = file.length() - (file.length()-in.available());
in.read(text, 0, (int) count);
out.write(text);
out.flush();
}
```

REFERENCES

Amounas, F., El-Kinani E.H. and Hajar, M. 2013. Novel Encryption Schemes Based on Catalan Numbers. *International Journal of Information & Network Security* 2(4): 339–347.

Aval, J.C. 2008. Multivariate Fuss-Catalan Numbers. *Discrete Mathematics* 308(20): 4660–4669.

Bacher, R. and Krattenthaler, C. 2011. Chromatic Statistics for Triangulations and Fuss-Catalan Complexes. *Electronic Journal of Combinatorics* 18(1): Art. No. P152.

Ballot, C. 2018. Lucasnomial Fuss-Catalan Numbers and Related Divisibility Questions. *Journal of Integer Sequences* 21(6): Art. No. 18.6.5.

Banerjee, M., Lee, J. and Choo, K. 2018. A Blockchain Future for Internet of Things Security: A Position Paper. *Digital Communications and Networks* 4(3): 149–160.

Boutros, A., Hesham, S., Georgey, B. and Abd El Ghany, M.A. 2017. Hardware Acceleration of Novel Chaos-based Image Encryption for IoT Applications. In: *IEEE International Conference on Microelectronics-ICM*, Lebanon, pp. 53–56.

Cachin, C. 2017. Blockchain, Cryptography, and Consensus. *Electronic Proceedings in Theoretical Computer Science* 261: 1–5.

Chandu, Y., Kumar, K.R., Prabhukhanolkar, N.V., Anish, A.N. and Rawal, S. 2017. Design and Implementation of Hybrid Encryption for Security IOT Data. In: *International Conference on Smart Technologies for Smart Nation*, Bengaluru, India, pp. 1228–1231.

Chaudhry, S. 2018. An Encryption-based Secure Framework for Data Transmission in IoT. In: *IEEE International Conference on Reliability INFOCOM Technologies and Optimization Trends and Future Directions*, Noida, pp. 743–747.

Chou, W.S., He, T.X. and Shiue, P.J. 2018. On the Primality of the Generalized Fuss-Catalan Numbers. *Journal of Integer Sequences* 21(2): Art. No. 18.2.1.

Collins, B., Nechita, I. and Zyczkowski, K. 2010. Random Graph States, Maximal Flow and Fuss-Catalan Distributions. *Journal of Physics A – Mathematical and Theoretical* 43(27): Art. No. 275303.

Di Francesco, P. 1998. New Integrable Lattice Models from Fuss-Catalan Algebras. *Nuclear Physics B* 532(3): 609–634.

Elkies, N.D. 2005. New Directions in Enumerative Chess Problem. *The Electronic Journal of Combinatorics* 11(2): 1–14.

Fernandez-Carames, T.M. and Fraga-Lamas P. 2018. A Review on the Use of Blockchain for the Internet of Things. *IEEE Access* 6: 32979–33001.

Higgins, P.M. 2008. *Number Story: From Counting to Cryptography*. Springer Science & Business Media, Germany.

Horak, P., Semaev, I. and Tuza, I.Z. 2015. An Application of Combinatorics in Cryptography. *Electronic Notes in Discrete Mathematics* 49: 31–35.

Huh, S., Cho, S. and Kim, S. 2017. Managing IoT Devices Using Blockchain Platform. In: *IEEE International Conference on Advanced Communication Technology*, Pyeongchang, South Korea, pp. 464–467.

Hussain, I., Negi, M.C. and Pandey N. 2017. A Secure IoT-Based Power Plant Control Using RSA and DES Encryption Techniques in Data Link Layer. In: *International Conference on INFOCOM Technologies and Unmanned Systems (Trends and Future Directions) (ICTUS)*, Dubai, UAE, pp. 464–470.

Hussein, A.B. 2019. On Representations of Fuss-Catalan Algebras. *Journal of Algebra* 519: 398–423.

Khan, M.A. and Salah, K. 2018. IoT Security: Review, Blockchain Solutions, and Open Challenges. *Future Generation Computer Systems* 82: 395–411.

Koshy, T. 2009. *Catalan Numbers with Applications*. Oxford University Press, New York.

Kshetri, N. 2017. Can Blockchain Strengthen the Internet of Things? *IT Professional* 19(4): 68–72.

Landau, Z.A. 2001. Fuss-Catalan Algebras and Chains of Intermediate Subfactors. *Pacific Journal of Mathematics* 197(2): 325–367.

Li, H.G., Zhang, F.G., Luo, P.R., Tian, H.B. and He, J.J. 2019. How to Retrieve the Encrypted Data on the Blockchain. *KSII Transactions on Internet and Information Systems* 13(11): 5560–5579.

Manimaran, A., Chandrasekaran, M., Gupta, A. and Porwal, R. 2016. Encryption and Decryption Using Algebraic Chess Notations. *International Journal of Pharmacy and Technology* 8: 22098–22105.

Panarello, A., Tapas, N., Merlino, G., Longo, F. and Puliafito, A. 2018. Blockchain and IoT Integration: A Systematic Survey. *Sensors* 18(8), Art. No. 2575.

Perez, S., Hernandez-Ramos, J.L., Skarmeta, A.F., et al. 2017. A Digital Envelope approach using Attribute-Based Encryption for Secure Data Exchange in IoT Scenarios. *In: IEEE Global Internet of Things Summit (GIOTS 2017)*, Switzerland, pp. 421–426.

Prajit, T.R. 2018. Chessboard Cryptalgorithm. In: *2018 Second International Conference on Advances in Electronics, Computers and Communications (ICAECC)*, Bangalore, pp. 1–6.

Qi, F. and Cerone, P. 2018. Some Properties of the Fuss-Catalan Numbers. *Mathematics* 6(12): Art. No. 277.

Qi, F., Shi, X. and Cerone, P. 2019. A Unified Generalization of the Catalan, Fuss, and Fuss-Catalan Numbers. *Mathematical and Computational Applications* 24(2): Art. No. 49.

Raikwar, M., Gligoroski, D. and Kralevska, K. 2019. SoK of Used Cryptography in Blockchain. *IEEE Access* 7: 148550–148575.

Saračević, M. and Adamovic, S. 2018. Applications of Catalan numbers and Lattice Path combinatorial problem in cryptography. *Acta Polytechnica Hungarica* 15(7): 91–110.

Saračević, M., Adamovic, S., Maček N., Elhoseny, M. and Sarhan, S. 2020. Cryptographic Keys Exchange Model for Smart City Applications. *IET Intelligent Transport Systems.* In Press, doi: 10.1049/iet-its.2019.0855.

Saračević, M., Adamović, S., Miškovic, V., Maček, N. and Šarac, M. 2019. A Novel Approach to Steganography based on the Properties of Catalan Numbers and Dyck Words. *Future Generation Computer Systems* 100: 186–197.

Saračević, M., Hadzic, M., Koricanin, E. 2017. Generating Catalan-Keys based on Dynamic Programming and Their Application in Steganography. *International Journal of Industrial Engineering and Management* 8(4): 219–227.

Saračević, M. and Koricanin, E. 2017. Encryption based on Ballot, Stack permutations and Balanced Parentheses using Catalan-keys. *Journal of Information Technology and Applications* 7(2): 69–77.

Saračević, M., Selimi, A. and Selimovic, F. 2018. Generation of Cryptographic Keys with Algorithm of Polygon Triangulation and Catalan Numbers. *Computer Science – AGH* 19(3): 243–256.

Singha, M., Kakkar, A. and Singh, M. 2015. Image Encryption Scheme Based on Knight's Tour Problem. *Procedia Computer Science* 70: 245–250.

Sivakumar, T. and Venkatesan, R. 2016.A New Image Encryption Method Based on Knight's Travel Path and True Random Number. *Journal of Information Science and Engineering* 32(1): 133–152.

Sudhakaran, P. and Malathy, C. 2019. Energy Efficient Distributed Lightweight Authentication and Encryption Technique for IoT Security. *International Journal of Communication Systems.* Art. No: e4198, doi: 10.1002/dac.4198.

Sung, B.Y., Kim, K.B. and Shin, K.W. 2018. An AES-GCM Authenticated Encryption Crypto-Core for IoT Security. In: *IEEE International Conference on Electronics Information and Emergency Communication*, Beijing, pp. 285–287.

Thamizhvanan, C., Priyadharshini, S., Punitha, R., ShamliPriya, R. and Thamizhselvi, V. 2017. Image Encryption and Decryption using Chess Cipher. *International Journal of Engineering Science and Computing* 7(4): 6385–6388.

Unde, A.S. and Deepthi, P.P. 2020. Design and Analysis of Compressive Sensing-based Lightweight Encryption Scheme for Multimedia IoT. *IEEE Transactions on Circuits and Systems* 67(1): 167–171.

Wan, S.H., Umer, T. and Bashir, A.K. 2020. Blockchain-enabled Multimedia in Industrial IoT. *Multimedia Tools and Applications.* doi: 10.1007/s11042-019-08541-w.

Zhai, S.P., Yang, Y., Li, J., Qiu, C. and Zhao, J.M. 2019. Research on the Application of Cryptography on the Blockchain. *Journal of Physics Conference Series* 1168: Art. No. 032077.

Zhu, X. and Fan, T. 2019. Research on Application of Blockchain and Identity-based Cryptography. *IOP Conference Series – Earth and Environmental Science* 252: Art. No. 042095.

8 Fog-Based Framework for Improving IoT/IoV Security and Privacy

M. Nahri and A. Boulmakoul
Hassan II University of Casablanca

L. Karim
Hassan 1st University

CONTENTS

8.1   INTRODUCTION

As humans build increasingly complex information and communication systems, this growth should be accompanied by strong security and privacy policies. In the Internet ecosystem, rules of security are always superposed on standards of computers and systems architectures, as well as network architectures. Systems such as firewalls, intrusion detection and prevention systems (IDS and IPS), hash and authentication, cryptography and public key infrastructure, Secure Socket Layer, and Transport Layer Security have been developed to be adapted to hardware and software protocols, as well as to network architectures.

However, Internet of Things (IoT) creates significant cybersecurity problems compared to conventional computer systems. Indeed, IoT data exchanges are ruled by a vast variety of protocols and communication technologies. Thus, security becomes deeply complex in such a context. Internet of vehicles (IoV) represents an instance of IoT in vehicular context, inheriting the security complexity of IoT, along with more constraints characterizing the vehicular environment such as high density and high velocity of its components. In fact, IoV tries to build an environment of data exchange between all objects involved in the road environment, centralizing the focus to the vehicle as a pivotal element. IoV ecosystem allows vehicle to vehicle (V2V), vehicle to infrastructure (V2I), and vehicle to everything (V2X) communication. In fact, more than Vanets systems (Tello-Oquendo et al. 2019), IoV tends to exploit the advents on vehicular sensing and communication capabilities and all types of existing network technologies such as LTE and 5G, as well as short-range communication, to create a complex context of data exchange satisfying user requirements . This concept has been promoted recently by the emergence of the software-defined network (SDN) (Tello-Oquendo et al. 2019) and fog computing concepts (Nahri et al. 2018). These advancements in technologies and architectures must be followed by advancements in security and privacy. Any attack targeting IoV environment can cost people lives, economic losses, and other catastrophic damages.

The variety, openness, and accessibility of IoV technologies make them more exposed to malicious attacks. Moreover, the complexity of IoV architectures enhances the complexity of security measures to be adopted. Furthermore, the security for IoV seems to be uncontrollable regarding the high velocity aspects of connected devices. IoV is likely to undergo all IoT attacks types and others with much more dangers. For instance, devices and networks are exposed to attacks such as DOS and channel inference, sybil, blackhole and wormhole, GPS deception, tampering, illusion, eavesdropping, replay, routing, and other (Sun et al. 2016). In general, the IoT architecture consists of three layers: perception/hardware layer, a network/communication layer, and a layer of interfaces/services. The attack vectors listed in the Open Web Application Security OWASP relate to these three layers. Therefore, the implementation of IoT security mitigation should encompass the security architecture of all IoT layers (Noor, Binti, and Hassan 2019). In the literature, authentication, encryption, trust management, secure routing, and other kinds of security measures, such as those based on blockchain technology, are considered for facing malicious attacks. However, any security solution must take into account the reality of IoV devices constrained by the computational power along with energy. Moreover, it must be more suitable for IoV architectures. Especially, the choice of the authentication mechanism and the cryptography algorithm presents a challenge.

Authentication and encryption management at low-level IoV components is facing many challenges. In fact, the application of public key infrastructure (Schukat and Cortijo 2015) to improve security in IoV is a subject of most recent research works. The U.S. National Institute of Standards and Technology considers the deployment of the vehicular public key infrastructure (VPKI) (Weil 2017) to secure vehicular communication in Vanets system. Rathore et al. (2019) proposed an authentication and key management system for IoT environment to secure high-velocity data in smart cities. Moreover, Kang et al. (2018) proposed a pseudonym affectation scheme for vehicles in an IoV environment. However, the proposed solutions in these works inevitably need cloud services or generally a central unit to complete required security functions.

On the other hand, emerging architectures for IoV propose centralizing IoV low-level device management through fog computing (Nahri et al. 2018). In addition to its benefits on reducing latency, the fog layer seems to be suitable for security and privacy management, especially for authentication management of edge devices, including vehicle's computing unit, considering its closeness to the latter. In this work, we present a fog-based security scheme comprising three essential points. The first affirms that a given fog node is the only handler of included devices on its managed geographical area. The second concerns permitting the interoperability between devices managed by different fog nodes. Finally, the third focuses on allowing the movement of a given device between fog nodes. One of the benefits of this approach is the improvement of the security manageability and the devices' control. Moreover, it gives more isolation and autonomy to the fog node; thus, if a given fog node is vulnerable, it cannot infect the entire system. Moreover, it is the role of the fog node to detect and remove the security issue occurring on its managed zone. Thus, fog-based security scheme guarantees the handover and collaboration between fog nodes, as well as gives autonomy to the fog node to manage the security of its own territory. Practically, every fog node contains a trust authority component allowing classic PKI services, such as registration, validation, and certification, as well as an interfog authority component promoting interfog real-time collaboration. In fact, every fog node collaborates with the surrounding fog nodes via this new component.

The rest of this chapter is organized as follows: in Section 8.2, we propose describing emergent technologies and architectures that shaped IoV. Moreover, we discuss IoT and IoV security challenges, potential threats, and proposed solutions in Section 8.3. Furthermore, in Section 8.4, we describe the proposed fog-based authentication scheme for IoV environment permitting interfog collaboration. Finally, open challenges and future works are discussed in the conclusion.

8.2 IOV-ENABLING TECHNOLOGIES AND ARCHITECTURES

This section explores enabling ITS architectures and technologies that promoted IoV. Thus, we are trying to give a clear vision about sensing, computation, and network technologies, as well as emerging architectures for IoV before discussing security issues.

8.2.1 IoV Technologies

IoT has revolutionized the automotive industry in recent years. Governments, researchers, and industrials invest heavily to improve smart vehicle performance, thus enhancing the comfort, safety, and efficiency of mobility.

8.2.1.1 Sensing Technologies

Effectively, four essential elements are taken into consideration for smart vehicles: sensing, computation, actuation, and connectivity (Liu, Niu, and Liu 2016). Sensing technologies can be divided into two categories. The first, qualified as interne, informs about vehicle state such as mechanical components state, speed, and position, among others. The second allows detecting the state of the surrounding environment, such as lane detection, distance detection, and environmental conditions sensing, among others. Indeed, positioning techniques, using systems such as GNSS and GPS, have greatly evolved. For instance, Broadcom chip enabling accuracy of 30 cm is now available for smartphones (Bricker 2017). Moreover, the power of detection of RADAR and LIDAR (Siegel, Erb, and Sarma 2018) technologies as well as integrated smart cameras allows recognizing distances, directions, lanes, and moving objects. All these technologies provide helpful information assisting drivers and permitting taking decisions and acting automatically and immediately. Thus, sensed data need to be gathered in a computing unit for treatment and external communication. Smart vehicles integrate an on-board unit (OBU) (Haroun and Mostefaoui 2017) equipped with the power of computing and communication. Indeed, OBU aggregates and gathers all data generated by sensors and performs some computation tasks at the vehicle level.

8.2.1.2   Intravehicle Communication

Intravehicle communication (Tuohy et al. 2015) is assisted by the advancements in short-range communication technologies, such as zigbee, Bluetooth, and NFC, among others. The described technologies favored significant advancement in smart vehicles having the power to decide and act in real-time to critical situations. Therefore, smart vehicles allow sensing, gathering, and treating information. On the other hand, communication with the exterior such as cloud, fog, and other vehicles or roadside units (RSUs) is allowed through long-range communication as well as short-range communication technologies.

8.2.1.3   V2X Communication

The vehicular environment is very dynamic. Moreover, applications such as traffic safety (Chang, Chen, and Su 2019) and control seem to be very critical for cities. Cities need to have a complete and dedicated communication environment to support vehicular connectivity, thus, favoring the deployment of vehicular collaborative applications for safety and traffic management. Vehicular ad-hoc networks (Vanets) were the first systems to support vehicular communication, allowing V2V and V2I communication. Based on wireless communication capabilities of connected vehicle and RSUs, Vanets have been a promising research area developed and adopted by many organizations and governments such as USDOT (Auer, Feese, and Lockwood 2016). The standardization of Vanets technologies has been the wish of these organizations, satisfying vehicular applications requirement in latency and efficiency. For example, applications such as FleetNet and CarTalk (Enkelmann 2003) have been deployed using Vanets and supporting traffic safety and management. Furthermore, Vanets have been accompanied by several standardizations, including IEEE 802.11p, WAVE and CALM (Hartenstein, Hannes, and Kenneth Laberteaux 2010), allocating a bandwidth of 75 MHz to vehicular communication.

Thus, IoV ecosystem is characterized by a wide range of communication technologies including Vanets technologies and other short-range and long-range communication technologies. Moreover, IoV allows deploying collaborative applications and performing efficient analytics by data centralization. Furthermore, other network technologies are proposed to support IoV such as 5G (Guan et al. 2019) and LTE-V (Molina-Masegosa and Gozalvez 2017) using nodes eNB instead of RSUs. These systems propose using cellular technologies, which have known a great advantage in creating an ecosystem for communicating vehicular data in a large and efficient manner.

8.2.2   IoV Architectures

Several architectures have been proposed to manage and exploit exchanged data in IoV. Alam, Saini, and El Saddik (2015) and Maglaras et al. (2016) discussed social IoV integrating Vanets to the cloud and considering IoV as a social ecosystem, in which vehicles form different relationships. Moreover, Chen et al. (2017)

and Jaballah, Conti, and Lal (2019) presented a software-defined IoV (SD-IoV) by applying SDN on IoV ecosystem to manage low-level network topology and logic (Misra and Saha 2019). In fact, SDN applied to IoV presents several advantages such as the automation of low-level configuration of the network and managing low-level network security. SD-IoV proposes an architecture of three layers for IoV: network layer encompassing physical network components like OBUs, switches, and RSUs; control layer comprising the logic of low-level network and commanding the network layer; and, finally, the application layer presenting a commanding interface exposed to users for executing their applications.

8.2.2.1 Edge–Fog–Cloud Architecture

Nahri et al. (2018) and Chen et al. (2017) proposed an architecture for IoV data management and analytics composed of three layers: the edge, the fog, and the cloud layers. Figure 8.1 shows an overview of this architecture. The edge layer contains physical and low-level devices such as in-vehicle computing unit and networking infrastructure. The fog layer, which is positioned between the edge and the cloud, performs services requiring low latency and ensures security and privacy management. The most interesting thing about fog computing layer is its ability to empower IoV with centralized management while resting close to connected devices. Finally, the cloud layer provides advanced services and delays tolerant services such as global data analytics.

Effectively, the fog layer will certainly play a pivotal role in security management and control in IoV due to its closeness to the edge devices. In fact, there is a major need to identify and authenticate low-level objects in IoV such as vehicles, pedestrians, and traffic signals to create a trusted information exchange. Moreover, edge devices are exposed to the risk of attacks due to public access,

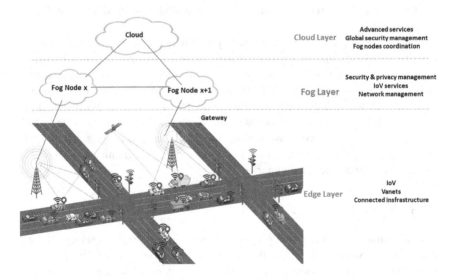

FIGURE 8.1 IoV edge–fog–cloud architecture.

wherein devices with a limited computational power are incapable of defending themselves. These ideas will be developed further in the next section.

8.3 IOV SECURITY: CHALLENGES, THREATS, AND COUNTERMEASURES

IoT security goal is to preserve confidentiality; ensure the safety of users, data, IoT device, and infrastructure; and ensure the availability of services provided by IoT ecosystem. However, given the heterogeneity of devices and protocols, as well as the number of IoT system nodes, security mechanisms are more difficult to enforce than a traditional system. Noor, Binti, and Hassan (2019) conducted surveys in IoT security published from 2017 to 2018. They highlight the current IoT security mechanisms, including authentication, encryption, trust management, secure routing protocols, and new technologies applied to IoT security. Current security challenges in the perception layer concern the detection of the abnormal sensor node, the choice cryptography algorithms, and the design of authentication management mechanism to be used, as well as preserving the anonymity of the data sender and protecting devices from vulnerabilities. IoV security inherits IoT security characteristics with much more complications. In fact, the vehicular environment is characterized by high dynamicity and dispersion geographic of its components. Moreover, the high dynamic topology of IoV networks makes it more exposed to attacks. Furthermore, security in IoV is critical. In fact, an attack on IoV can cause significant damages affecting people's lives and economic losses. Identifying and managing malicious attacks is at the core of IoV design and implementation.

8.3.1 FORMAL METHODS AND IoT SECURITY

Connected objects create significant cybersecurity problems compared to conventional computer systems. From a technical viewpoint, connected objects are very accessible by their wireless connection in a public network. Often without supervision, it can slow down the detection of attacks. From an economic viewpoint, stakeholders have little interest in the security of connected objects, they will not bear the cost of a possible attack, while they will have to bear the additional costs related to security. The advent of smart objects and their increasing use in IoT applications creates more security challenges. In IoT, security is paramount but presents many difficulties to achieve. On one hand, the usual trade-off between highly secure operation and efficient deployment is more compelling than ever. Considerations in terms of cost analysis are also increasingly required. Therefore, the designers of IoT applications not only need tools to assess the possible risks and study the countermeasures but also need to evaluate their costs. Various approaches have been proposed in the literature. Bodei et al. (2005, 2016) have developed a methodology based on the formal calculation process (Huttel, Sangiorgi, and Walker 2002). Their work is based on a specification language that describes IoT systems. Designers can model the architecture of a system, the dynamic behavior of its smart connected objects, and their interactions via formal expressions. The formal specifications can infer quantitative measures on the evolution of the developed systems. The quantitative assessment is used to calculate the cost of any security countermeasure, in terms of time and energy.

8.3.2 IoT and IoV Security Classification

A clear vision about intervention fields of any encountered danger as well as of any countermeasure improves the reliability and the precision of any proposed security solution. Thus, here, we suggest classifying (Sharma et al. 2018; Gulzar and Abbas 2019) IoV and IoT security into several intervention fields presented below.

8.3.2.1 Authentication

Every element has to prove its identity before accessing or using any service or performing any operation. This identity can be registered before a trusted authority or it can be created and attributed at the start of the connection demand.

8.3.2.2 Authorization

Authorization concerns access control management that consists of specifying rights and privileges for an element to access given services or information or performing given functions.

8.3.2.3 Confidentiality

Confidentiality consists of protecting data. Data is reachable and readable only to those authorized for access. Thus, the source and the destination are the only sides to know the content of any transmitted information, and of course some information about the source and the destination are known by the transmission tools, such as network equipment. Encryption is one of the tools used to preserve confidentiality.

8.3.2.4 Integrity

Integrity is about accuracy, authenticity, consistency, and completeness of data. Thus, data should be unmodifiable during transfer and unalterable by an unallocated element.

8.3.2.5 Privacy

Privacy is the ability to protect data from access, extraction, or exploitation from third parties without prior knowledge of holders, except for a trusted authority that can access data to manage access control management.

8.3.2.6 Data Availability

Data availability focuses on ensuring the accessibility of required data whenever its needed by the authorized parties.

8.3.2.7   Quality of Service (QoS)

QoS concerns reducing data losses and latency when communicating and processing data. Attackers can influence QoS by shutting down some equipment or by perturbing transmission channels. This field also manages the performance of all involved technical solutions tending to ensure data communication and data processing. We mention that any solution aimed at managing above security fields must not have a big influence on QoS.

These fields are the most interesting points to envision when designing any security solution. In addition to the reliability of the solution, these solutions must preserve the QoS of the system. This point is critical and we consider it further in the proposed solution in Section 8.4 by delegating security management to the fog layer promising to reduce the latency. In the next two subsections, we describe susceptible IoV attacks and proposed solutions for them, projecting the classification described in this part.

8.3.3   IoV ATTACKS

IoV devices can encounter several types of attacks (Bhushan and Sahoo 2018). As explained in the Section 8.3.2, attacks can affect confidentiality, integrity, availability, authenticity, nonrepudiation, QoS, and privacy (Sun et al. 2016; Sicari et al. 2015; Sakiz and Sen 2017). Most common attacks in IoV environment are discussed below.

8.3.3.1   Denial of Service (DOS)

In this kind of attack, attackers try to affect network and services by targeting services directly or by overloading the system by fake requests. Thus, user requests cannot be handled by the attacked service. In addition, DOS attacks can occur from several locations, representing distributed denial of service attacks, making it difficult to detect and identify attackers. Moreover, another type of attack that can be a part of DOS is called channel interference, which includes launching wireless signals interfering with the allocated channel and perturbing the communication between IoV devices. DOS attacks influence particularly the availability and the QoS.

8.3.3.2   Sybil Attack

Sybil attack occurs when a single device pretends to have multiple identities (Hamdan, Hudaib, and Awajan 2019). Thus, many fake virtual active devices in the network communicate false information about traffic and positions. This can have disastrous consequences on traffic management and safety. Moreover, sybil attacks can facilitate other kinds of attacks without identifying their source. This kind of attacks create a great security risk impacting authentication and authorization, integrity, and QoS.

8.3.3.3   GPS Deception

This attack consists of capturing the GPS signal and trying to provide IoV devices with a fake position. Thus, the attacker spoofs the identity of the GPS satellite distributing false position to demanders, creating a big issue by misinforming traffic and safety applications, which are based on GPS position. The integrity of data and QoS are the most impacted by GPS deception.

8.3.3.4 Masquerading

In masquerading, the attacker assumes the identity of another vehicle or device; thus, it uses a false identity to gain more privileges and communicate fake messages. This attack can have a significant impact on data access management. For instance, it can be used to influence emergency and traffic management. This attack creates many concerns related to authentication, authorization, and integrity.

8.3.3.5 Eavesdropping

Eavesdropping consists of listening to the traffic between the sender and the receiver. Thus, the attacker can access data exchanged in the network and collect information about the sender, receiver, and content of the message. Indeed, an attacker can gather traffic data communicated by following specific road users. In addition, the attack is performed in the middle of the network making it hard to detect. This kind of attack impacts privacy and confidentiality.

8.3.3.6 Routing Attack

In this type of attack, the attacker tries to reroute packets from their destination to another wrong one. The attacker disrupts the normal mechanism of the network, changing its operational logic or dropping some transmitted packets. This leads to an incompleteness of exchanged data having a big influence on the efficiency of the network system. Thus, routing attack influences the QoS and the integrity of data.

In the next subsection, we present the most used countermeasures to manage these attacks.

8.3.4 IoV Attacks Countermeasures

Authentication remains the most common security technique used, while trust management is gaining popularity due to its ability to prevent or detect malicious nodes. Moreover, secure routing in Vanet and IoV is a dynamic research field. Recently, researchers are using blockchain and SDN technologies to bring more security to the vehicular environment (Muthanna et al. 2019).Kim (2019) presented the design of a secure network of enhanced IoV using the blockchain governance game. Kim (2019) and Kim (2018) proposed a model based on stochastic game theory to find the best strategies to prevent network malfunction caused by an attacker.

On the other hand, recent encryption researches for IoT have focused on lightweight cryptography algorithms allowing low-power consumption and less computational requirement respecting device constraints (Singh et al. 2017). Nain et al. (2017) and Mahmood, Ning, and Ghafoor (2017) focused on implementing low-energy consumption and lightweight encryption in the physical and network environments. RSA (Rivest–Shamir–Adleman) and ECC (elliptic curve cryptography) algorithms are the most used for IoT environment. However, ECC seems to be the most suitable for IoT and IoV environment regarding its minimal resource consumption. Furthermore, the trend is to apply public key infrastructure to IoT and IoV environments. However, these solutions face several challenges both in the design and deployment.

8.4 FOG-BASED APPROACH FOR IOV SECURITY

In this section, we present an approach based on fog layer, exploiting its advantages in low-level devices to improve security and privacy in an IoV environment while taking into consideration mobility constraints.

8.4.1 RELATED WORKS

As mentioned in the above Section 8.2.2.1, the fog layer can play a pivotal role in the security of edge–fog–cloud architecture. Indeed, several studies have been conducted in the fog paradigm for managing security in IoT as well as IoV architectures. Moreover, work suggested in (Liu, Niu, and Liu 2016) presents a pseudonym scheme for managing IoV identifiers and improving privacy, especially addressing the location privacy issue. This work recommends creating and distributing pseudonyms with more precision and according to the context, as well as reducing pseudonym management overhead through passing by the fog nodes, which are distributed geographically and close to vehicles. This approach presents several advantages compared to centralized pseudonym management affecting performance and efficiency. According to Kang et al. (2018), all vehicles should be registered in a central authority before going on the road. Work done by the U.S. National Institute of Standards and Technology (Weil 2017) presents a VPKI and certification authority model for improving security and preserving privacy. The proposed security architecture is dedicated to vehicular ad-hoc network systems, allowing registration before a central certificate authority (CA) server and accompanied by a system for intrusion detection. Moreover, several works which are mainly based on SDN, fog, and blockchain paradigms are proposed for enhancing and managing security. For instance, authors in (Kahvazadeh et al. 2017) propose to improve security management in critical infrastructure based on master/slave strategy, fog-cloud, and SDN. Moreover, Muthanna et al. (2019) proposed a hierarchical fog-based SDN and blockchain architecture increasing security for IoT ecosystem.

However, most of these works pass through the cloud or a central security server for registration or performing partial security tasks, knowing that all security tasks are supposed to be executed in a minimal duration. Moreover, in IoV, most of the proposed architectures do not take consider IoV and its overall context. Of course, the vehicle is at the heart of IoV, but each vehicle can make a trusted connection with all surrounding connected objects such as pedestrians, traffic signals, and other roadside objects. Therefore, there is a need to think about a holistic approach of distributing identifiers, certificates, and security keys while respecting the latency tolerated by sensible IoV services. In the next part, we outline the design of an architecture based on the fog paradigm and PKI architecture, improving security and privacy while taking into account all types of IoV devices and promising minimal latency.

8.4.2 FOG-BASED SECURITY ARCHITECTURE: OVERVIEW

IoV architecture based on edge-fog-cloud layers presents several advantages due to the proximity of the fog, with the computational power, to the devices and network equipment. This closeness attributes to the fog layer additional responsibilities related to low-level device management, network management, as well as security and privacy management. Indeed, the global architecture is composed of one cloud node, several geographically distributed fog nodes, and edge devices. Moreover, every fog node is responsible for managing devices belonging to or positioned on its managed geographical area.

8.4.2.1 Fog Node Architecture

In Section 8.2.2, we have presented SDN as emerging solution for network management and discussed its applicability for managing IoV low-level network. SDN concept proposes three levels for network management: data plan, control plan, and management plan. The control plan or the SDN controller is the most interesting level between them. Effectively, gathering SDN controller and security management in the same entity will certainly have a great benefit for both services. For instance, it allows blocking malicious devices and disseminating information about them in real time. Another point that can be of great value on security is equipping fog nodes with a system for intrusion prevention and detection, as well as a system for behavior analysis. Furthermore, a credential management system at the fog level is necessary for registering and authenticating every device before accessing any IoV service. Figure 8.2 shows an overview of theses fog

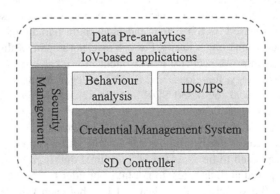

FIGURE 8.2   Architecture of the fog layer.

components. Finally, we mention that fog security system must have more autonomous functionalities while collaborating with the overall IoV architecture.

8.4.2.2 Edge Node Architecture

Edge devices present the endpoints of the proposed security architecture. Here, we describe the OBU architecture that must be respected by low-level IoV devices. In fact, security components in OBU have to be in synergy with security components of higher levels, such as fog security management component, while preserving the independence of intravehicle functionalities, such as sensing and actuation, as well as the independence of internal applications, such as a self-driving system. Figure 8.3 shows the IoV device architecture. The security entity within the device is positioned between external communication interfaces and IoV applications. Therefore, before accessing any IoV service or establishing any peer-to-peer communication, the device needs to prove its safety. We mention that the security entity, which we describe here, is closely linked to the security component positioned at the fog. Therefore, the security schema can be modeled as a client–server system, in which the client is the security entity of the device and the server side is the fog security component.

Security components of both fog node and device node are ideal for implementing security systems such as PKI and credential management system. PKI infrastructure according to standard X.509 (Schukat and Cortijo 2015) is based on three components: registration authority (RA), CA, and validation authority (VA). The idea is to give the fog nodes, which are geographically distributed, the autonomy to manage the security of their devices while respecting their mobility. Thus, the collaboration between fog nodes is primordial. For this reason, we add interfog authority (inter-fog-A) as a new component to the PKI architecture, ensuring the collaboration between fog nodes.

8.4.3 INTERFOG-A FUNCTIONALITIES AND SCENARIOS

Certainly, a big advantage will be gained if we give total autonomy to the fog node to manage the security of devices included in its territory. However, it seems to be difficult to control OBUs with high velocity and moving permanently between fog

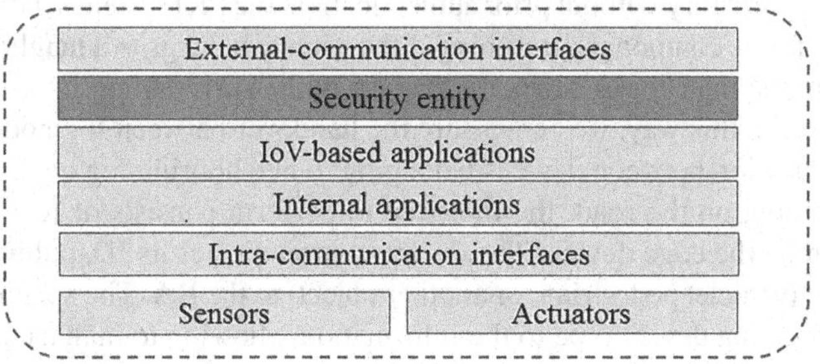

FIGURE 8.3 Architecture of the IoV device (OBU).

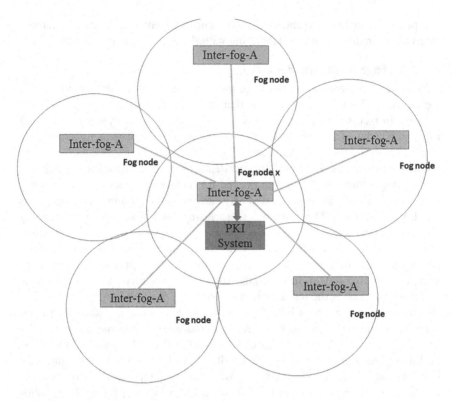

FIGURE 8.4   Inter-fog-A component.

nodes. To face this constraint, we added a new component that we call inter-fog-A to the PKI system, ensuring the communication between PKI systems of different fog nodes. Exactly, inter-fog-A of a given fog node is in permanent communication with similar components of surrounding fog nodes. The schema in Figure 8.4 shows the communication between interfog components.

Effectively, in a given fog node, the inter-fog-A stands besides the PKI system responding to their requests in terms of collaboration. For instance, if a vehicle enters the fog territory without prior authentication, the PKI system sends a request to the inter-fog-A component searching if the vehicle belongs to a neighboring fog node. Then, the inter-fog-A sends the request to the surrounding inter-fog-As for this purpose. In this way, we can ensure the handover between fog nodes as well as communication between devices belonging to neighboring fog nodes.

Before going on the road, the first step to perform consists of requesting for registration by the edge device. The device communicates its ID, public key, and device type (vehicle, pedestrian, or another object) to the RA. The specificity here consists of adding device type to the information, allowing to manage authorizations. Then the RA passes a command to the certification authority for generating a digital certificate and communicating it to the device, as shown in Figure 8.5.

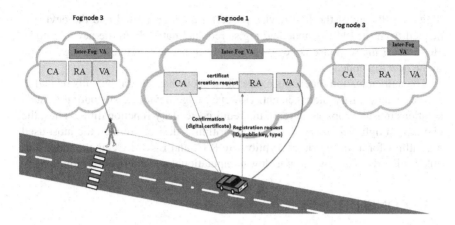

FIGURE 8.5   Registration and Inter-fog-As collaboration.

After the registration, the device can establish a trusted connection with fog services as well as with other similar devices at the edge.

8.4.3.1   Establish a Trusted Connection between an Edge Device and Fog Services

Before establishing any connection, the device must be registered. The VA performs this task and allows the device to connect to the required service.

8.4.3.2   Establish a Trusted Connection between Edge Devices

8.4.3.2.1   Devices Managed by the Same Fog Node

Clearly, when two devices at the edge need to communicate with each other, the verification of security information passes through the VA at the fog. Then, registered devices can exchange keys and establish a trusted connection. This process can be easily executed when devices belong to the same fog node; however, it is more complicated when devices are managed by different nodes.

8.4.3.2.2   Devices Managed by Neighboring Fog Nodes

When a device requests connection to another device belonging to neighboring fog node, the VA of the second fog node forwards the request to the inter-fog-A, searching if the first device belongs to a neighboring fog node. Subsequently, if the response is affirmative information and the digital certificate of the first device requester are transmitted, a trusted connection between the two devices is established.

8.4.3.3   Handover Mechanism

The same approach is applied when a device is moved from one fog node to a neighboring fog node. Indeed, the inter-fog-A searches if the device belongs to a neighboring fog node; if the response is affirmative, then information and the digital certificate of the first device requester are transmitted and the device is deleted from the old fog node and registered automatically in the new fog node. However, if the response is negative, the device needs to request for registration in the new fog node.

The authentication mechanism described above has several benefits related to privacy and security management. However, edge devices contributing to such solutions must be capable of realizing required security functionalities. Thus, the choice of a lightweight cryptography algorithm is desired. In fact, the most used algorithms for asymmetric encryption are RSA and ECC. ECC is the most suitable for IoT devices regarding its low computational and energy consumption.

8.5 CONCLUSION

Technologies, security classification, attacks, and countermeasures in IoV environment have been discussed in this chapter. Moreover, a fog-based authentication scheme is highlighted. The development and the test of the framework are planned in the near future. Another security measurement needs to be considered besides authentication and cryptography management systems, such as IDS based on machine learning, to detect every abnormal behavior of devices and other solutions for securing network infrastructure and communication channels. Finally, even though IoV constitutes a promising environment to improve road safety efficiency and comfort, it must be acceptable ethically, socially, and legally. Any proposed security solution for IoV needs to be aligned with these requirements. Thus, more than the technical challenges that we discussed in this chapter, other types of challenges have to be considered, for example, business challenges.

ACKNOWLEDGMENT

This work was partially funded by the Digital Development Agency (ADD) and the National Center for Scientific and Technical Research (CNRST) in partnership with the Ministry of Industry, Commerce and Green and Digital Economy (MICEVN) and the Ministry of National Education, Professional Training, Higher Education and Scientific Research (MENFPESRC) # AL KHAWARIZMI Program # *Intelligent & Resilient Urban Network Defender: A distributed real-time reactive intelligent transportation system for urban traffic congestion symbolic control and monitoring.*

REFERENCES

Alam, Kazi Masudul, Mukesh Saini, and Abdulmotaleb El Saddik. 2015. "Toward Social Internet of Vehicles: Concept, Architecture, and Applications." *IEEE Access* 3 (c): 343–357. doi:10.1109/ACCESS.2015.2416657.

Auer, Ashley, Shelley Feese, and Stephen Lockwood. 2016. "History of Intelligent Transportation Systems." Department of Transportation. Intelligent Transportation Systems Joint Program Office FHWA-JPO-1: 1–56.

Bhushan, Bharat, and Gadadhar Sahoo. 2018. "Recent Advances in Attacks, Technical Challenges, Vulnerabilities and Their Countermeasures in Wireless Sensor Networks." *Wireless Personal Communications* 98 (2): 2037–2077. doi:10.1007/s11277-017-4962-0.

Bodei, Chiara, Mikael Buchholtz, Pierpaolo Degano, Flemming Nielson, and Hanne Riis Nielson. 2005. "Static Validation of Security Protocols." *Journal of Computer Security* 13 (3): 347–390. doi:10.3233/JCS-2005-13302.

Bodei, Chiara, Pierpaolo Degano, Gian Luigi Ferrari, and Letterio Galletta. 2016. "A Step towards Checking Security in IoT." *Electronic Proceedings in Theoretical Computer Science, EPTCS* 223 (Ice): 128–142. doi:10.4204/EPTCS.223.9.

Bricker, Jonathan. 2017. "Super – Accurate GPS Coming to Smartphones in 2018." *IEEE Spectrum* 54 (11): 10–11.

Chang, Wan Jung, Liang Bi Chen, and Ke Yu Su. 2019. "DeepCrash: A Deep Learning-based Internet of Vehicles System for Head-on and Single-Vehicle Accident Detection with Emergency Notification." *IEEE Access* 7: 148163–148175. doi:10.1109/ACCESS.2019.2946468.

Chen, Jiacheng, Haibo Zhou, Ning Zhang, Wenchao Xu, Quan Yu, Lin Gui, and Xuemin Shen. 2017. "Service-Oriented Dynamic Connection Management for Software-Defined Internet of Vehicles." *IEEE Transactions on Intelligent Transportation Systems* 18 (10): 2826–2837. doi:10.1109/TITS.2017.2705978.

Enkelmann, Wilfried. 2003. "FleetNet – Applications for Inter-Vehicle Communication." In *IEEE Intelligent Vehicles Symposium, Proceedings*, 162–167. doi:10.1109/IVS.2003.1212902.

Guan, Ke, Danping He, Bo Ai, David W. Matolak, Qi Wang, Zhangdui Zhong, and Thomas Kurner. 2019. "5-GHz Obstructed Vehicle-to-Vehicle Channel Characterization for Internet of Intelligent Vehicles." *IEEE Internet of Things Journal* 6 (1): 100–110. doi:10.1109/JIOT.2018.2872437.

Gulzar, Maria, and Ghulam Abbas. 2019. "Internet of Things Security: A Survey and Taxonomy." In *2019 International Conference on Engineering and Emerging Technologies, ICEET 2019*. IEEE, 1–6. doi:10.1109/CEET1.2019.8711834.

Hamdan, Salam, Amjad Hudaib, and Arafat Awajan. 2019. "Detecting Sybil Attacks in Vehicular Ad Hoc Networks." *International Journal of Parallel, Emergent and Distributed Systems*. doi:10.1080/17445760.2019.1617865.

Haroun, Amir, and Ahmed Mostefaoui. 2017. "Data Fusion in Automotive Applications." *Personal and Ubiquitous Computing* 21 (3): 443–455. doi:10.1007/s00779-017-1008-2.

Hartenstein, Hannes, and Kenneth Laberteaux. 2010. *VANET: Vehicular Applications and Inter-Networking Technologies*. Chichester, UK: Wiley.

Huttel, Hans, Davide Sangiorgi, and David Walker. 2002. "The π-Calculus: A Theory of Mobile Processes." *The Bulletin of Symbolic Logic* 8 (4): 530. doi:10.2307/797961.

Jaballah, Wafa Ben, Mauro Conti, and Chhagan Lal. 2019. "A Survey on Software-Defined VANETs: Benefits, Challenges, and Future Directions." *ArXiv Preprint ArXiv:1904.04577*. http://arxiv.org/abs/1904.04577.

Kahvazadeh, Sarang, Vitor B Souza, Xavi Masip-Bruin, Eva Marín-Tordera, Jordi Garcia, and Rodrigo Diaz. 2017. "An SDN-Based Architecture for Security Provisioning in Fog-to-Cloud (F2C) Computing Systems." In *Proceedings of 2017 Future Technologies Conference (FTC): 29-30 November 2017, Vancouver, Canada*: 732–738. The Science and Information (SAI) Organization.

Kang, Jiawen, Rong Yu, Xumin Huang, and Yan Zhang. 2018. "Privacy-Preserved Pseudonym Scheme for Fog Computing Supported Internet of Vehicles." *IEEE Transactions on Intelligent Transportation Systems* 19 (8): 2627–2637. doi:10.1109/TITS.2017.2764095.

Kim, Song-Kyoo. 2018. "The Trailer of Blockchain Governance Game." *ArXiv Preprint ArXiv:1807.05581*, 1–13. http://arxiv.org/abs/1807.05581.

Kim, Song-Kyoo. 2019. "Enhanced IoV Security Network by Using Blockchain Governance Game." *ArXiv Preprint ArXiv: 1904.11340.*

Liu, Yazhi, Jianwei Niu, and Xiting Liu. 2016. "Comprehensive Tempo-Spatial Data Collection in Crowd Sensing Using a Heterogeneous Sensing Vehicle Selection Method." *Personal and Ubiquitous Computing* 20 (3): 397–411. doi:10.1007/s00779-016-0932-x.

Maglaras, Leandros, Ali Al-Bayatti, Ying He, Isabel Wagner, and Helge Janicke. 2016. "Social Internet of Vehicles for Smart Cities." *Journal of Sensor and Actuator Networks* 5 (1): 3. doi:10.3390/jsan5010003.

Mahmood, Zahid, Huansheng Ning, and Ata Ullah Ghafoor. 2017. "A Polynomial Subset-Based Efficient Multi-Party Key Management System for Lightweight Device Networks." *Sensors* 17 (4). doi:10.3390/s17040670.

Misra, Sudip, and Niloy Saha. 2019. "Detour: Dynamic Task Offloading in Software-Defined Fog for IoT Applications." *IEEE Journal on Selected Areas in Communications* 37 (5): 1159–1166. doi:10.1109/JSAC.2019.2906793.

Mohamad Noor, Mardiana Binti, and Wan Haslina Hassan. 2019. "Current Research on Internet of Things (IoT) Security: A Survey." *Computer Networks* 148: 283–294. doi:10.1016/j.comnet.2018.11.025.

Molina-Masegosa, Rafael, and Javier Gozalvez. 2017. "LTE-V for Sidelink 5G V2X Vehicular Communications: A New 5G Technology for Short-Range Vehicle-to-Everything Communications." *IEEE Vehicular Technology Magazine* 12 (4): 30–39. doi:10.1109/MVT.2017.2752798.

Muthanna, Ammar, Abdelhamied A. Ateya, Abdukodir Khakimov, Irina Gudkova, Abdelrahman Abuarqoub, Konstantin Samouylov, and Andrey Koucheryavy. 2019. "Secure and Reliable IoT Networks Using Fog Computing with Software-Defined Networking and Blockchain." *Journal of Sensor and Actuator Networks* 8 (1). doi:10.3390/jsan8010015.

Nahri, Mohamed, Azedine Boulmakoul, Lamia Karim, and Ahmed Lbath. 2018. "IoV Distributed Architecture for Real-Time Traffic Data Analytics." *Procedia Computer Science* 130: 480–487. doi:10.1016/j.procs.2018.04.055.

Nain, Ajay Kumar, Jagadish Bandaru, Mohammed Abdullah Zubair, and Rajalakshmi Pachamuthu. 2017. "A Secure Phase-Encrypted IEEE 802.15.4 Transceiver Design." *IEEE Transactions on Computers* 66 (8): 1421–1427. doi:10.1109/TC.2017.2672752.

Rathore, M. Mazhar, Yaser Jararweh, Muhammad Raheel, and Anand Paul. 2019. "Securing High-Velocity Data: Authentication and Key Management Model for Smart City Communication." *2019 Fourth International Conference on Fog and Mobile Edge Computing (FMEC)*. IEEE, 181–188. doi:10.1109/fmec.2019.8795312.

Sakiz, Fatih, and Sevil Sen. 2017. "A Survey of Attacks and Detection Mechanisms on Intelligent Transportation Systems: VANETs and IoV." *Ad Hoc Networks* 61: 33–50. doi:10.1016/j.adhoc.2017.03.006.

Schukat, Michael, and Pablo Cortijo. 2015. "Public Key Infrastructures and Digital Certificates for the Internet of Things." *2015 26th Irish Signals and Systems Conference, ISSC 2015*, no. ii. doi:10.1109/ISSC.2015.7163785.

Sharma, Mini, Aditya Tandon, Subhashini Narayan, and Bharat Bhushan. 2018. "Classification and Analysis of Security Attacks in WSNs and IEEE 802.15.4 Standards : A Survey." *Proceedings -2017 3rd International Conference on Advances in Computing, Communication and Automation (Fall), ICACCA 2017*, January 2018, 1–5. doi:10.1109/ICACCAF.2017.8344727.

Sicari, Sabrina, Alessandra Rizzardi, Luigi Alfredo Grieco, and Alberto Coen-Porisini. 2015. "Security, Privacy and Trust in Internet of Things: The Road Ahead." *Computer Networks* 76: 146–164. doi:10.1016/j.comnet.2014.11.008.

Siegel, Joshua E., Dylan C. Erb, and Sanjay E. Sarma. 2018. "A Survey of the Connected Vehicle Landscape—Architectures, Enabling Technologies, Applications, and Development Areas." *IEEE Transactions on Intelligent Transportation Systems* 19 (8): 2391–2406. doi:10.1109/TITS.2017.2749459.

Singh, Saurabh, Pradip Kumar Sharma, Seo Yeon Moon, and Jong Hyuk Park. 2017. "Advanced Lightweight Encryption Algorithms for IoT Devices: Survey, Challenges and Solutions." *Journal of Ambient Intelligence and Humanized Computing.* doi:10.1007/s12652-017-0494-4.

Sun, Yunchuan, Lei Wu, Shizhong Wu, Shoupeng Li, Tao Zhang, Li Zhang, Junfeng Xu, and Yongping Xiong. 2016. "Security and Privacy in the Internet of Vehicles." *Proceedings -2015 International Conference on Identification, Information, and Knowledge in the Internet of Things, IIKI 2015*, 116–121. doi:10.1109/IIKI.2015.33.

Tello-Oquendo, Luis, Shih Chun Lin, Ian F. Akyildiz, and Vicent Pla. 2019. "Software-Defined Architecture for QoS-Aware IoT Deployments in 5G Systems." *Ad Hoc Networks* 93: 101911. doi:10.1016/j.adhoc.2019.101911.

Tuohy, Shane, Martin Glavin, Ciarán Hughes, Edward Jones, Mohan Trivedi, and Liam Kilmartin. 2015. "Intra-Vehicle Networks: A Review." *IEEE Transactions on Intelligent Transportation Systems* 16 (2): 534–545. doi:10.1109/tits.2014.2320605.

Weil, T. 2017. "VPKI Hits the Highway: Secure Communication for the Connected Vehicle Program." *IT Professional* 19 (1): 59–63.

9 An Overview of Blockchain and its Applications in the Modern Digital Age

Reinaldo Padilha França, Ana Carolina Borges Monteiro, Rangel Arthur, and Yuzo Iano
University of Campinas (UNICAMP)

CONTENTS

9.1 INTRODUCTION

Bitcoin was the first complete implementation of a blockchain, which originated in 2008 and was made available as open source in 2009. It is a payment system (digital peer-to-peer asset) with no points of failure. Blockchain is one of the emerging technologies that has been gaining prominence in the world's technological scenario. It is a public ledger that registers a virtual currency operation (the most known is Bitcoin), which makes it transparent, unchanging, and reliable [1].

The blockchain concept was created in 2008 to allow the development of Bitcoin. It is basically a set of rules that determine the functioning of the virtual currency. There are indications that Bitcoin was developed by Satoshi Nakamoto. The currency continued to be developed and traded, and the community itself formed around it began to work on solving problems in the code. It is a chain of blocks, hence the name, that is part of a collective registration system. This means that the information is not saved in a specific place and instead of being saved on a single computer, all blockchain information is distributed among the various computers connected to it [2].

The blockchain gave Bitcoin a fixed collection of mechanical rules so that operations could happen between users privately without the need for intermediaries. As Bitcoin gained fame, other digital coins started their own blockchain drafting. Each successful new elaboration of this collaborative technology has taken others to realize it, increasing the interest in blockchain technology for various applications in various industries. The blockchain technology registers information such as the number of currencies traded, who sent it, who received it, the moment that this transaction was effected, and where the book is registered [3,4].

One of the advantages of blockchain is that it allows customers to authenticate employing encryption keys, which means that there is no storage of their personal data. Thus, they decide whether they want to share information with different platforms. Another highlight is the lower propensity to interrupt the service. This is because the network works in a distributed environment, that is, there is less chance of a generalized shutdown affecting the entire system. This shows that transparency is one of the main blockchain predicates [5].

With all these advantages and still being an incorruptible technology, its technological potential has been reaching other applications that are not restricted to cryptocurrencies, attracting the interest of banks, companies, and governments. Blockchain uses decentralization to store information more securely. It has the ability to enable tracking on certain transactions. This is possible because its blocks are spread over thousands of computers around the world and, thus, a change can only be made when accepted by all systems belonging to a given network [6].

Through blockchain, data will never be erased or changed; transactions and interactions will be done with reliable rules; there will be proof of time, ownership, and rights; and there will be resistance against a single point of failure and censorship, just as there will be transparency and selective privacy. Based on this growing trend of process automation, there is a need to register and verify transactions done by machines connected to the Internet, such as replacement, maintenance, purchase, and sale, among others [7,8].

In addition, it automates this system of operations in a decentralized and secure manner, creating a way to deal more securely with the Internet of Things (IoT), allowing its development and ensuring the reliability of its future transactions.

The IoT market has developed remarkably in recent years, and with it, the number of invasions in homes, businesses, transport systems, and smart cities [9,10].

This is because, in addition to tracking, blockchain breaks with the current model of centralized data storage and bets on decentralization. With blockchain, we now have technology capable of encrypting, tracking, and certifying any information. Blockchain can avoid a vulnerable device from transmitting false information throughout the system or, in the case of a network overload, it can identify the point where something went wrong [11].

The goal of blockchain is to offer a model that adds confidence to unreliable environments and to reduce business interruptions by offering transparent access to information available in the chain. Within this context, blockchain can be used by companies to provide reliable, decentralized, encrypted, and uneditable databases. With blockchain, a broader view of information security is viewed, rather than traditional endpoint protection tools. This vision is broader and includes the security of the user's identity, transactions, and communication infrastructure through transparent processes [12,13].

This vision is increasingly necessary in today's connected world. In a world in which digital businesses and enterprises continually reinforce their defenses against security breaches, fraud, and intruders, blockchain can be an alternative to digital transactions.

This chapter aims to provide a scientific contribution related to the discussion and overview of blockchain technologies and how it impacts IoT, their applications in the current era, as well as the technology's potential for security and trust issues in IoT.

Therefore, this chapter intends to provide a modern review and an overview of blockchain and their applications, addressing its evolution and fundamental concepts, showing its relationship as well as debating its success, exposing a concise bibliographic scope, and particularizing and characterizing the potential of the technology.

The chapter is organized as follows: Section 9.2 argues blockchain concepts. Section 9.3 discusses blockchain technologies for security. Section 9.4 describes IoT and its security. Section 9.5 presents how both blockchain and IoT technologies interact. Section 9.6 presents future trends for blockchain, and finally, in Section 9.7, the conclusions of the research are presented.

9.2   BLOCKCHAIN CONCEPTS

Blockchain is an emerging technology that has been gaining great prominence in the world's technological scenario. Although initially created to enable secure transactions between virtual currency, due to its technological potential, the technology is not restricted to cryptocurrencies, and has been reaching other applications and attracting the interest of banks, companies, and governments, among others [14,15].

This technology is a kind of database where all information about Bitcoin transactions is stored, which is accessible to all users. In this sense, it is possible to access this database through a computer, or a mobile device, and see a negotiation that took place between two people, one in the United States and another in Brazil, for example. It can also be considered as a public ledger that registers a virtual currency operation so that the record is reliable and immutable [16].

For a better understanding, it is possible to use the simple analogy of "pages" of this "ledger" that are saved in various "libraries" around the globe; for this reason, erasing the knowledge stored in it is a difficult task. Blockchain technology is a chain of blocks that is part of a collective registration system. This implies that the information is not saved in a single place. Instead of being saved on a specific computer, all information is distributed among the various computers connected to it [17].

The blockchain registers information such as the number of virtual currency transacted, that is, who sent it, who received it, the moment that this transaction was effected, and wherein the book is registered. Hence, transparency is one of the essential features of blockchain technology. Although the details about who is involved in this transaction are impossible to know because everything is encrypted, it is known that the transaction occurred and it is recorded on blockchain technology [18].

Blockchain technology stores information for a group of operations in blocks, marking each block with a date and time stamp. Every time period, usually minutes on the blockchain, a new operation block is built, which is linked to the antecedent block. It is impossible to undo or change a transaction after it is entered into the system [19].

The blocks form a chain of blocks, which are dependent on each other, formed by a network of interconnected computers, a peer-to-peer network. The blockchain network is established by miners who perform checking and registration operations on the block. This establishes a great technology for information record that needs to be trusted, as in the case of a virtual currency transaction, as well as for exchanging information between IoT devices. In this sense, a set of transactions is placed within each of these blocks, which are locked by a strong encryption layer. On the other hand, the blockchain technology is public, implying that anyone can check and audit the movements recorded on it [20].

In the virtual financial transaction environment, the so-called "miners" lend processing and computing power to the network, are encouraged to continue contributing to make the network sustainable and more secure, and, in turn, receive a reward in the form of digital currencies. Each of the transactions carried out has a unique code, that is, a digital signature. The security of the blocks is in the specific digital signature, called a proof of work or hash. This signature is similar to the fingerprint of a block and assists to make the process more secure as everything is encrypted [21].

Any transaction made by the blockchain can only be validated when an entire "block" is filled with transactions. In other words, using the context of a virtual financial transaction environment, this is the only way to enable digital currency leave the hands of one person and go to another. This ensures that each currency reaches the right destination, one currency is not used more than once, as well as previous transactions are not changed without compromising the entire chain. Blockchain technology is considered inviolable because any attempt to change the record of a past transaction is easily identified by the computers on the network that are responsible for validating the process using a consensus algorithm [22].

9.2.1 CONSENSUS ALGORITHM

A consensus algorithm can be established as a mechanism by which each blockchain network reaches an agreement. Public (decentralized) blockchain networks are elaborated as distributed systems, which have no dependency on a central authority; the computers on the network must agree to validate the transactions. The consensus algorithms ensure that all transactions take place reliably and that the rules of the protocol are being followed, ensuring that each currency is spent only once [23].

Basically, the protocols are primary rules for using the blockchain network, and this algorithm acts as a mechanism by which these rules are followed. The blockchain network works in the sense that all participants and all different parts of the system in the network must follow the rules of the protocol, which defines how the system should work [24].

Thus, the algorithm dictates the system and what steps it needs to take to agree with those rules defined by the protocol, which produces the expected result. In the virtual financial transaction environment, what determines the validity of blocks and transactions is the consensus algorithm of a blockchain. Considering Bitcoin as a protocol, PoS (proof of stake) and PoW (proof of work) act as consensus algorithms [25].

The PoW algorithm was the first consensus algorithm developed. It is applied in Bitcoin and countless other cryptocurrencies. The PoW algorithm is a fundamental part of the mining operation. PoW mining covers countless initiatives to solve the puzzles, so the more computing power the more attempts per second. It consists of defining mining in the ordering of the blocks [26].

In the case of PoW, powerful machines are needed at an industrial level to carry out mining. It is for these cryptographic breaks and block creation that miners are rewarded. In PoW, virtual currencies are made available as blocks are closed and created, and have a mining limit in the case of Bitcoin. Once this limit is reached, the mining will stop and the remuneration will come from the fees charged by the network [27].

Considering that PoW offers protection against DDoS that reduces the total amount of the miner's amount, these blockchain algorithms offer a good deal of difficulty for hackers to use the system that requires a lot of power and computational effort. Therefore, miners who have high computational power are more likely to find a valid solution to the mathematical problem of the next block. PoW ensures that miners can only validate new transaction blocks and add them to the blockchain if the distributed network of nodes reaches consensus and agrees that the solution to the mathematical problem presented is a valid proof of the effort employed in the process [28].

The PoS algorithm was created in 2011 as an option for PoW. Even if PoW and PoS share the same goals, they have relevant differences and peculiarities, especially when it comes to validating new blocks. PoS replaces the PoW mining mechanism using each participant's allocated capital to validate new blocks [29].

This algorithm consists of the "proof of participation" philosophy, where miners are called "nodes" and need to have a large amount of coins in their possession for the blockchain to "draw them" so that they can solve the calculations and close the blocks, thus receiving the network bonus. Unlike PoW, coins from the PoS network are precreated and made available on the network before the blocks are created. Thus, there is no mining, but a forge. The bonus is already made with the fees of each block. As a blockchain using this system requires little performance from the machines that make up the network, it becomes a more environmentally friendly option for those interested [29].

PoS deals with the main disadvantages of the PoW algorithm, where each block is validated before the network adds another block to the blockchain record. This validator for each block (forger) is defined by the investment of the cryptocurrency in question and not by the computational power designated to the process. Each PoS system can employ the algorithm in different forms; blockchain is preserved by a pseudorandom election that judges the wealth of the age and the node of the coins (how long the coins are stored) together with a randomization factor [29].

LPoS (leased proof of stake) follows the same line as PoS; the difference is leasing. This system facilitates the maintenance of nodes through the contribution of network users. Users, who do not have a node, "lend" their coins to one for it to have more strength within the network, so that it does not necessarily need its own coins to forge. With this, the users themselves strengthen the network. In addition, to continue with the decentralization proposal, the user can cancel the lease at any time and can perform the procedure on the same node as many times and for as long as it exists. In this way, the proposal becomes more attractive and aims to bring more and more users to the network [30,31].

PoC (proof of capacity) requires storage capacity from the miner. Despite being the least known, it consumes the least energy and performance on the user's machines. It follows the same PoS draw concept, but what gives the miner strength is the machine's HD storage capacity. The more space, the more strength, and the more chance of being drawn [32].

This is how dPoW (delayed PoW) exists which uses the hash rate of another blockchain to promote the second layer of protection for a given blockchain. The dPoS (delegated PoS) allows participants to elect a representative to ensure the security of the network. The PoA (proof of authority) causes transactions to be validated by approved accounts. Proof of reputation, which depends on the reputation of the participants to keep the network safe, is similar to PoA, causing the network to penalize participants who try to attack it. Several other consensus algorithms are being used and developed with the focus on increasing the security of virtual currency transactions [32].

In the virtual financial transaction environment, consensus algorithms are primordial to preserving the security and integrity of a cryptocurrency network.

They enable ways for a network of nodes to reach a consensus regarding which version of the virtual currency is the real one. With these algorithms, a culture is created to correct failures, thereby increasing security and participation in the network [32].

9.2.2 HASH FUNCTION

Another factor that contributes to the security of this process is hashes. Each block has its own hash, a specific cryptographic signature. Similar to the security of the information exchanged, the link between the blocks is made by a hash, wherein each block has its own hash and the hash of a precedent block. As a result, a chain is established that links various blocks of information together [33].

Overall, to access the information contained in a block, it will be necessary to decipher the encryption of its hash and of the previous block. Because everything is connected like a chain, this process would have to be done successively with practically no end. In addition, to defraud blockchain, it would be necessary to change the data recorded on each of the various computers connected to the network. In this sense, any attempt to defraud the transaction history requires enormous computational power, which makes this attempt to change unfeasible. Considering that if a member of the network suspects fraud, the information is simply discarded. Therefore, defrauding the system is practically impossible. Thus, the information recorded in the blocks is reliable, immutable, and transparent as long as the majority of the network remains honest [34].

Each block creates a constant rhythm, like the beat of a heart or a song, having the maximum capacity. In applications of Bitcoin, new blocks are added to the network every approximately 10 minutes, considering the context of virtual financial transactions. Therefore, at this time, various transactions for the sale or purchase of virtual currency between users can be added and verified. Therefore, after a complete block is filled and verified, the cryptocurrency leaves the virtual wallet of the user who sold it and passes to the wallet of the one who bought it [35].

The security that the blockchain system offers is so relevant that companies and even government institutions are showing interest in using this technology. This is because the idea allows not only to protect data but also to share it with without giving up control over that information. In short, blockchain technology is a public distributed ledger, with the ability to register all virtual currency transactions in a blockchain, wherein anyone can participate. Due to decentralization, blockchain is becoming more and more comprehensive and bringing even more possibilities of use, because with the immutable registration proposal, speed in sending, receiving data, and network confirmations, the technology is finding several applications [36].

9.3 BLOCKCHAIN TECHNOLOGIES FOR SECURITY

Blockchain benefits information security. The information unit in technology is called a transaction, which can be anything from music, money, financial assets, property, and even data and information between devices, considering an IoT universe. The technology can be defined as a digital public register in which transactions are carried out, verified, and kept permanently. Its main function is to validate transactions. The data is saved within cryptographic blocks connected hierarchically, creating a chain of blocks [37].

It is possible to track and verify all transactions already made. Each time a transaction is made, it is marked with the information of who and when it was made. These data and the transactions are transmitted to a worldwide network and linked to the previous one, if there is a consensus between the participating computers. Once a transaction is certified by that network and saved, it can never be modified again. It is as if each block carries its own digital signature plus that of the previous block, which generates the current characteristic. This process, which is fundamental for security, allows the validation of all transactions in an interdependent manner [38].

Blockchain has three clear and main characteristics, which are decentralization, which is a key point as the distribution of transaction information between multiple computers is the basis of the technology. It is inherent and enables security and independence as there is no administrator or superior that has all the power and responsibility, and the control is in the hands of users collectively [39].

It enables tracking as long as it is possible to have an anonymous, decentralized Internet with guaranteed privacy protection. In blockchain, the system consists of two parts: a peer-to-peer (P2P) network and a decentralized database. This P2P network has users who share tasks, work, or files without the need for a central server (which brings a significant cost reduction), with all participants having equal privileges and influence on the environment. Each computer in the network is a node, and, whenever new data enters the system, it is received by all nodes. This information is encrypted and there is no way to track who added it, it is only possible to verify its validity. As a security measure, the method makes the distributed registration of information to decentralize the process. Thus, when one node leaves the network, the others already have a copy of all the information shared. Similarly, if new nodes enter it, the remaining create copies of the information for them [40].

Cryptography, considering that when a transaction is carried out, it receives a unique code, a digital signature, as described in Section 9.2. This stamp, with date and time, is verified by the chain participants themselves, and the operation is only incorporated into the blockchain (in the form of a new block) after it has been approved. This verification is an important step in preventing fraud. When a new valid block is found by the network, it gains a PoW, a hash, which is a code composed of encrypted numbers that work as a "protocol" implying that the transaction is valid. PoS serving as a protocol certifies that that user owns that information. After the block is created and validated by the network, it is added to the blockchain network [41].

In addition to the activity's digital signature, each block has its own encrypted code (hashes). The blocks carry the hash of the previous block, which is the link that keeps them connected. From this, a chain is created that protects the information by gathering transactions that have not yet been placed in a block and then calculating the hash to form the link between them. These calculations are quite complex and, therefore, made by high-performance computers. Furthermore, it is perfectly possible to trace the origin of any block from the fingerprint of just one of the parts [42].

9.3.1 ROLE OF THE HASH FUNCTION IN CRYPTOGRAPHY

The hash function is an algorithm used by Bitcoin and other cryptocurrencies, which receives an input of any length and creates a fixed-length output to transform a large amount of information into a fixed-size hexadecimal numeric sequence. In the case of cryptocurrency, each hash is created with the aid of a double-SHA-256 algorithm, which creates a random number of 512 bits (or 64 bytes). The number, like 128 or 256, usually refers to the length of the output, that is, SHA256 will produce a 256-bit output [43].

This function, in addition to protecting information, allows it to be shared without losing control over it. The reason is the complexity of the operation. For the hash to be accepted by the blockchain, the computer must generate a value that has the first 17 digits composed of the number zero. This happens only in a 1.4×1020 hash creation operations. Another way to understand hash functions is compression. A large entry is essentially compressed into a very short string representation of that entry. Blockchain employs hash functions everywhere in the block chain, whereas the data on the blockchain is "hashes" in each block. The hash value of the precedents block is applied to calculate the hash value of the current block, generating a link between the blocks [44].

That is also the point that makes Bitcoin so secure. Each block receives a hash based on the previous block. Therefore, if someone modifies a block already added to the blockchain to create a fake transaction, their hash will be modified. As each block is made from information from the previous block, a modification to the blockchain would affect the entire chain. As a result, users' ability to identify fraud and keep the system free from failures is increased [44].

Blockchain only accepts new hashes. Because each hash must have a certain number of zeros in the first digits, creation takes more time as there is no way to program the hash function to create keys according to the user's preference and the data is always being modified. Each attempt consumes processing power, energy, and, consequently, money. The easiest way to detect if the entry has been changed is to compare the message digest from two previous versions. If they match, it is certain that the person holding the purchase and sale title, for example, is actually the real owner of the house [45].

Another property of hashes is that they are unique. It is very easy to calculate a message digest, but it is almost impossible to discover the entry, meaning it would take approximately billions of years to do so. Thus, it is immutable because, if there is a change, it will be detected and rejected by the other nodes; and thus, hash functions play an important role [46].

9.3.2 SECURITY AS DECENTRALIZATION

In a centralized database, which is currently used by companies, all it takes is one "door" for an attacker to be able to enter the server and gain access to the data often without leaving a trace. With blockchain, an attacker would have to have control of 51% of the nodes in the chain, that is, he/she would have to invade several machines to validate a change. Moreover, all movements would be traceable. Blockchain decentralizes data and stores it more securely. Thus, it makes it possible to track information, with cybersecurity being one of the main results. In addition, the blocks can be stored on several computers around the world. Even though they are geographically distant, they are interconnected, and when a change is made to the blocks, it is only accepted if the systems that make up the network allow it [47].

Therefore, from the moment a computer (or terminal) is invaded, other systems will try to prevent the attack. Consequently, the invasion is eliminated by blockchain technology. This means that if an attacker succeeds in entering a single machine from a certain company, the other computers that make up the block will act, invalidating the action. Keeping this information safe is crucial to keeping the company's market value and sensitive information safe as it is used as a basis for strategic decision-making [48].

9.3.3 PUBLIC AND PRIVATE BLOCKCHAIN

Blockchain is a type of distributed ledger technology (DLT). Not every DLT is a blockchain, often it is just a computer network sharing the same database. Often these networks are small and centralized, which is the opposite of blockchain. It can be divided into two categories, public and private. The difference is between the groups granted to participate in the network, execute the consensus protocol to certify the transactions, and preserve the shared record [49].

In public networks, anyone can participate; in general, there is some incentive mechanism to encourage new members. An open and huge blockchain (like Bitcoin transactions) requires a demand for complex data processing with high computational capacity. A private or permissioned blockchain is restricted to invited or authorized people; they authorize specific people on the network to authenticate blockchain operations through an access layer. The entry needs to be validated by the person who started the chain or by rules predetermined by him/her. After an entity joins the network, it becomes part of the decentralized maintenance of the blockchain. As they do not require a consensual process, they are less disruptive due to dependence on an internal authority [50].

When considering the role of analytical intelligence for blockchain, it is possible to distinguish two categories of related data, that is, "data at rest," static data that already exists in the immutable database of a blockchain; and "data in motion," data that is being produced whenever a transaction is created on the blockchain [51–53].

9.3.4 ANALYTICAL INTELLIGENCE

Exporting static blockchain data to an analytical intelligence platform requires various analysis of the characteristics of transactions, their segmentation, prediction of future events, trend analysis, and identification of the relationships between the blockchain and several other data sources. Reflecting on the emergence of streaming analytics, blockchain technology offers analytical initiatives enabling the identification, close to real time, which results in an opportunity to take immediate action addressing activity on the blockchain operations [51–53].

Analytical models produced using static data can be leveraged to data in motion ensuring the authenticity and integrity of a blockchain. Real-time blockchain analytics can identify fraudulent activity and block any suspicious transactions at the time they are happening [51–53].

9.3.5 ADVANTAGES OF BLOCKCHAIN SECURITY

The greatest virtues of blockchain, as already discussed, are the distribution and decentralization of the network, security of data immutability, and transparency. Thus, the transaction is registered on all computers on the network and anyone can see it, and at the same time it is private because it is necessary to identify the transaction data and the parties involved [51–53].

Cybersecurity is one of the advantages that most excite blockchain enthusiasts. After all, in addition to enabling information tracking, it focuses on decentralization to store data more securely than other technologies we know. As described in Section 9.2, the blocks are "stored" on thousands of computers around the globe. Although geographically distant, they are interconnected. Therefore, if a computer is possibly hacked, others will notice this movement and try to avoid the attack. Consequently, it hampers the invasion of a system with this technology [52,53].

The security, privacy, and reliability of the technology are related to the fact that each user simultaneously processes the operations performed within the system, which is encrypted and only allows access for previously registered users. It is less subject to attacks and security breaches against private data than conventional systems, which generates greater reliability and verification possibilities, without depending on a specific and unique central authority. The efficiency of the technology is related to being a completely automated system based on computer code programming, and it has the benefit of being efficiently fast. It is possible to transfer data and other intangible assets in a matter of a few minutes, making transactions much more efficient [52,53].

The costs of each transaction are reduced as there is no central verifying authority, roles, excessive regulations, or too much human capital involved. This is a benefit for all users which increases throughput and makes blockchain technology cheaper. The system is transparent because transactions registered on a blockchain system are public and accessible by its users. Operations are recorded in chronological order to guarantee the validity and verification of each step practiced within the system, as in an accounting ledger. Thus, users will have access to the information as long as they have access to the system [52–54].

As more data is added to the blockchain, it acquires greater security; support over each new block is produced based on the shared precision of the preceding block. The security of the technology is improved by the transactions made by users and "things" in real-time; although it is not a foolproof method against poor security and bad data practices, it has the ability to prevent fraud [52–54].

Blockchain works like a database, a registration book that guarantees the authenticity and integrity of this transaction, starting from the impossibility of any type of alteration. In this sense, blockchain and information security are two closely aligned topics. Applying these technologies, it is possible to guarantee that a company will operate in a more protected manner and in line with the digital transformation that the world is going through [52–54].

9.4 SECURITY AND TRUST ISSUES IN IOT

IoT is a term that defines the set of devices that use wireless communication interfaces to collect, send, and receive digital data. The devices can "see, hear and capture motion" with high precision, as well as send such records to software over the Internet. IoT has already reached several devices due to the low cost of processing power, popularization of the wireless Internet, and low cost of sensors, making it more accessible to both governments and businesses and home users [52–54].

In the home environment, televisions can execute voice commands due to sensors and processors that analyze spoken messages in an instant. In the same sense, sensors act side by side with thermostats to set the temperature of business environments automatically, resulting in a reduction in cooling costs, among many other application possibilities. With the advent of smartphones and Big Data technology, IoT has become a commercial asset. Provided that through it, companies are allowed to collect a much larger number of records for analysis. Consequently, commercial strategies are defined with high precision. Making it easier for managers to find those factors that most influence the costs and profitability of the enterprise, and enabling them to remodel their routines strategically [52–54].

However, vulnerable connected devices pose a serious problem as existing vulnerabilities can be investigated by invaders to create botnets. The IoT era offers many products and services that enable the development of smart homes, from lights turning on, doors opening automatically, connected toys, recording devices, IP cameras, among many other equipment, and all types of devices imaginable. Among other applications, even vehicles without drivers, and crop information in real-time, through precision farming, are in the danger of existing security vulnerabilities [52–54].

9.4.1 IoT SECURITY ASPECTS

The main objective of using IoT in the corporate environment is to increase the level of business automation and increase the amount of information available for business analysis. However, for this to be possible, IoT gadgets need to capture data continuously. However, the more devices connected to the network, the greater the exposure to risks, such as theft or hijacking of data, alteration of information, or even the connected environment. This is due to security breaches on both sides, manufacturing, and use. Therefore, from the viewpoint of information security, it is necessary to consider the possible security loopholes that industries generally overlook [52–54].

As they are focused on associating new features and instituting devices easier to use and connect, manufacturers overlook an essential factor, device security. Similarly, insecure web interface, insecure network, lack of encryption, weak or insufficient authentication, insecure mobile and cloud interface, and security settings that make software unsafe are flaws that should not exist. Manufacturers must take meticulous care in all stages of implementation of their products and services, that is, from conception, development, and implementation to release [53–55].

IoT network security tends to have more challenges than traditional network security as there is a wider range of device features, communication protocols, and standards, which can present more complex problems. The main resources needed are traditional security tools, such as antivirus and antimalware, in addition to firewalls and detection systems and intrusion prevention [53–55].

Authentication of an IoT device is permitted as embedded machine-based sensors without any human intervention. For this, it is necessary to have special rules that allow these authentications with security and confidence without disturbing the processes [53–55].

Encryption relates to a wide variety of hardware and IoT devices profiles limited to have standard encryption protocols and processes. Also, all IoT encryption is required to be complemented by encryption key lifecycle management processes. Analytical security in IoT relates to solutions that monitor, aggregate, collect, and normalize data with the addition of refined machine learning techniques, Big Data, and artificial intelligence (AI). All of this provides predictive modeling for anomaly detection, in addition to reducing the number of false positives. Analytical security on IoT is necessary for detecting specific intrusions in the environment of connected "things," concerning complex attacks that may not be pointed by traditional network security solutions, as firewalls [53–55].

9.4.2 IoT COMMUNICATION SECURITY

In the same sense, the security of the APIs involved in the communication between different systems is essential to protect the integrity of the data exchanged between the devices and the backend software. One of the biggest problems of IoT concerns insecurity in relation to access interfaces, whether for information theft or DDoS attacks. as tools are needed that can detect threats and attacks against specific APIs (Application Programming Interface), where the objective is to certify that only authorized developers, applications, and devices, are communicating with these APIs [53–55].

Taking into account personal user information stored on the device or in the environment to which the device is connected, it is necessary to use passwords that cannot be changed and easy to guess through the use of brute force technique, or are still publicly available, and even backdoors in firmware or client software that allow unauthorized access to systems, taking advantage of these vulnerable passwords, which make these devices insecure, inappropriate, or unauthorized. Lack of encryption or access control for sensitive data that is within the ecosystem, including data at rest, in transit, or processing; what affect security issues in web, cloud, mobile, or backend APIs in an ecosystem of insecure interfaces, which are considered outside the devices what allow both devices and certain related components to be compromised [53–55].

What compromises the authenticity, confidentiality, or availability of information or allows unauthorized control remotely are unsafe and unnecessary network services running on the device itself, especially those exposed to the Internet. Devices or systems with unsafe default settings or without the possibility of making the system more secure by applying restrictions based on configuration changes [53–55].

The lack of firmware validation on the device, the scarcity of mechanisms to avoid going back, the absence of a simple system to update the device safely, the scarcity of security when sending (unencrypted traffic), and the absence of notifications about security changes due to updates. In the same way lack of security support on devices released for production, which include asset management, update management, secure disarming, systems monitoring, and response capabilities, are critical points that compromise the confidentiality, authenticity, or availability of information [53–55].

Similar to the monitoring and the adoption of safer standards to protect wireless networks, impacting the entire network infrastructure to search for failures and intruders, which is related to the creation of a hidden Wi-Fi network only for IoT equipment to isolate them from the rest of the infrastructure, along with the definition of rules for access and control of the devices to delimit the number of sensitive areas of the infrastructure that they can access, as well as the implementation of a firewall system, and the planning of the drafting of new equipment with a focus on security from the first steps [53–55].

The use of hardware components or third-party software from a compromised supply chain, as well as the use of unsafe software components/libraries and/or obsolete, taking into account the inclusion of insecure customization of the operating system platform, are factors that can compromise the device. The lack of measures allow the strengthening of IoT devices from a physical viewpoint, which means that cybercriminals can have access to confidential information that could be useful to take local control of the device in a future remote attack [53–55].

IoT devices can pose a major security issue for companies and activities that need to keep their data and information safe from attack. When it is no longer used or complied with, such measures mentioned above, for example, complex passwords or the creation of rules for accessing a given system, such equipment ends up becoming the gateway to the corporate network, allowing the visualization of internal data. Like internal users of these devices, they must have their responsibility. They should pay attention to purchasing products from companies that value security, while avoiding taking actions both online and offline that make them vulnerable [53–55].

Despite the risks, when adopting the IoT, infrastructure planning for receiving such equipment is important without expanding the existing level of vulnerabilities. Thus, the benefits of these gadgets tend to be taken full advantage of, creating increasingly effective and intelligent workflows [53–55].

9.5   BLOCKCHAIN AND IOT

IoT applications now require enhanced security as the integrated sensors and devices transmit information between themselves and over the Internet. Blockchain guarantees the protection of these communications, preventing the IoT devices from being compromised by cyber-attacks and users' behavioral patterns from being revealed. Unlike current IoT systems, whose architecture is based on a centralized server, which makes them vulnerable to failures at a single point in the network, blockchain functions as a distributed ledger of transactions and communication between the "nodes" of the IoT networks. By storing the list of transactions encrypted on several servers participating in the network, blockchain increases security and reliability [53–55].

Blockchain has become a great ally of companies in this regard by offering a standardized method that speeds up data exchange, allowing the execution of processes between IoT devices without intermediaries and in a secure manner. The current IoT ecosystem is completely connector-based and relies on a central server to identify the network node, which can be a smart device or object. The blockchain's differential lies in eliminating the need for this central agent to approve and validate transactions. Thus, the blockchain makes these devices communicate without the need for a central unit to identify them. Another important aspect is that the network can handle multiple devices without the need for additional resources [53–55].

9.5.1   BLOCKCHAIN FOR IoT SECURITY

Blockchain principles that make IoT more secure are basically the possibility of anonymity, enhanced security features, and the decentralization of transactions. Consequently, IoT requires a secure and private system that is lightweight, scalable, and distributed. Blockchain technology has the potential to meet the challenges imposed by IoT precisely because of its distribution, security, and privacy [53–55].

Blockchain provides the parties with a management platform to analyze a large amount of data. Because the IoT has the problem of organizing and analyzing this huge amount of data coming from related devices, blockchain ensures that the information is accepted and released only to trusted parties. This technology consists of a chain of blocks that aims at decentralization as a form of security. Because there is no central server validating requests to allow the exchange of information, users themselves participate in the validation of all transactions on the network [53–55].

In a distributed blockchain IoT network, the devices are in a point-to-point mesh network, being able to authenticate transactions and execute them based on predetermined rules, without the need for a central server. It is based on a decentralized platform (a Dapp), and its security does not depend on a single controller but on the network structure. The combination of both technologies can provide the creation of a reliable vehicle history platform based on a device that connects to the diagnostic port of a used car, extracting information, and recording this data on a blockchain. Thus, there is an immutable record of the history of a used car. No batch of used cars or duplicate ex-owners can alter the data, or even disconnect the device, which would mean a suspicious blank period in the records [53–55].

In the context of a smart device, for example, a code connected to the device can be linked to the Internet and generate alerts when cookies are ready or if the washer has finished the washing process. Energy efficiency and control of these devices when a person is away from home are among other benefits. In blockchain, the encryption of these devices ensures that only the owner of the devices is allowed to transfer information. Still, in the context of agriculture, a farm-to-plate tracking system for high-quality food, for example, how long has this steak been refrigerated at a certain temperature, how far has it traveled today, and so on. Using blockchain nodes mounted on boxes and trucks, it allows the creation of an untouchable record of the passage of an item through the transport infrastructure [53–55].

The sensors provide companies with end-to-end visibility of their supply chain, providing condition of supplies from data on the location as they are transported worldwide. With blockchain, it is possible to have the ability to store, manage, transfer, and protect this information against attacks and fraud [53–55].

Thus, the application of blockchain in IoT increases the guarantees of uniqueness and the "nonobstruction" of information due to the generation of a hash key that can prove the veracity of the sender using a digital signature. This signature generation process works from the fact that blockchain adds a private key of the sender (symmetric encryption) to the original data, hashing the encrypted information, and re-encrypting it with the recipient's public key. Through this process, the data is digitally signed by the issuer (self-signed certificate) [53–55].

Blockchain technology helps companies improve security through decentralized interaction and data exchange by optimizing the reliability of devices, applications, and platforms. By merging blockchain and IoT, companies have greater encryption, and the decentralized control offered by the blockchain are highly effective alternatives to replace traditional security mechanisms [53–55].

Hence, the use of IoT enhances the use of solutions based on blockchain. IoT sensors can generate the necessary evidence that a certain fact happened, guaranteeing the veracity of each transaction registered on the blockchain. IoT can be defined as the communication between machines and monitoring and control systems via the Internet, which allows from cars to smart devices or industrial machines, as well as to have the ability to share data to perform certain tasks. IoT has as a basis for the functioning data networks and sensors as it makes communication between "things" possible [53–55].

Blockchain allows IoT ecosystems to break the traditional paradigm of centralized networks based on connectors, where devices rely on a central server in the cloud to identify and authenticate individual devices. In addition, centralized networks are difficult to establish in many industrial sectors, such as large farms, where IoT nodes will expand into huge areas with scarce connectivity networks. Blockchain will allow the creation of more secure network meshes, in which IoT devices reliably interconnect, preventing threats such as impersonation and device spoofing [53–55].

However, in this environment, it is worth noting that it is not absurd to consider that IoT devices and sensors can be compromised to transmit wrong information to a blockchain. Still, the blockchain's distributed architecture can help solve many of the problems brought about by the increased use of IoT devices, such as monitoring the data measured by the sensors and preventing duplication with other malicious data; the sensors used can exchange data directly with each other instead of using a third-party system to establish digital trust; implementation and operating costs are reduced as there is no need for intermediaries; device identification, authentication, and data transfer used in IoT implementations are ensured; the feature of the blockchain concerning distributed registration eliminates the danger of failures of any link within the ecosystem, preventing the data on the devices from being manipulated. In case of problems, it allows the recording of transaction histories, which is easily accessible and cannot be manipulated [53–55].

9.5.2 Blockchain Benefits for IoT Security

The main advantages of integrating blockchain and IoT technologies are greater privacy for users, decentralization of records, more reliability for the IT infrastructure, and reduced operating costs. In the industrial environment, there is a level of complexity in the industry that helps to increase inefficiency, fraud, and waste of inputs related to the fact that companies generally need to rely on intermediaries to execute some processes. This lack of transparency in the supply chain forces companies to rely on third parties or paper records. With the addition of blockchain along with IoT, greater control of who can access the network is achieved. They provide a shared and immutable platform for all supply chain actors to track assets, saving time and resources and reducing fraud [53–55].

Blockchain can mitigate the risk of compromising IoT devices through a switch, improving the scalability of IoT implementations. In principle, blockchain allows the protection of IoT networks by forming consensus groups on suspicious behavior. Individuals and organizations using IoT should look for multilayered security with end-to-end protection, preventing potential intrusions and compromises on the network through a gateway to the endpoint. It is also essential to create more secure network meshes, in which IoT devices interconnect reliably, making it possible to avoid threats such as impersonation and device spoofing.

These factors strengthen security even more in the same way as blockchain; hence, transactions accomplished by various sources can be managed through a transparent and immutable record, in which physical assets and data are tracked across the supply chain. In case of an incorrect decision or system overload, the blockchain registry must be able to recognize and identify the point at which something went wrong [53–55].

Blockchain technology consists of a verifiable and transparent system that changes the way people exchange assets and value, share data, and enforce contracts. It avoids the risks of a centralized data system by storing data across a given network. Therefore, attackers cannot exploit vulnerabilities as they do on a centralized network. Encryption is used by security methods that adopt the use of public and private keys [53–55].

Blockchain technology is a secure and shared record of transactions distributed across a computer network. Companies can use blockchain technology to enable a new class of applications to act as a common data layer. Making data and business processes shared among countless organizations, reducing the risk of fraud, creating new revenue streams, and eliminating waste. Blockchain has been defined as synonymous with reliable and decentralized transactions [53–55].

9.6 FUTURE TRENDS

Blockchain is considered to be one of the most relevant and disruptive technologies that have emerged in recent years. Despite this, the chain records of blocks are still in an embryonic stage if we consider their full application potential. One of the main trends in the use of blockchain is to ensure that the data collected by the devices that power the IoT are not altered. At the same time, pondering the implementation of a computer and hardware infrastructure for receiving a ready and regulated blockchain cloud platform for legal transactions. In this ecosystem model, there are private and public keys working together, a process that provides flexibility in cases of emergency and privacy [56,57].

Distributed cloud storage will be another blockchain application that companies can benefit from. Most companies choose to use cloud services offered by technology giants for storing and accessing their data as many of the major providers are still vulnerable to external attacks. The integration of this technology for cloud computing can be used by companies providing reliable, decentralized, encrypted, and noneditable databases. Native blockchain encryption can also be used to protect networks of interconnected devices from external interference [58–60].

IoT, AI, and other technologies with blockchain to unlock the next level of value that these networks should bring to the first users. Blockchain networks, as organizations integrate their existing systems and business processes into these solutions, in turn, will trigger the link between blockchain solutions, probably at all levels of technology [61,62].

The tokenization of an asset is nothing more than transforming a contract, a property, a work of art, or even a part of a company into a digital fraction called a token. Using blockchain, the digital contract is issued and the document represents a real asset, having a market value, and can be negotiated quickly with less bureaucracy and legal validation [63,64].

In the same way that an incorruptible "data chain" was created, used in the financial market, especially for the transition of cryptocurrencies through the network. It can be used for predictive analysis, supply chain logistics, integration with IoT, for trade without financial intermediaries, and even in the health sector, with the transition of medical records in a safe and decentralized manner [65–67].

9.7 CONCLUSIONS

Blockchain can be simply understood as an online, public, and decentralized database created to make the distribution of information transparent and reliable, without the need for an external and centralizing agent to validate the process.

The great advantage of this technology is in the encrypted registration of information, with "fingerprints" that generate a dynamic of secure and reliable chaining, besides being based on a decentralized structure and always registering any changes that occur in the chain, making information exchange even safer. These advantages enable the emergence of blockchain applications that will change the dynamics of the business sector in the coming years.

Blockchain has numerous benefits and applications. It eliminates intermediaries and transactions take place in real time. In addition to providing security, blockchain reduces the risk of fraud as the data is auditable and verifiable.

As already described, the blockchain applicability is huge, mainly linked to IoT, as it prevents double expenses and fraud, still considering the potential in a standardized way for software and device to identity each data sent and received, records its current state, and validates that a particular transaction is legitimate.

REFERENCES

1. Nakamoto, S. (2019). *Bitcoin: a peer-to-peer electronic cash system*. Manubot.
2. Ulrich, F. (2017). *Bitcoin: a moeda na era digital*. LVM Editora.
3. Narayanan, A., Bonneau, J., Felten, E., Miller, A., & Goldfeder, S. (2016). *Bitcoin and cryptocurrency technologies: a comprehensive introduction*. Princeton University Press.
4. Crosby, M., Pattanayak, P., Verma, S., & Kalyanaraman, V. (2016). Blockchain technology: beyond bitcoin. *Applied Innovation*, 2(6–10), 71.
5. Vishwakarma, P., Khan, Z., & Jain, T. (2018). A brief study on the advantages of blockchain and distributed ledger in financial transaction processing. *International Engineering Management & Applied Science*, 7, 76–79.
6. Yuan, Y., & Wang, F. Y. (2018). Blockchain and cryptocurrencies: model, techniques, and applications. *IEEE Transactions on Systems, Man, and Cybernetics: Systems*, 48(9), 1421–1428.
7. Padilha, R. F. (2018). Proposta de um método complementar de compressão de dados por meio da metodologia de eventos discretos aplicada em um baixo nível de abstração [Proposal of a complementary method of data compression by discrete event methodology applied at a low level of abstraction]. [Master's thesis, University

of Campinas (UNICAMP), Campinas-São Paulo, Brazil] Repositório da Produção Científica e Intelectual da Unicamp.

8. Zheng, Z., Xie, S., Dai, H. N., Chen, X., & Wang, H. (2018). Blockchain challenges and opportunities: a survey. *International Journal of Web and Grid Services*, 14(4), 352–375.

9. França, R. P., Peluso, M., Monteiro, A. C. B., Iano, Y., Arthur, R., & Estrela, V. V. (2018, October). Development of a kernel: a deeper look at the architecture of an operating system. In *Brazilian Technology Symposium* (pp. 103–114). Springer, Cham, Switzerland.

10. França, R. P., Iano, Y., Monteiro, A. C. B., & Arthur, R. (2020). A proposal of improvement for transmission channels in cloud environments using the CBEDE methodology. In *Modern principles, practices, and algorithms for cloud security* (pp. 184–202). IGI Global, Pensilvânia, EUA.

11. Karame, G. O., & Androulaki, E. (2016). *Bitcoin and blockchain security*. Artech House, Norwood, MA, EUA.

12. França, R. P., Iano, Y., Monteiro, A. C. B., & Arthur, R. (2020). Intelligent applications of WSN in the world: a technological and literary background. In *Handbook of Wireless Sensor Networks: Issues and Challenges in Current Scenario's* (pp. 13–34). Springer, Cham, Switzerland

13. Bahga, A., & Madisetti, V. (2017). *Blockchain applications: a hands-on approach*. Vpt.

14. Wattenhofer, R. (2016). *The science of the blockchain*. Inverted Forest Publishing.

15. Drescher, D. (2017). *Blockchain basics* (Vol. 276). Berkeley, CA: Apress.

16. Hofmann, E., Strewe, U. M., & Bosia, N. (2017). *Supply chain finance and blockchain technology: the case of reverse securitization*. Springer, Switzerland.

17. Bashir, I. (2017). *Mastering blockchain*. Packt Publishing Ltd, Birmingham, United Kingdom.

18. Khan, M. A., & Salah, K. (2018). IoT security: review, blockchain solutions, and open challenges. *Future Generation Computer Systems*, 82, 395–411.

19. Fernández-Caramés, T. M., & Fraga-Lamas, P. (2018). A review on the use of blockchain for the Internet of Things. *IEEE Access*, 6, 32979–33001.

20. Conoscenti, M., Vetro, A., & De Martin, J. C. (2016, November). Blockchain for the Internet of Things: a systematic literature review. In *2016 IEEE/ACS 13th International Conference of Computer Systems and Applications (AICCSA)* (pp. 1–6). IEEE, Agadir, Morocco.

21. Garbow, Z. A., Kulack, F. A., & Paterson, K. G. (2017). U.S. Patent No. 9,830,634. Washington, DC: U.S. Patent and Trademark Office.

22. Fanning, K., & Centers, D. P. (2016). Blockchain and its coming impact on financial services. *Journal of Corporate Accounting & Finance*, 27(5), 53–57.

23. Mingxiao, D., Xiaofeng, M., Zhe, Z., Xiangwei, W., & Qijun, C. (2017, October). A review on consensus algorithm of blockchain. In *2017 IEEE International Conference on Systems, Man, and Cybernetics (SMC)* (pp. 2567–2572). IEEE, Banff, Canada.

24. Nguyen, G. T., & Kim, K. (2018). A survey about consensus algorithms used in blockchain. *Journal of Information Processing Systems*, 14(1).

25. Bach, L. M., Mihaljevic, B., & Zagar, M. (2018, May). Comparative analysis of blockchain consensus algorithms. In *2018 41st International Convention on Information and Communication Technology, Electronics and Microelectronics (MIPRO)* (pp. 1545–1550). IEEE, Opatija, Croatia.

26. Baliga, A. (2017). Understanding blockchain consensus models. *Persistent*, 2017(4), 1–14.

27. Mingxiao, D., Xiaofeng, M., Zhe, Z., Xiangwei, W., & Qijun, C. (2017, October). A review on consensus algorithm of blockchain. In *2017 IEEE International*

Conference on Systems, Man, and Cybernetics (SMC) (pp. 2567–2572). IEEE, Banff, Canada.

28. Corchado, J. M. (2019, June). Blockchain technology: a review of the current challenges of cryptocurrency. In *Blockchain and applications: International Congress* (Vol. 1010, p. 153). Springer, Switzerland.

29. Gupta, S., & Sadoghi, M. (2019). Blockchain transaction processing. In S. Sakr, A. Zomaya (eds.), Springer International Publishing AG, part of Springer Nature 2018, Encyclopedia of Big Data Technologies, doi: 10.1007/978-3-319-63962-8_333-1

30. Salimitari, M., & Chatterjee, M. (2018). An overview of blockchain and consensus protocols for IoT networks. arXiv preprint arXiv:1809.05613.

31. Król, M., Sonnino, A., Al-Bassam, M., Tasioupolos, A., & Psaras, I. (2019, May). Proof-of-prestige: a useful work reward system for unverifiable tasks. In *2019 IEEE International Conference on Blockchain and Cryptocurrency (ICBC)* (pp. 293–301). IEEE, Seoul, South Korea.

32. Mattila, J. (2016). *The blockchain phenomenon*. Berkeley Roundtable of the International Economy.

33. Jain, A., & Mangal, N. (2018, December). Application of stream cipher and hash function in network security. In *International Conference on Computer Networks, Big Data and IoT* (pp. 429–435). Springer, Cham, Switzerland

34. Sharma, S. R. (2017). Blockchain technology review and its scope. *International Research Journal of Engineering and Technology (IRJET)*, 4.

35. Seebacher, S., & Schüritz, R. (2017, May). Blockchain technology as an enabler of service systems: a structured literature review. In *International Conference on Exploring Services Science* (pp. 12–23). Springer, Cham, Switzerland

36. Mansfield-Devine, S. (2017). Beyond Bitcoin: using blockchain technology to provide assurance in the commercial world. *Computer Fraud & Security*, 2017(5), 14–18.

37. Lin, I. C., & Liao, T. C. (2017). A survey of blockchain security issues and challenges. *IJ Network Security*, 19(5), 653–659.

38. Halpin, H., & Piekarska, M. (2017, April). Introduction to security and privacy on the blockchain. In *2017 IEEE European Symposium on Security and Privacy Workshops (EuroS&PW)* (pp. 1–3). IEEE, Paris, France.

39. Mahmood, B. B., Muazzam, M., Mumtaz, N., & Shah, S. H. (2019). A technical review on blockchain technologies: applications, security issues & challenges. *International Journal of Computing and Communication Networks*, 1(1), 26–34.

40. Li, X., Jiang, P., Chen, T., Luo, X., & Wen, Q. (2020). A survey on the security of blockchain systems. *Future Generation Computer Systems*, 107, 841–853. Elsevier.41. Stinson, D. R., & Paterson, M. (2018). *Cryptography: theory and practice*. CRC press, Boca Raton, FL, EUA.

42. Goldreich, O. (2019). On the foundations of cryptography. In *Providing sound foundations for cryptography: on the work of Shafi Goldwasser and Silvio Micali* (pp. 411–496).

43. Swathi, E., Vivek, G., & Rani, G. S. (2016). Role of hash function in cryptography. *International Journal of Advanced Research in Science (IJAERS)*.

44. Alotaibi, M., Al-hendi, D., Alroithy, B., AlGhamdi, M., & Gutub, A. (2019). Secure mobile computing authentication utilizing hash, cryptography and steganography combination. *Journal of Information Security and Cybercrimes Research (JISCR)*, 2(1).

45. Maseleno, A., Othman, M., Deepalakshmi, P., Shankar, K., & Ilayaraja, M. (2019). Hash function based optimal blockchain model for the Internet of Things (IoT). In *Handbook of multimedia information security: techniques and applications* (pp. 289–300). Springer, Cham, Switzerland.

46. Buldas, A., Saarepera, M., & Pearce, J. (2018). U.S. Patent Application No. 15/877,495.
47. Mylrea, M., & Gourisetti, S. N. G. (2017, September). Blockchain for smart grid resilience: exchanging distributed energy at speed, scale and security. In *2017 Resilience Week (RWS)* (pp. 18–23). IEEE, Wilmington, DE, USA.
48. Taylor, P. J., Dargahi, T., Dehghantanha, A., Parizi, R. M., & Choo, K. K. R. (2019). *A systematic literature review of blockchain cybersecurity.* Digital Communications and Networks.
49. Guegan, D. (2017). Public blockchain versus private blockchain. Archive ouverte en Sciences de l'Homme et de la Société
50. Gabison, G. (2016). Policy considerations for the blockchain technology public and private applications. *Science and Technology Law Review*, 19, 327.
51. Lai, R., & Chuen, D. L. K. (2018). Blockchain – from public to private. In *Handbook of blockchain, digital finance, and inclusion* (Vol. 2, pp. 145–177). Academic Press, Cambridge, MA, EUA.
52. Katragadda, R. B., Ramirez, J., Kumar, G. K., Karipineni, C., Vellanki, S., & Kolachalam, S. (2020). U.S. Patent No. 10,542,046. Washington, DC: U.S. Patent and Trademark Office.
53. Ølnes, S., & Jansen, A. (2018, May). Blockchain technology as infrastructure in public sector: an analytical framework. In *Proceedings of the 19th Annual International Conference on Digital Government Research: Governance in the Data Age* (pp. 1–10). Delft, Netherlands.
54. Jeschke, S., Brecher, C., Meisen, T., Özdemir, D., & Eschert, T. (2017). Industrial internet of things and cyber manufacturing systems. In *Industrial Internet of Things* (pp. 3–19). Springer, Cham.
55. Srivastava, G., Parizi, R. M., & Dehghantanha, A. (2020). The future of blockchain technology in healthcare internet of things security. In *Blockchain cybersecurity, trust and privacy* (pp. 161–184). Springer, Cham, Switzerland.
56. Lao, L., Li, Z., Hou, S., Xiao, B., Guo, S., & Yang, Y. (2020). A survey of IoT applications in blockchain systems: architecture, consensus, and traffic modeling. *ACM Computing Surveys (CSUR)*, 53(1), 1–32.
57. Gupta, S., Malhotra, V., & Singh, S. N. (2020). Securing IoT-driven remote healthcare data through blockchain. In *Advances in data and information sciences* (pp. 47–56). Springer, Singapore.
58. Xie, S., Zheng, Z., Chen, W., Wu, J., Dai, H. N., & Imran, M. (2020). Blockchain for cloud exchange: a survey. *Computers & Electrical Engineering*, 81, 106526.
59. Wang, H., Qin, H., Zhao, M., Wei, X., Shen, H., & Susilo, W. (2020). Blockchain-based fair payment smart contract for public cloud storage auditing. *Information Sciences*, 519, 348–362.
60. Wei, P., Wang, D., Zhao, Y., Tyagi, S. K. S., & Kumar, N. (2020). Blockchain data-based cloud data integrity protection mechanism. *Future Generation Computer Systems*, 102, 902–911.
61. Parker, B., & Bach, C. (2020). The synthesis of blockchain, artificial intelligence and Internet of Things. *European Journal of Engineering Research and Science*, 5(5), 588–593.
62. Abraham, M., Aithal, H., & Mohan, K. (2020). Real-time smart contracts for IoT using blockchain and collaborative intelligence-based dynamic pricing for the next generation smart toll application. arXiv preprint arXiv:2002.12654.
63. Narayan, R., & Tidström, A. (2020). Tokenizing coopetition in a blockchain for a transition to circular economy. *Journal of Cleaner Production*, 263, 121437.

64. Rodríguez Bolívar, M. P., & Ølnes, S. (2020, January). Introduction to the Minitrack on Blockchain, DLT, Tokenization, and Digital Government. In *Proceedings of the 53rd Hawaii International Conference on System Sciences*, Maui, Hawaii.

65. Wamba, S. F., Queiroz, M. M., & Trinchera, L. (2020). Dynamics between blockchain adoption determinants and supply chain performance: an empirical investigation. *International Journal of Production Economics*, 107791. doi:10.1016/j.ijpe.2020.107791.

66. Somin, S., Gordon, G., Pentland, A., Shmueli, E., & Altshuler, Y. (2020). ERC20 transactions over ethereum blockchain: network analysis and predictions. arXiv preprint arXiv:2004.08201.

67. Zehir, S., & Zehir, M. (2020). Internet Of Things in blockchain ecosystem from organizational and business management perspectives. In *Digital business strategies in blockchain ecosystems* (pp. 47–62). Springer, Cham, Switzerland.

10 Blockchain as a Lifesaver of IoT
Applications, Security, and Privacy Services and Challenges

Sukriti Goyal and Nikhil Sharma
HMR Institute of Technology & Management

Ila Kaushik
Krishna Institute of Engineering & Technology

Bharat Bhushan
School of Engineering and Technology, Sharda University.

Abhijeet Kumar
HMR Institute of Technology & Management

CONTENTS

10.1 INTRODUCTION

Today, human beings live with technology at their fingertips or in their hands. Technology has developed by giving trendy and fun tools in our palms, and the progress in technology is not just limited to our homes or in our palms but at work, in hospitals, in schools, at airports, and many other places. Technological advancements have helped us to instate so much more by giving more knowledge, security of data, and low optimization of power and latency. Nowadays, technology is not just an auxiliary but a requirement when it comes to data security. Among the list of several technologies, here we introduce the Internet of Things (IoT).

Over the last few years, humans have observed the capacity of IoT to supply amazing services across various fields ranging from agriculture to healthcare, from intelligent transportation to smart cities, from social media to smart business, and from smart homes to smart industries [1]. In very simple terms, IoT is the concept of "taking all the things (like gadgets) in the world and linking them to the Internet" [2]. IoT builds up a network of embedded gadgets including sensors and connection operations [3]. The different gadgets are seamlessly interrelated by IoT with distinct range of capabilities in the human and device pivoted systems to achieve the developing needs of the above-mentioned fields. It is becoming one of the major technologies of the fourth industrial revolution because it has given rise to new or fresh values in the linked smart world by storing data in huge amounts, uploading data into clouds, and then processing the uploaded data in intelligent networks.

IoT is a necessity in our lives. It helps us live and work smarter not harder. From offering smart gadgets to smart homes, IoT is necessary for businesses as well. IoT gives businesses a real-time insight into everything from the service of machines to supply chain and logistic operations, as well as a look into how their machines or systems work. Although it is clear that IoT is one the major technologies of everyday life, it has security challenges that need to be resolves. At present, it is not only humans with their computer systems but also there are "things" that connect with the Internet without human interference of humans, and the Internet is continuously communicated by these "things," for example, a vehicle sending messages to the mechanic to inform about its level of engine oil. Hence, it is not incorrect to say that IoT is an amazing technology in many fields and ways, but unluckily, developers have not developed it yet to overcome many issues such as security to make it completely safe and secure. However, blockchain is a gamechanger technology for IoT.

The technology of blockchain has laid the foundation of a new kind of Internet by permitting digital data to be disseminated but not copied. In brief, a blockchain is a time-stamped sequence of inconvertible records of information that are maintained by a group of computer systems not belonging to any individual system. Each of these blocks of information is safe and secure and is bound to each other using cryptographic principles. The network of blockchain does not have any type of middlemost authorization. As it is a contributed and inconvertible ledger, the data is open for anyone and everyone to see. Therefore, blockchain can strengthen the IoT in several ways, such as by securing the connection and configuring IoT. Hence, it can also be known as the "life saver of IoT." With the recent rise in the interest in IoT and blockchain technologies, many new applications and platforms have been introduced. The main focus of this chapter is on blockchain technology, how blockchain acts as a solution for

IoT concerning the concept of security and privacy, and what is the significance of blockchain-based IoT. This chapter is organized as follows. After a detailed introduction, Section 10.2 presents the history of IoT and how it evolved from the time of pulse dial phone. As the network of IoT cannot processed individually, it requires some elements to process efficiently. The fundamental elements and different architectures of the IoT, mainly three-layered and five-layered architecture, are illustrated as the technology of IoT has no integrated architecture. As we know, each object has two aspects, advantageous and disadvantageous, similarly, the IoT network along with its advantages has some issues which might be dangerous for the users; therefore, in this section, these issues are introduced. Every problem has a solution, therefore, the problems associated with the IoT also has a solution, that is, blockchain. How the blockchain technology works as a problem-solver of IoT is described in Section 10.3. In Section 10.4, blockchain is explained in detail along with a brief introduction to the architecture of the blockchain technology, working systems of mining, and features of blockchain. The main goal of this section is to familiarize the readers with the technology of blockchain as well as its primary theories. Further, an idea of merging both the blockchain and the IoT technology is proposed in Section 10.5 followed by the conclusion in Section 10.6.

10.2   HISTORY AND EVOLUTION OF INTERNET OF THINGS

The Internet isn't that primitive compared to the World Wide Web. The architecture of the TCIP/IP network that we are familiar with existed in 1974. However, recently, Internet is taking command of almost every facet of our life, whether it is completely accepted by us or not. What was once a pipe dream has turned into a reality, and the credit goes to the IoT. Now, it's the time to know about the history of IoT.

From Figure 10.1, one thing is clear that the term "Internet of Things" is not a new term. It is continuously used for many years. In 1926, the primary look at the IoT came from Nicola Tesla. In 1998, Google was integrated, and InTouch, a project that evolved at MIT, was put into application by Professor Hiroshi Ishii and Scott Brave who declared: "We then present in InTouch, which applies 'synchronized disseminated physical objects' to develop a 'tangible telephone' for long distance communication." The actual technology of IoT was introduced by Mark Weiser in 1998. Mark developed a water fountain that was marvelous as well as delectable to each person who saw it. It increased and decreased according to the trends of pricing and the volume of stock on the NYSE.

In 1999, Kevin Ashton spoke of IoT who was the executive director of the Auto-ID Center at that time and business week in the same year was the time of the next huge proclamation about the term IoT. Then, the technology of IoT started to grow continuously and the future of IoT is now. Today, the world completely relies upon linked medical gadgets, linked cars, linked smartphones, and even linked homes or smart homes. Currently, companies and organizations are struggling to get their own networks of IoT online and newer enrollments are

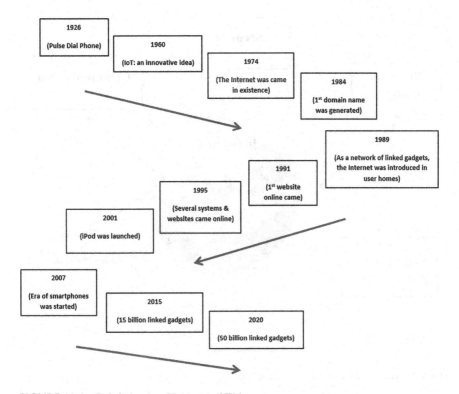

FIGURE 10.1   Brief history of Internet of Things.

being brought daily to head up IoT networks in companies and organizations from small to large.

10.2.1   ELEMENTS OF INTERNET OF THINGS

In short, the technology of IoT is the idea linking any gadget to the Internet and to other linked gadgets. It is a huge system of linked things and humans – all of which store and transmit data about the way they are used and their surroundings [4]. There are few basic necessary elements of IoT that are needed to deploy an IoT network as presented in Figure 10.2.

10.2.1.1   Sensors/Gadgets

In IoT, gadgets or sensors are the major physical objects linked to the network. They help in storing minute information from the environment. All of this stored information can have several degrees of involutions ranging from a basic temperature invigilating sensor. A gadget can have a number a sensors that are tied together to do more than just sense things, and one of the best examples is mobile phones that have many sensors such as and camera, GPS, but it does not easily sense things.

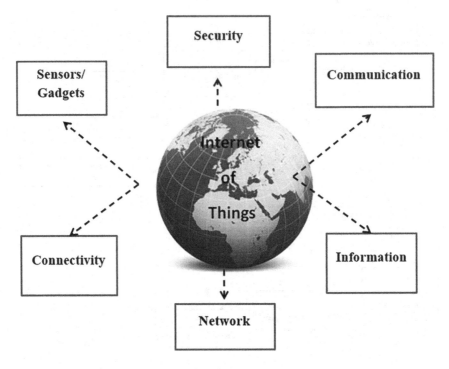

FIGURE 10.2 Elements of Internet of Things.

10.2.1.2 Security

Security is a delicate element of the IoT network, yet it is too often ignored in the production of networks. Approximately, on a daily basis, penetrability in IoT is being exploited with nasty intentions, yet the huge majority of them can be stopped easily. A safe network begins with rooting out the penetrability in IoT gadgets and equipping these gadgets in such a way that they can find and prevent nasty attacks and recover from them.

10.2.1.3 Communication

IoT gadgets need a medium for transferring the data perceived at the gadget level to a service based on cloud for further processing. This is where the great value ingrained in IoT is formed which needs either Wide Area Network (WAN) communications or Wireless LAN-based communications (Wi-Fi). Moreover, other abilities may also be required relying on the requirement of short-range communications.

10.2.1.4 Information

It is an essential element in IoT networks. The useful and important information is delivered to the end user that may be a commercial, a consumer and an orga-nizational user. In addition, it may be another gadget in the Machine to Machine

(M2M) workflow. The main motive in a consumer use case is to give the data in an easy and efficient manner.

10.2.1.5 Network

The components of IoT are linked together by networks using several wire-oriented and wireless standards, technologies, and protocols to give extensive connectivity.

10.2.1.6 Connectivity

The IoT network requires a means for transporting the stored data to the cloud. Through multiple means of transports and communications like Wi-Fi, WAN, cellular networks, Bluetooth, satellite networks, low-power WAN, and many others, the sensors can be linked to the infrastructure of cloud. Every choice user opts for has some characteristics and trade-offs between optimization of power, bandwidth, and range. Therefore, it is important in the network of IoT to opt the best connectivity choice.

10.2.2 Architecture of Internet of Thing

There is no individual integrated architecture of IoT which can satisfy all its requirements. Some of the important requirements are illustrated in Table 10.1.

TABLE 10.1
Requirements for the Architecture of IoT [5]

Requirements	Description
a. Availability and QoS	Minimum fault staying and delays in the network.
b. Concurrent storage of data	It supports the storage and investigation, and controls a large number of data from various sensors or actuators.
c. Communication and connectivity	Provides best connectivity of network and support for flexible protocols between the sensors or actuators and the cloud architecture for communications.
d. Gadget management	Enables the management and updates of remote or automated gadgets.
e. Data security	End-to-end encryption of data and its invigilating.
f. Described APIs	Each level should have a described or defined API that permits convenient unification with current applications and unification with other IoT solutions.
g. Composed, platform-independent, and flexible	Each level should permit characteristics, hardware, or cloud architecture to be sourced from distinct providers.
h. Optimized handling of data	Reduction of raw data and increase the useful and actionable data.
i. Scalability	Using the similar infrastructure, scale each individual component in the network.

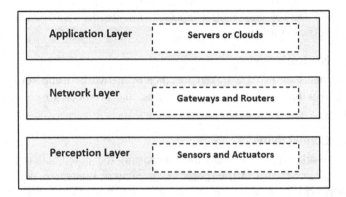

FIGURE 10.3   Three-layered architecture of IoT.

While it is impossible to cover all the possibilities and permutations, it's time to discuss the different architectures of IoT which give a greater understanding of the basic structure and major functional layers in an end-to-end stack of IoT [6]. Two architectures are considered and described as follows.

10.2.2.1   Three-Layered IoT Architecture

In the IoT, the most core architecture is the three-layered architecture, as illustrated in Figure 10.3. It was proposed in the early phases of research in the field of IoT. It consists of three layers, which are known as the application layer, the network layer, and the perception layer.

- *Perception Layer*: This is the physical layer in the architecture, which has sensors for sensing and gathering the data about the surrounding environment. Also, it percepts other physical parameters and finds out other smart things in the surrounding.
- *Network Layer*: This layer is responsible for linking to other smart objects, network gadgets, and servers. Its characteristics are also used for delivering and implementing data of sensors.
- *Application Layer*: This layer transmits application-specific services to the user. It describes several applications in which IoT can be deployed such as smart cities, smart electricity, smart healthcare, and smart homes.

10.2.2.2   Five-Layered IoT Architecture

As in the above architecture, the three-layered architecture of IoT describes the core concept of IoT. But it is not enough for research on IoT because research often targets accurate aspects of the IoT. So, here is the five-layered architecture of IoT which includes the business layer and the processing layer. Thus, all five layers are business, application, processing, transport, and perception layer, as shown in Figure 10.4. The task of the application and the perception layers in this architecture is similar to the three-layered architecture. The functioning of the remaining three layers is:

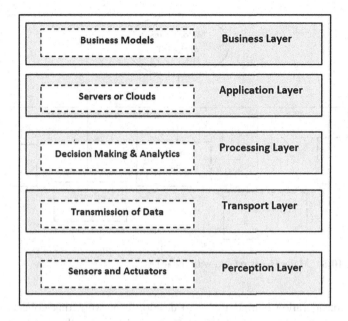

FIGURE 10.4 Five-layered architecture of IoT.

- *Transport Layer*: The data of sensors is transmitted by the transport layer from the perception layer to the processing layer and vice versa through networks such as RFID, LAN, 3G, NFC, and Bluetooth.
- *Processing Layer*: This layer is known as the middleware layer. A huge amount of data that comes from the transport layer is gathered, examined, and processed by the processing layer. In addition, a distinct set of services can be maintained and supplied by it to the lower layers. It employs multiple technologies such as big data processing resources, databases, and cloud computing.
- *Business Layer*: The complete network of IoT including, applications, profit, and business models as well as privacy of users are managed by the business layer.

10.2.3 MAJOR ISSUES OF INTERNET OF THINGS

IoT is a bond of gadgets and services that permit sharing or swapping of data. These gadgets range from home appliances to production machineries that can be implanted with several software and mediums of electronic connectivity. To maintain objects in an easier manner, everything and anything can be coordinated to link ion the Internet and become a portion of the web. Thus, the network of IoT spans a huge number of gadgets, including people and their communications via the Internet.

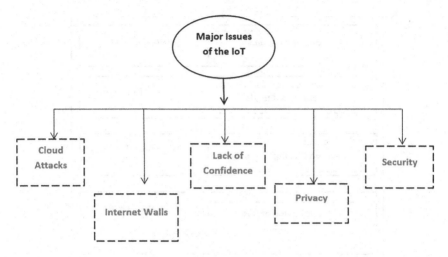

FIGURE 10.5 Major issues related with the IoT.

The consideration behind IoT is to develop a network that collects all the data needed by people without having a straight hand in gathering it. Currently, the influence of IoT on our lives is rising exponentially, and according to many researches, in forthcoming years, the influence of IoT on the world will be overwhelming. Thus, many questions have risen simultaneously on the potential threats and dangers of inculcating IoT into all dimensions of human lives. The major issues related with the IoT, as depicted in Figure 10.5, are explained in the subsections below.

10.2.3.1 Cloud Attacks

As the systems of cloud have huge stocks of data to implement the IoT, it is highly likely that the next probable risks to the IoT would be cloud systems.

10.2.3.2 Internet Walls

The danger of losing crucial information via hacks is a hazardous proposition not only for organizations but also for nations via cross-border invasions, and there is a high probability that these kinds of invasions will propel nations to form Internet walls that will limit IoT operation to specific regions. Furthermore, nations will be pushed to save their economic interests as governments cannot execute intensively in a global network of online industries and organizations, which finally deals the amazing concept of the IoT as hurdles clog the unregulated transfer of information that many companies and enterprises claim.

10.2.3.3 Lack of Confidence

It is the most common and major issue in the IoT networks. According to some researches, 54% of users obtain an average of four gadgets of IoT, but after that, only 14% take into account that they are aware of the security of IoT gadgets.

Moreover, 65% of users are petrified about a hacker controlling their gadgets, while on the other side, around 60% are irritated of either their professional or personal data being leaked.

10.2.3.4 Privacy

Another major issue with IoT is user privacy. Hacking is not only considered a disruption of security, but breaching of user's privacy is also considered as disruption of security. In the networks of IoT, there is a deficiency of privacy due to which users are largely unsatisfied with it. Because of the increase in awareness of the extent of cyber-control in users, they have started considering their confidentiality more weightily and demand that complete monitoring over their information should remain with them only rather than any other third party.

10.2.3.5 Security

Security risks are one of the greatest risks to the IoT. As the regulations of security are still not completely developed, it is possible that the networks of cloud will be the first to be compromised. The current growth in ransomwares is considered a serious risk of security, in which the service suppliers and cloud vendors will be the major goals. If cloud network is compromised, interconnectivity can also serve as a deficit to several companies.

Apart from these issues there are many other problems associated with the IoT that require a solution, which will be discussed in the coming sections. Thus, it is evident that to drive efficiency and industrial flexibility and to fulfill the demands of their customers, organizations all over the world are boarding onto digital transformation businesses that are IoT-driven. There might be threats to these businesses, however, if managed properly, organizations could be assured, and the path to the victory of IoT and potential will be smooth.

10.3 BLOCKCHAIN AS A SOLUTION FOR IOT

Blockchain is a cryptographically safe and disseminated ledger that permits the safe transmission of data between authorized parties. The conventional networks of IoT rely on a concentrated structure in which data is transferred from the gadget to the cloud where data is implemented using analytics and then further transferred back to the gadgets of IoT. With millions or billions of gadgets set to connect networks of IoT in the forthcoming years, this kind of concentrated network has very bounded flexibility, discloses millions of weak points that deal with the secrecy of network, and will become implausibly costly as well as slow if unauthorized third-parties have to frequently monitor and verify each and every micro transaction between the gadgets.

The systems of blockchain consist of smart contracts that permit gadgets to operate safely and independently by forming agreements that are only performed upon fulfillment of particular requirements. Smart contracts not only permit greater computerization, flexibility, and inexpensive transmissions but also stop cancellation by unauthorized parties that want to use the information for their own advantage. It is very tough to deal with the security of network as the data is transmitted across a decentralized, cryptographically secured system. Ultimately with a concentrated system, the threat of a single point of failing leads to disabling an entire system is a real probability. This threat is reduced by a decentralized network of blockchain with billions of separate nodes that transmit data on the basis of P2P (peer-to-peer) network to keep the remaining part of the IoT system running fluently.

10.4 MORE ABOUT BLOCKCHAIN

A blockchain is a disseminated, safe, and secured ledger that helps collect and track the resources without requiring a concentrated believed authorization. In a peer-to-peer (P2P) network where disseminated conclusions are made by majority of authorities instead of by an individual concentrated authority, blockchain permits two parties to interact and share resources [7]. It is practically safe against invaders who try to command the system by compromising the central commander. The technology of blockchain has drawn terrific interest from both industry as well as academia. Various applications that are trendy and running the research of networking are expanded recently by this technology. IoT [8], storage of cloud [9], and healthcare [10] are included in these applications. In general, the technology has endorsed its efficiency in any application that recently needs a centric ledger. The controlling of network as well as security services inclusive of secrecy, authenticity, confidentiality, and verification are some applications among the network of blockchains. These services are supplied by authenticated or verified third-party using incapacitated disseminated scenarios, which conclude that for present applications, security is a primary dare. On the other hand, the guarantee of security solves many conventional problems, including giving a completely disseminated, consensus as well as practicably secure solution is provided by the blockchain. The distinctness between the conventional access monitoring and the technology of blockchain-based access monitoring is described in Figure 10.6.

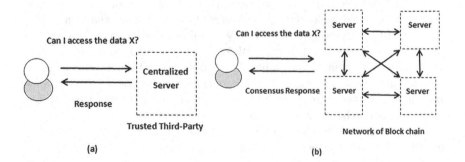

FIGURE 10.6 (a) Conventional centralized access control. (b) Blockchain-based access control.

10.4.1 BACKGROUND OF BLOCKCHAIN

Here, first, a concise introduction to the architecture of the technology of blockchain is presented. Then, the systems of mining or block building are discussed. Along with a proportion of distinct open-source blockchain executions, the amazing features of this technology are also introduced. The main goal of this part is to familiarize the readers with blockchain as well as its major theories.

10.4.1.1 Architecture of Blockchain

As shown in Figure 10.7, a network of nodes and database constitute a blockchain. The database of a blockchain is a disseminated, defect-endurable, attach-only, and shared that manages the record of data in blocks. Although the blocks of blockchain are permeable by all its users, they cannot be modified or removed by them. In the chain, the blocks are linked in such a manner that each block has a hash value of its preceding block, and a timestamp implying the composition time of that block is included in each block of the chain. The blockchain network constitutes the nodes that handle the chain of blocks in a disseminated and P2P manner. Although all of the nodes have access to the blocks, they cannot monitor them fully.

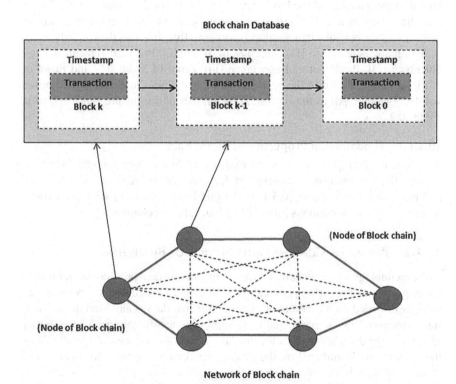

FIGURE 10.7   Basic architecture of blockchain.

The technology of blockchain permits the interacting parties to communicate in the absence of a verified and trusted third-party. The communications between the parties are entered in the database of blockchain with the required security. The "transaction" of the user is broadcasted to the blockchain network when a user communicates with another user of blockchain. When the communications are permissible and create a new block of permissible or acceptable transactions by merging various acceptable transactions (or mining) in the chain, only the various nodes in the blockchain network are verified. If the newly created block is found to be recognizable, it is added in the database of the blockchain and cannot be modified or removed in the future, otherwise the block collapses.

10.4.1.2 Mining a Block in a Blockchain

The mechanism of constructing blocks that will be linked to the database of a blockchain is known as "mining." In some of the applications of blockchain, like Bitcoin, the miner who constructs the first recognizable block is rewarded by the system; for financial applications, the reward is basically given in terms of money. The process of mining is one of the delicate concepts in the technology of block-chain. The nodes in the network are permitted to construct new blocks which will be verified by other blocks. If the new block is discovered as acceptable, it is linked to the database of the blockchain, and the nodes that construct blocks are known as "mining nodes." Further, mining nodes run to confirm the transactions and construct a new block as soon as possible so that they win the reward.

Minimum Block Hash [11], Measure of Trust [12], Practical Byzantine Fault Tolerance [13], PoI Proof of Importance [14], Proof of Stake [15] and Proof of Work [16] are included in the various current techniques to decide which miner will be rewarded. All of these primary techniques of mining are summarized in Table 10.2.

10.4.1.3 Fundamental Properties of Blockchain

The fundamental properties of blockchain includes cryptographic security, decentralized consensus, disseminated feature, and nonrepudiation guarantee and trustless system. Table 10.3 lists the problems associated with the various schemes along with solutions provided by blockchain technology.

10.4.2 PRIVACY OR CONFIDENTIALITY SERVICES OF BLOCKCHAIN

In the technology of blockchain, privacy is the major concept. A consumer knows the importance of securing personal data which is recorded on its servers or network or recorded on its servers or system. Personal data is any kind of information associated with a recognized or recognizable natural person. The authority to monitor the data, apply rules for the data as well as sources made available by the network are introduced by the privacy services to the user. In other terms, monitoring data is allowed for the resource or data owner by privacy services. Basically, this is done by allowing the owner describe its access control list (ACL) itself. In this, the needs of giving data privacy, its significance for the existing

TABLE 10.2

Comparison of Various Distinct Techniques of Mining [17,18]

Mining Techniques	Minimum Block Hash	MoT	PBFT	PoI	PoS	PoW
Implementation(s)	Not implemented	Not implemented	Hyperledger	NEM	Ethereum	Bitcoin
Randomness	Randomized selection of blockchain	No randomness	No randomness	No randomness	Randomized selection of blockchain	No randomness
Required resources	None	Trustworthiness	None	Significance of node	Stake or wealth	High computation power
Reward miner...?	Yes	Yes (trusted)	No	Yes	No	Yes

TABLE 10.3

Problems Associated with Various Schemes along with Solutions Provided by Blockchain Technology

Schemes	Problem to Be Resolved	Solution by Blockchain
Cryptographic security	Algorithms for security should be proved in such a way that they become extremely tough to break.	The technology of blockchain uses elliptic curve cryptography that is very tough to break; moreover, the decentralized consensus and the trustless system of blockchain make it even more tough to break.
Decentralized consensus	Centralized decisions by one controller can make the controller a single point of failure.	Decisions of blockchain are instated by decentralized consensus, majority votes, and nodes agreement.
Disseminated feature	Existing applications need disseminated monitoring and techniques of security as they are disseminated by feature, but most of the recent practical solutions of security are centric in nature and incapacitated for these applications.	As the blockchain technology is disseminated by feature, security services based on blockchain can be executed in a disseminated manner.
Nonrepudiation guarantee	In the system, users can refuse their communications.	Signatures of transactions and blocks in addition to permanent databases are used by the blockchain so that transactions cannot be refused in the future.
Trustless system	If the party is compromised, security given by the third party can inflict privacy as well as secrecy threats.	The technology of blockchain inflicts a trust of majority votes, which is impermeable to compromise unless invaders have command over the entire network.

applications, the conventional approaches for confidentiality, and the disputes recently faced in ensuring privacy are summarized. Further, how the technology of blockchain can be utilized to give confidentiality is provided, as well as a few current privacy systems based on the blockchain technology is introduced.

10.4.2.1 Data Privacy and Access Control List

Data privacy requires that all private and susceptive data stay private and access to that data can be monitored by only the owners of the data. When and who can access a particular group of information is assured by the ACL by describing a specific group of rules. To understand the issue of confidentiality, recognize the consumers in organizations such as Google, Facebook, government, and banks where personal data is given by each and every user. Hence, these organizations have a collection of huge amounts of sensitive and private data of users that should not be made open to access by anyone. In this way, each single user has either little or no control over the data and any kind of access to his/her data, which can lead to violations data privacy. Confidentiality troubles arise whenever users' information is recorded, stored, consumed, dissipated, or removed. In simple terms, privacy gives the information pace and stability. To stop data loss or leakage, various federal laws have been formulated, for example, the healthcare data confidentiality laws [19]. Therefore, confidentiality/privacy is a primary consideration for developers of systems as well as applications.

10.4.2.2 Significance of Confidentiality in Existing Applications

In the period of systems of networking and cloud computing where many consumers use a similar network or physical storage, the confidentiality of information is a major issue. The collection and computations of applications are transferred to the cloud networks by the developers, followed by the granting of data privacy. In addition, healthcare, automotive, IoT, and various other applications of networking execute and collect huge amounts of information by basically consuming the technology of cloud computing. For most the mentioned applications that are included with private data or knowledge about location, confidentiality is a major requirement. In circumstances using several cloud networks and internetworking among them, the subject of confidentiality is enlarged.

10.4.2.3 Conventional Approaches for Privacy of Data

Basically, data confidentiality can be ensured by handling over the definitions of ACL to the owners of the data and applying approaches of encryption to stop any other party or any other individual from accessing the data. Thus, the organizations that are not permitted by the ACL do not have the right of accessing the data. To provide privacy, structure and execution approaches are the most active subject for research, and many approaches have been introduced so far. For example, a way to provide data confidentiality is homomorphic encryption [20], which permits the computation and implementation of the encrypted information and gives the encrypted outputs.

Another facet of privacy is securing the identity of the user. The techniques of data anonymization and distinctive confidentiality secure the user's identity and make it harder to connect the data to its owner. For example, T-closeness is a technique that monitors the dissemination of critical data [21]. K-anonymity needs the critical data to be same to at least k-1 other entries and a way to unnamed the datasets [22]. L-diversity is another approach (and an explication of the K-anonymity technique) which guarantees that the critical data is collected in "diverse enough" plausible positions [23].

10.4.2.4 Challenges of the Conventional Approaches

Irrespective of the various existing endeavors, a potential service of data confidentiality is yet defying. Potential, flexibility, ownership of data, and deficiency of systematical data lifespan are included in the problem areas. In this subsection, these issues are summarized [24].

10.4.2.4.1 Data Ownership and Control

The subject of who owns the data and who can alter it are sensitive in the concept of privacy. Basically, the party that concludes the access monitoring rules for the data is known as the owner of the data. However, unfortunately, the conventional approaches illustrated in the previous subsection still do not provide an answer to the question of ownership.

10.4.2.4.2 Technique of Systematic Data Lifespan

To describe the lifespan of the data, a base for data privacy needs to be created. This base should recognize the stages, describe their needs of confidentiality, and permit scalability in the lifespan transpositions. The gaining, transmitting, and removing of the information as well as the sources embraced in the network can be included in these stages. Thus, in most of the introduced approaches of confidentiality, an orderly technique is still unavailable.

10.4.2.4.3 Potential and Flexibility

There are many approaches of data secrecy that depend on complicated cryptographic algorithms; thus, they are incapacitated and rigorous to scale with huge applications. Existing research is attempting to decrease the complications and increase the potential of these cryptographic approaches [25]. However, in most circumstances, the introduced techniques are not practical, and most of the algorithms lose flexibility with huge amount of data implementing in the recent systems.

10.4.2.5 Data Privacy Approaches Based on Blockchain

Some of the disputes proposed in the preceding subsection can be resolved by guarantees of decentralized end-to-end data privacy provided by blockchain. Particularly, it can be a solution to the problem of ownership of the data and transposition the rights of access dynamically, whenever required. The approaches based on blockchain are still complicated because they rely on cryptographic algorithms. The concept behind the idealistic confidentiality of data based on blockchain is to construct a layer of blockchain over the layer of data collection; for example, let the owner of the data describe the required ACL through smart contracts, then release the ACL, and thus data becomes blockchain transactions. In this form, the data is not owned by companies like Google or Facebook as occurs in the conventional approaches. In blockchain, they will be a portion of the system of blockchain and will only be able to implement the data when the ACL published by the owner of the data permits them. Thus, this kind of blockchain is known as permissioned blockchain. In this part, to provide privacy, some of the existing techniques that optimize the technology of blockchain are discussed.

10.4.2.5.1 Fair Access Technique

Smart contracts are utilized by Fair Access to define the policies of access monitoring and develop authentication decisions. To define the tokens of authentication, the network consumes the blockchain transaction. Then, the transmitter uses these tokens to authenticate the receiver in accessing sections of the data of

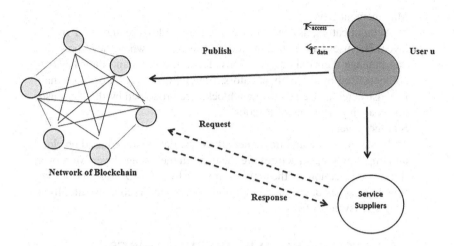

FIGURE 10.8   Elements of Zyskind's system.

the transmitter. Registration of resource, access permission, request access, and reverse access are Fair Access operations.

10.4.2.5.2   Zyskind's Technique

A decentralized technique of data confidentiality that assures the monitoring of user over its data and consumes the blocks of blockchain to collect the data and the ACL has been introduced by Zyskind, Pentland and Nathan. As shown in Figure 10.8, Zyskind's system comprises three major elements, namely, user(s), service suppliers/ providers, and the network of blockchain [26].

In general, two types of transactions are accepted by the network of block-chain in this technique: T_{access} and T_{data}. The transaction T_{access} is applied for the monitoring and management processing on the data such as altering the rights of access and defining the access control list; and transaction T_{data} is applied for storage and restitution of data.

Here, T_{access} = User u allows service s access, and T_{data} = Access to data of user u.

10.4.2.5.3   Data Sharing Based on Blockchain

To supply secrecy for the medical entries in cloud, another technique is introduced, that is, data sharing based on blockchain, also known as blockchain-based data sharing [27]. A sorted structure of blockchain that is flexible and suitable for lightweight interaction networks is used by this technique. Basically, the network constitutes three layers: the user layer, the management layer, and the storage layer. All these layers are described below in brief.

- User Layer

    An individual user or parties desire to either store or access their information and services are included in the user layer.

- Management Layer

    Authenticators, issuers, and consensus nodes are included in this layer. The users are authenticated by the issuers when they come first and manage their data registrations. Then, the users and their keys are authenticated by the authenticators. Finally, the network of blockchain is constructed as well as the new blocks are processed by the consensus nodes as in processing of Bitcoin.

- Storage Layer

    To safely record and implement the data, data storage based on cloud and implementing structures are included in the storage layer. Moreover, the communications in the network are safe by authorization and encryption approaches based on recognition, which are practicably safe, lightweight, easy, and efficient.

10.5 BLOCKCHAIN-BASED INTERNET OF THINGS

We are well aware that security is a primary consideration in the technology of IoT. Blockchain can be applied to track the data of sensors and remove the repetition of data with nasty content. Hence, rather than preferring a third-party for setting up trust, sensors can be used to swap the information through a blockchain. The merging of both the technologies (i.e., IoT and blockchain) would not only eliminate technical troubles and inability but also provide a support for P2P communication. For businesses and fulfilling the target of optimized cost, an integration of blockchain and IoT is compatible. In general, the technology of blockchain can be exploited by the technology of IoT in four ways, namely, decrease in cost, accelerated exchange of data, scaled security, and trust building [28]. This will also enable the development of novel value business models, efficient surrounding, decrease threat, operating speed, structure completeness, confidentiality, low-transaction costs, trusted system, remote services, antiimitative, certification authentication, and micro-transactions [29]. For allowing exchange of messages, smart contracts can be energized by gadgets, which further structures agreement between two organizations, allowing autonomous operating of smart gadgets without any centralized organization [30]. However, in blockchain, the major subject is the issue of flexibility and the high throughput of transactions in the millions or billions of gadgets in the system, which constantly produce a massive amount of content. For resolving this issue, Byzantine fault tolerant and proof of work blockchains are exploited by the hybrid architecture of IoT. The major advantages of a blockchain are its permissioned feature where each and every member of the network has rights of access to the system; secureness of network where consensus is needed from all the members of the network; and distributed nature which is a shared network of entries among users on a business system.

However, for developing IoT solutions, blockchain is a better solution; in some cases of IoT, it is not always the better solution to apply. It can only be applied when the features of the applications require the characteristics provided by it

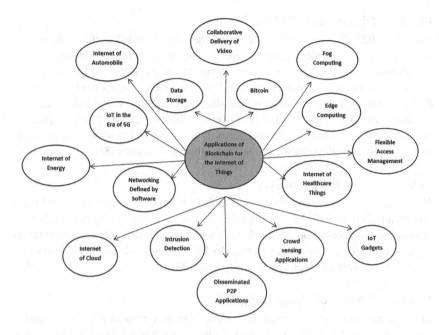

FIGURE 10.9 Blockchain applications for the IoT.

because without any motive applying it will only increase cost. Thus, before finalizing the design, the application must be completely understood.

10.5.1 APPLICATIONS OF BLOCKCHAIN FOR THE IoT

It can be clearly seen in Figure 10.9 that the technology of blockchain can be dominantly used in almost all applications of IoT. Also, note that it is not mandatory to apply blockchain in some circumstances. For example, there is no requirement of using the technology when the entities of IoT trust a third-party organization. Some of the applications of blockchain for the IoT are illustrated below.

10.5.1.1 Intrusion Detection

In IoT, many approaches based on machine learning for executing systems of intrusion detection have been introduced [31]. The concept of optimizing the technology of blockchain to enhance the shared systems of intrusion detection and to ensure safe swapping of alerts between the shared nodes of network is introduced. In a system of intrusion detection, the practicability of the technology of blockchain is described [32]. Upcoming systems of intrusion detection must be relied on for shared interaction among disseminated intrusion detection systems [33], as well as claiming massive data sharing among individuals and authenticated computation. The technology of blockchain is used for considerations of confidentiality that are uplifted by the sharing of data and to press inner invasions. Hence, the application of authorized third party, also considered as a minute point of failure, can be deferred.

10.5.1.2 Disseminated P2P Applications

For IoT, in P2P disseminated applications, the gadgets of IoT collaborate as well as self-manage for modern applications, namely, sending files, cooperative movies, transmitting messages, and broadcasting information through the use of networks of sensor. An authenticated stimulus scheme dependent on blockchain for P2P disseminated surroundings is introduced [34] to encourage consumers for collaboration. A pricing scheme is introduced by this approach which permits intermediary nodes to get a reward from transactions of blockchain to intercept selfish users, as well as shield against the collusion invasions [35].

10.5.1.3 IoT in the Era of 5G

5G will allow a completely linked and mobile community for millions of linked things in the era of IoT [36]. An approach of confidentiality protection and information exchange based on the blockchain technology is discussed. Dependent on the concept of including blocks to the blockchain, each and every fresh block is linked to the chain of blocks by its hash value [37].

10.5.1.4 Internet of Things Gadgets

Through nasty codes in malware, invaders scrutinize to obtain the information from gadgets of IoT, particularly on the open-source platform of Android [38]. A system of malware finding based on the consortium kind blockchain (also known as, CB-MDEE) is proposed by optimizing statistical examination mechanism. Based on blockchain, an approach of firmware update is introduced to secure the embedded gadgets in the IoT [39]. In this approach, the embedded gadgets have two distinct process cases, which are reply from a node of authentication to a node of request and reply from a node of response to a node of request [40].

10.5.1.5 Crowdsensing Applications

The paradigm of integrating mobile crowdsensing is a major category of mobile application in IoT, for example, application of geographical sensing [41]. Based on blockchain cryptocurrency, an amazing stimulus procedure for protecting confidentiality in application of crowdsensing is proposed.

10.5.1.6 Edge Computing

The technology of edge computing is a profusely apparent stage that allows data recording and computing between end users and center of data of the conventional cloud computing [42]. It is observed that the gadgets of fog can interact without each other and without any third-party interference. To provide the facility of interactions between gadgets of IoT and nodes of fog network, the approach of blockchain can be applied [43]. For sourcing the computations of gadgets of fog, an appropriate technique of payment is introduced. This approach (based on the Bitcoin) focuses some characteristics of security, such as accountability, appropriateness, and wholeness.

10.5.1.7   Data Storage

For systems of data collection based on the technology of IoT, data storage can do agreement with different sources of data. The major issues concerning data storage in IoT are how to exchange and secure crucial information. A confidential keyword search known as "Searchain" for decentralized collection of data is introduced [44], which is based on blockchain. Two main components such as a chain of blocks of all the sorted blocks and nodes of transaction in a P2P architecture are included in the structure of the Searchain. In addition, the structure of Searchain can also provide user confidentiality and accountability.

10.5.2   Basic Concept of Security and Privacy in IoT Based on Blockchain

The concept of security and confidentiality has a special position in the technology. It includes a huge number of features, some of which are described below.

10.5.2.1   Trust

A payment approach based on the technology of blockchain employed in a remote area setting is supposed to have an intermediate associativity to the central system of a bank [45]. A disseminated trust is established with the application of a two-layered structure, where the bank authenticates a group of chosen villagers to act as miners who in turn authenticate transactions among villagers with tokens as well as the bank. A method where every user rises and keeps its own chain of transactions is proposed [46]. This scheme supplies a disseminated trust, without the requirement of any kind of gatekeeper, while being vigilant against cyber invasions.

10.5.2.2   Certification

Recently, an approach of transitively closed undirected graph certification that can endorse systems of identity management based on the technology of blockchain has been introduced [47]. The difference between this approach and other competing approaches of certification is that an additional ability of removing or adding edges and nodes is provided by their proposals.

10.5.2.3   Preservation of Confidentiality

A system that merges the storage of blockchain as well as off-blockchain is introduced to form a platform of confidential data management concentrated on confidentiality [48]. In particular, this system prevents the confidentiality problems, such as ownership and transparency of data and verification [49].

10.5.3   Challenges of Blockchain

Although there are many efficient advantages of blockchain, it has some challenges that restrict its practicality for most of the applications [50]. Some of the challenges are depicted in Figure 10.10 are explained in the subsections below.

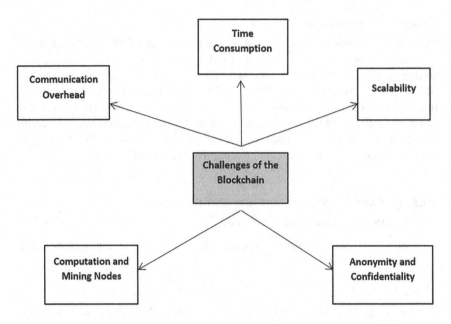

FIGURE 10.10 Blockchain challenges.

10.5.3.1 Communication Overhead

The existing applications require continuous transpositions in the ACLs and the production data because these applications are highly dynamic in nature. Hence, the nodes are forced by this to transmit continuous transactions to update the ACL or alter the data of production [51]. On the other hand, the technology of blockchain is a system of P2P dissemination, which implies that a prominent overhead will be included in terms of the abilities of system implementation and the traffic of network [52]. Hence, the overhead added in the network of blockchain is a prominent and thinkable challenge for blockchain. In blockchain for applications of security, additional challenges are brought by the collection and implementation overheads.

10.5.3.2 Time Consumption

To provide services of data safety, fast abilities of implementation are required, and specifically in the existing systems, where milliseconds can cost billions of dollars. Moreover, in blockchain, mining and getting consensus still consume more time. By forming conclusions from the logs of local blockchain without needing any disseminated consensus, the introduced techniques solve the problem [53]. Many obligations have been made to solve the issues of time in Bitcoin in platforms of "Hyperledger and Ethereum." However, the time utilized in mining is two or three seconds compared to the requirement of milliseconds. Moreover, structuring approaches of encryptions and confidentiality over the blockchain technology escalates the issue of time consumption as these approaches are complicated and consume a huge amount of time [54]. Therefore, to employ the technology of blockchain for real-time applications, faster approaches of implementation and mining are required.

10.5.3.3 Scalability

The motive of the technology of blockchain is to scale much better than the conventional centralized techniques. It is observed that, as the counting of the users and the nodes of networking increments then, the blockchain starts performing poorly which is not advantageous for the system. It is considered as a primary challenge in blockchain, specifically regarding the applications of network security, where millions of consumers need to be served and the system scales up fast [55]. The platforms of Hyperledger and Ethereum have their own obligations for the feature of scalability. Thus, it is shown by performance tests that both of the platforms still suffer from the issues of scalability.

10.5.3.4 Anonymity and Confidentiality

Giving pseudo-user anonymity is one of the major characteristics as well as benefits of blockchain. As public blockchains are open publicly and there is large probability of disclosing the data of consumer to the invaders, the above-mentioned feature is sensitive for data security [56]. In the case of Bitcoin, the issue of anonymity is resolved by applying the public key of consumer as the recognition of consumer. Thus, this gives pseudo-anonymity; to provide completely anonymized techniques, more researches are required that fulfill the needs of application of confidentiality.

10.5.3.5 Computation and Mining Nodes

In many of the existing applications, the network nodes do not have high abilities of computation, which implies that the client of blockchain needs to be as simple as possible to meet the needs of low computation ability. On the other hand, generally, the services of confidentiality need prominent computations in encryption, decryption, and signature [57]. Furthermore, the issue of mining would be solved by permitting the nodes of application to be the clients of blockchain, as well as by proposing appropriate nodes of mining that are included only to implement mining for most of the introduced approaches. Further, the needed high power of computation for the nodes adds to the charge of the system, which is not beneficial for the system [58]. Hence, a much better technique would include reducing the requirements of computation for mining, and some convenient cryptographic techniques should be designed to decrease the requirements of computation for encrypting and signing the data.

10.6 CONCLUSION

IoT and blockchain are already on their way to reconstructing our digital world, featured by the drastic transpositions in the structure of the existing network. The IoT has converted the things around into smart things and making them capable of interacting with each other, thus storing a huge amount of data by frequently capturing the physical world, for examining and implementing some smart operations based on the same. However, as there is always a threat of losing of confidential data in the network, blockchain is proposed to solve the issues of data security in IoT networks. In this chapter, an idea of merging IoT and blockchain is presented because their conjunction would produce a better outcome in every possible field. It has the potential to provide security, privacy, and efficiency to the field it had been applied to. This chapter presents a detailed description of both the technologies, the issues of IoT, and how blockchain acts as a lifesaver for IoT. Further, it illustrates their combination, their application, and the concept of security and privacy in IoT based on blockchain. Finally, the challenges of blockchain have been discussed to intrigue researchers to contribute to the field.

REFERENCES

1. Pilkington, M. (2015). Blockchain Technology: Principles and Applications. *Research Handbook on Digital Transformations*, 225–253. doi:10.4337/9781784717766.00019
2. Jaidka, H., Sharma, N., & Singh, R. (2020). Evolution of IoT to IIoT: Applications & Challenges. *SSRN Electronic Journal*. doi: 10.2139/ssrn.3603739
3. CoinMarketCap.Com, "Crypto Currency Market Capitalization," [online] Available: https://coinmarketcap.com/currencies/ (accessed August 15, 2017).
4. Beltran, V., & Skarmeta, A. F. (2016). An overview on delegated authorization for CoAP: Authentication and authorization for Constrained Environments (ACE). *2016 IEEE 3rd World Forum on Internet of Things (WF-IoT)*. doi:10.1109/wf-iot.2016.7845482
5. Lee, J. (2018). BIDaaS: Blockchain Based ID as a Service. *IEEE Access*, 6, 2274–2278.
6. Li, X., Jiang, P., Chen, T., Luo, X., & Wen, Q. (2018). A Survey on the Security of Blockchain Systems. *Future Generation Computing Systems*, 107, 841–853.
7. Christidis, K., & Devetsikiotis, M. (2016). Blockchains and Smart Contracts for the Internet of Things. *IEEE Access*, 4, 2292–2303.
8. Arora, A., Kaur, A., Bhushan, B., & Saini, H. (2019). Security Concerns and Future Trends of Internet of Things. In *2019 2nd International Conference on Intelligent Computing, Instrumentation and Control Technologies (ICICICT)*. doi: 10.1109/icicict46008.2019.8993222
9. Zhang, Y., Xu, C., Lin, X., & Shen, X. (2019). Blockchain-Based Public Integrity Verification for Cloud Storage against Procrastinating Auditors. *IEEE Transactions on Cloud Computing*. doi: 10.1109/TCC.2019.2908400
10. Mettler, M. (2016). Blockchain Technology in Healthcare: The Revolution Starts Here. *2016 IEEE 18th International Conference on e-Health Networking, Applications and Services (Healthcom)*, 1–3.
11. Paul, G., Sarkar, P., & Mukherjee, S. (2014). Towards a more democratic mining in bitcoins. In *Information Systems Security Lecture Notes in Computer Science*, 185–203. doi:10.1007/978-3-319-13841-1_11.
12. Darwish, M.A., Yafi, E., Ghamdi, M.A., & Almasri, A.H. (2020). Decentralizing Privacy Implementation at Cloud Storage Using Blockchain-Based Hybrid Algorithm. *Arabian Journal for Science and Engineering*, 45, 1–10.
13. Castro, M., & Liskov, B. (2002). Practical Byzantine Fault Tolerance and Proactive Recovery. *ACM Transactions on Computer Systems (TOCS)*, 20, 398–461.

14. Wikipedia, "NEM (Cryptocurrency)," [online] Available: https://en.wikipedia.org/wiki/NEM_(cryptocurrency)#Proof-ofimportance (accessed February 13, 2018).

15. Bitcoinwiki, "Proof of Stake," [online] Available: https://en.bitcoin.it/wiki/Proof_of_Stake (accessed February 13, 2018).

16. Zhang, Q., Leng, Y., & Fan, L. (2018). Blockchain-based P2P File Sharing Incentive. *IACR Cryptol.* ePrint Arch., 2018, 1152.

17. Boireau, O. (2018). Securing the Blockchain against Hackers. *Network Security*, 2018, 8–11.

18. Ferrag, M.A., Maglaras, L.A., Janicke, H., & Jiang, J. (2017). Authentication Protocols for Internet of Things: A Comprehensive Survey. *Security and Communication Networks*, 2017, 6562953:1–6562953:41.

19. Saini, H., Bhushan, B., Arora, A., & Kaur, A. (2019). Security Vulnerabilities in Information Communication Technology: Blockchain to the Rescue (A Survey on Blockchain Technology). In *2019 2nd International Conference on Intelligent Computing, Instrumentation and Control Technologies (ICICICT)*. doi: 10.1109/icicict46008.2019.8993229

20. Yi, X., Paulet, R., & Bertino, E. (2014). *Homomorphic Encryption and Applications.* SpringerBriefs in Computer Science, 27–46. doi:10.1007/978-3-319-12229-8_2

21. Li, N., Li, T., & Venkatasubramanian, S. (2007). t-Closeness: Privacy Beyond k-Anonymity and l-Diversity. In *2007 IEEE 23rd International Conference on Data Engineering*, 106–115.

22. Byun, J., Kamra, A., Bertino, E., & Li, N. (2007). Efficient k-Anonymization using clustering techniques. *Advances in Databases: Concepts, Systems and Applications Lecture Notes in Computer Science,* 188–200. doi:10.1007/978-3-540-71703-4_18.

23. Machanavajjhala, A., Gehrke, J., Kifer, D., & Venkitasubramaniam, M. (2006). L-diversity: Privacy beyond k-anonymity. *22nd International Conference on Data Engineering (ICDE'06)*. doi:10.1109/icde.2006.1.

24. Bertino, E. (2015). Big Data – Security and Privacy. In *2015 IEEE International Congress on Big Data*, 757–761.

25. Varshney, T., Sharma, N., Kaushik, I., & Bhushan, B. (2019). Authentication & Encryption Based Security Services in Blockchain Technology. In *2019 International Conference on Computing, Communication, and Intelligent Systems (ICCCIS)*. doi: 10.1109/icccis48478.2019.8974500

26. Zyskind, G., Nathan, O., & Pentland, A. (2015). Decentralizing Privacy: Using Blockchain to Protect Personal Data. In *2015 IEEE Security and Privacy Workshops*, 180–184.

27. Xia, Q., Sifah, E.B., Smahi, A., Amofa, S., & Zhang, X. (2017). BBDS: Blockchain-Based Data Sharing for Electronic Medical Records in Cloud Environments. *Information*, 8, 44.

28. S. Gopal, "Blockchain for the Internet of Things," *Tata Consultancy Services White Paper*. Available: https://www.tcs.com/content/dam/tcs/pdf/technologies/internet-of-things/abstract/blockchain-for-iot.pdf

29. Kshetri, N. (2017). Blockchain's Roles in Strengthening Cybersecurity and Protecting Privacy. *Telecommunications Policy*, 41, 1027–1038.

30. IBM Research Editorial Staff (2018), "What is Blockchain?," Available: https://www.ibm.com/downloads/cas/K54GJQJY (accessed September 21, 2020)

31. Alexopoulos, N., Vasilomanolakis, E., Ivánkó, N.R., & Mühlhäuser, M. (2017). *Towards Blockchain-Based Collaborative Intrusion Detection Systems. Critical Information Infrastructures Security Lecture Notes in Computer Science,* 107–118. doi:10.1007/978-3-319-99843-5_10.

32. Meng, W., Tischhauser, E., Wang, Q., Wang, Y., & Han, J. (2018). When Intrusion Detection Meets Blockchain Technology: A Review. *IEEE Access*, 6, 10179–10188.

33. Cruz, T., Rosa, L., Proença, J., Maglaras, L.A., Aubigny, M., Lev, L., Jiang, J., & Simões, P. (2016). A Cybersecurity Detection Framework for Supervisory Control and Data Acquisition Systems. *IEEE Transactions on Industrial Informatics*, 12, 2236–2246.

34. He, Y., Li, H., Cheng, X., Liu, Y., Yang, C., & Sun, L. (2018). A Blockchain Based Truthful Incentive Mechanism for Distributed P2P Applications. *IEEE Access*, 6, 27324–27335.

35. Gu, J., Sun, B., Du, X., Wang, J., Zhuang, Y., & Wang, Z. (2018). Consortium Blockchain-Based Malware Detection in Mobile Devices. *IEEE Access*, 6, 12118–12128.

36. Arora, S., Sharma, N., Bhushan, B., Kaushik, I., & Ahmad, A. (2020). Evolution of 5G Wireless Network in IoT. In *2020 IEEE 9th International Conference on Communication Systems and Network Technologies (CSNT)*. doi: 10.1109/csnt48778.2020.9115773

37. Ahmad, A., Bhushan, B., Sharma, N., Kaushik, I., & Arora, S. (2020). Importunity & Evolution of IoT for 5G. In *2020 IEEE 9th International Conference on Communication Systems and Network Technologies (CSNT)*. doi: 10.1109/csnt48778.2020.9115768

38. Hu, Y., Manzoor, A., Ekparinya, P., Liyanage, M., Thilakarathna, K., Jourjon, G., & Seneviratne, A. (2019). A Delay-Tolerant Payment Scheme Based on the Ethereum Blockchain. *IEEE Access*, 7, 33159–33172.

39. Ouaddah, A., Elkalam, A.A., & Ouahman, A.A. (2017). Towards a Novel Privacy-Preserving Access Control Model Based on Blockchain Technology in IoT. *Advances in Intelligent Systems and Computing*, 520, 523–533.

40. Tiwari, R., Sharma, N., Kaushik, I., Tiwari, A., & Bhushan, B. (2019). Evolution of IoT & Data Analytics using Deep Learning. In *2019 International Conference on Computing, Communication, and Intelligent Systems (ICCCIS)*. doi: 10.1109/icccis48478.2019.8974481

41. Eyal, I., Gencer, A.E., Sirer, E.G., & Renesse, R.V. (2016). *Bitcoin-NG: A Scalable Blockchain Protocol*. In *13th {USENIX} symposium on networked systems design and implementation ({NSDI} 16)*, 45–59.

42. Malavolta, G., Moreno-Sanchez, P., Schneidewind, C., Kate, A., & Maffei, M. (2019). *Anonymous Multi-Hop Locks for Blockchain Scalability and Interoperability*. In *Proceedings 2019 Network and Distributed System Security Symposium*. doi:10.14722/ndss.2019.23330.

43. Goyal, S., Sharma, N., Kaushik, I., Bhushan, B., & Kumar, A. (2020). Precedence & Issues of IoT based on Edge Computing. In *2020 IEEE 9th International Conference on Communication Systems and Network Technologies (CSNT)*. doi: 10.1109/csnt48778.2020.9115789

44. Jiang, P.L., Guo, F., Liang, K., Lai, J., & Wen, Q. (2020). Searchain: Blockchain-based Private Keyword Search in Decentralized Storage. *Future Generation Computer Systems*, 107, 781–792.

45. Gupta, S., Sinha, S., & Bhushan, B. (2020). Emergence of Blockchain Technology: Fundamentals, Working and Its Various Implementations. *SSRN Electronic Journal*. doi: 10.2139/ssrn.3569577

46. Otte, P., Vos, M.D., & Pouwelse, J.A. (2020). TrustChain: A Sybil-resistant scalable blockchain. *Future Generation Computer Systems*, 107, 770–780.

47. Lin, C., He, D., Huang, X., Khan, M.K., & Choo, K.R. (2018). A New Transitively Closed Undirected Graph Authentication Scheme for Blockchain-Based Identity Management Systems. *IEEE Access*, 6, 28203–28212.

48. Dinh, T.T., Wang, J., Chen, G., Liu, R., Ooi, B.C., & Tan, K. (2017). BLOCKBENCH: A Framework for Analyzing Private Blockchains. In *Proceedings of the 2017 ACM International Conference on Management of Data – SIGMOD '17*. doi:10.1145/3035918.3064033.

49. Chen, Y., Li, Q., & Wang, H. (2018). Towards Trusted Social Networks with Blockchain Technology. ArXiv, abs/1801.02796.

50. Varshney, T., Sharma, N., Kaushik, I., & Bhushan, B. (2019). Architectural Model of Security Threats & Their Countermeasures in IoT. In *2019 International Conference on Computing, Communication, and Intelligent Systems (ICCCIS)*. doi: 10.1109/icccis48478.2019.8974544

51. Soni, S., & Bhushan, B. (2019). A Comprehensive Survey on Blockchain: Working, Security Analysis, Privacy Threats and Potential Applications. In *2019 2nd International Conference on Intelligent Computing, Instrumentation and Control Technologies (ICICICT)*. doi: 10.1109/icicict46008.2019.8993210

52. Conoscenti, M., Vetrò, A., & Martin, J.C. (2016). Blockchain for the Internet of Things: A Systematic Literature Review. In *2016 IEEE/ACS 13th International Conference of Computer Systems and Applications (AICCSA)*, 1–6.

53. Rustagi, A., Manchanda, C., & Sharma, N. (2020). IoE: A Boon & Threat to the Mankind. In *2020 IEEE 9th International Conference on Communication Systems and Network Technologies (CSNT)*. doi: 10.1109/csnt48778.2020.9115748

54. HealthIT (2015, April), "Guide to Privacy and Security of Electronic Health Information," [online] Available: https://www.healthit.gov/sites/default/files/pdf/privacy/privacy-andsecurity-guide.pdf (accessed February 13, 2018).

55. Jadon, S., Choudhary, A., Saini, H., Dua, U., Sharma, N., & Kaushik, I. (2020). Comfy Smart Home using IoT. *SSRN Electronic Journal*. doi: 10.2139/ssrn.3565908

56. Kreuter, B., Shelat, A., Mood, B., & Butler, K.R. (2013). *PCF: A Portable Circuit Format for Scalable Two-Party Secure Computation*. USENIX Security Symposium. Available: https://www.usenix.org/conference/usenixsecurity13/technical-sessions/paper/kreuter (accessed September 21, 2020).

57. Manchanda, C., Sharma, N., Rathi, R., Bhushan, B., & Grover, M. (2020). Neoteric Security and Privacy Sanctuary Technologies in Smart Cities. In *2020 IEEE 9th International Conference on Communication Systems and Network Technologies (CSNT)*. doi: 10.1109/csnt48778.2020.9115780

58. Hassan, M.U., Rehmani, M.H., & Chen, J. (2019). Privacy preservation in blockchain based IoT systems: Integration issues, prospects, challenges, and future research directions. *Future Generation Computing Systems*, 97, 512–529.

11 Business Operations and Service Management within Blockchain in the Internet of Things

Keith Sherringham
A.C.N. 629 733 633 Pty. Ltd.

Bhuvan Unhelkar
University of South Florida Sarasota–Manatee

CONTENTS

11.1 INTRODUCTION

From the connectivity of napkin holders and the automated ordering of stock levels, to the management of food in domestic or commercial fridges and freezers, or the self coordination of drones in search and rescue operations; the Internet of Things (IoT) is increasing its roles within society. The IoT is more than the connectivity of a series of devices with information sharing. The IoT brings new business opportunities, new uses of technology, and innovative ways to solve business challengers. As with other business transformative technologies, the IoT is part of a complex interdependent system of people, processes, information, and technology combined to deliver services and products. The IoT uses a series of technologies, including blockchains [1], that are brought together to solve the design, development, management, operation, and servicing of technology across businesses.

As the IoT becomes an integrated component of the core platform (the convergence of the Internet, mobile computing, cloud computing, social media, big data, and cloud-based services, with real-time decision-making, machine learning, and artificial intelligence) that underpins society, the IoT will both require and assist in the required service management, regulatory, audit, compliance, privacy, security, safety, independence, skilling, training, processes, service models, and more. Blockchains are increasingly playing a role within the IoT, the management and operation of the IoT, and the business services that use the IoT. Whether it is the transformation of knowledge workers around automation and the provision of knowledge worker services from the cloud [2] or the use of blockchains within set transaction processing [3], it is the business integration and service management of the technology that is required.

Security for the IoT and for the resultant products and services is pivotal to the implementation, uptake, management, and success of the IoT; whether supplied by blockchain and/or a combination of technologies [4] based on architecture [5,6]. From mitigating brand damage to ensuring regulatory compliance and avoiding fines and from prosecution prevention to providing a competitive advantage to a business, security of the IoT is integral to the establishment, operation, and successful adoption of the IoT. The security of and for the IoT requires an integrated end-to-end approach from both business and technology, across multiple business silos and businesses.

This chapter reflects the business experience of the authors in management and implementation of technology and its transformation through to the assurance and operations of services in support of the IoT, including addressing the role of blockchains in information sharing and role-based access within the IoT from a business perspective. The chapter starts with a business view on the provision of security for ICT, the role of blockchains in service management provision is considered, including the use of layered operations and integrated management. The role of blockchains in information exchange and role-based access are also addressed. With a foundation established from the business principles, the operations and service management of security are addressed, including the role of blockchains alongside the role of artificial intelligence (AI). The chapter then considers the emerging role of blockchains within automation. Finally, the chapter is summarized from an implementation and operational perspective.

11.2   PROVISION AND MANAGEMENT OF IOT SECURITY

Security for the IoT requires an integrated end-to-end solution across business and technology, of which information sharing and authentication by blockchains both device to device and person to device with AI forms a part. Blockchain in their own right or with the IoT plays a role in addressing the following considerations, but blockchain is not an answer in their own right or by themselves.

11.2.1   Geopolitical Considerations

For the security of the IoT, a range of geopolitical considerations are of impact to a business, including the jurisdictions in which the IoT operates (i.e., an IoT may operate across both well and less regulated regimes with corresponding risk profiles for a business). Often related to overall sovereign risk, considerations for security include:

- *Political Stability*: Regimes of political stability are conducive to secure and stable operations.
- *Political Interference*: Regimes that enable corruption, fear of loss of job for being independent, and environments open to political pressure and interference tend to be less secure. For some regimes, the ability to interrogate complete ledger records across devices (as blockchains provide) using AI has its advantages over the traditional consolidated databases with greater controls.
- *Infrastructure Stability*: The stability of the infrastructure in support of the IoT (e.g., power supplies and physical security). Stable environments and well-managed environments enable security in the IoT.

11.2.2   IoT Security from an Integrated Service

The main components of the IoT are shown in Figure 11.1 (excluding the business use and derived services), including the devices that make up the IoT and their

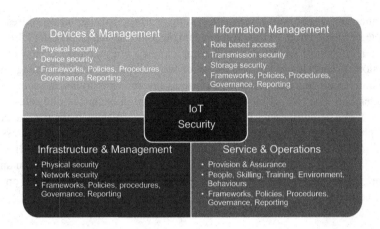

FIGURE 11.1   Elements for security in the Internet of Things.

management, the infrastructure and its management that connects the devices, the management of the information from the devices, and the overall management and operations of the IoT (both the technology and business) for which security applies.

For the main components (Figure 11.1), key elements of security for the IoT are summarized.

11.2.2.1   Devices and Management

Security considerations for the management of devices includes, but is not limited to:

- *Physical Security*: The physical environment in which the devices for the IoT are located and the related security threat posed. This is technology agnostic. As the business criticality of the device and its information increases, so does the physical security of the environment with corresponding increases in governance and management. Blockchains are an emerging contribution to the information exchange as part of the physical security.
- *Device Security*: This is the security of the device and its vulnerability. From ease of opening the device to accessing the hardware and software operating on the device that is used to access information or change the information provided by the device. As the devices and information become more critical to a business, the associated security increases with corresponding increases in governance and management. Again, this requirement is technology agnostic.
- *Frameworks, Policies, Procedures, Governance, Reporting*: Managing the devices, device security, and the device physical security requires the necessary governance with reporting on what is occurring as well as policies (standards) and procedures. It is the proactive management that assures the security of the IoT and which is required by the business. Frameworks to support the use of blockchain technologies are required.

11.2.2.2   Infrastructure and Management

Security considerations for infrastructure and its management includes, but is not limited to:

- *Physical Security*: For the infrastructure (mainly the networking that connects the devices of the IoT) that supports the operation and management of the IoT, the security of the physical environment needs to be secure and kept secure. As the business criticality increases, so does the physical security required, with corresponding increases in governance and management. This is also technology agnostic.
- *Network Security*: The security of the network on which the IoT operates including the hardware and software of the network with the performance, monitoring, and management is a critical element of the IoT. AI and blockchains can assist.
- *Frameworks, Policies, Procedures, Governance, Reporting*: Managing the infrastructure in support of the IoT and its security requires the necessary governance with reporting on what is occurring and policies (standards) and procedures. It is the proactive management that assures the security of the IoT and which is required by the business. Blockchains can be used with regulators for selective reporting and operations, and AI is expected to be used for routine reporting and exceptions management.

11.2.2.3   Information Management

Security considerations for information and management includes, but is not limited to:

- *Transmission Security*: The security in the transmission of information between devices in the IoT and to final storage. Again, there are many aspects to this element from differences in devices, to cross-jurisdictional standards, to the security of the network on which information is transmitted. The blockchain technologies form the basis as the standards emerge.
- *Storage Security*: The security of storage of the information in the IoT and from the IoT and for its support and management, either temporarily or finally, either on devices in the IoT or stored on external devices. This includes the management and access to the information.
- *Role Based Access*: Whether accessed by machines or people, role-based access to the information within the IoT is required with necessary permissions. This information includes the data itself, metadata, derived information, corresponding processing [7], as well as the information regarding what makes the IoT. Maintaining the currency of this role-based access is part of the security.
- *Frameworks, Policies, Procedures, Governance, Reporting*: Managing the information generated by the IoT as well as the information on the

IoT necessary to operate the IoT have their own security. This requires the necessary governance with reporting on what is occurring, as well as policies (standards) and procedures. It is the proactive management that assures the security of the IoT and which is required by the business.

11.2.2.4 Service and Operations

Security considerations for service and operations includes, but is not limited to:

- *Provision & Assurance*: For security of the IoT, the service has to be provided managing the issues of budgets, resources, performance management, vendor management, people, scheduling, and many other common business tasks. With the IoT, like other technology services, the security is achieved by assurance of the service provision, assurance of performance standards, standardization of service, and service improvement.
- *People, Training, Skilling, Environment, Behaviors*: Security for the IoT and the related operations and services is achieved when the right people, with required training, with the correct skills, working in the conducive environment, and showing the necessary behaviors deliver the defined outcomes.
- *Frameworks, Policies, Procedures, Governance, Reporting*: The governance with reporting, policies (standards) and procedures, for managing the service in support of the IoT is required. Proactive management is required to assure the security of the IoT, as well as the necessary services and operations required by the business.

As with other technologies, and the integration and adoption of the technologies, blockchains have a role to play and can be used, but it is the business management and operations that remain.

11.2.3 IoT Security from Integrated Management

With blockchain playing a role within integrated management for the security of the IoT, an emerging role for blockchain is in facilitating the management of the interdependencies required (Figure 11.2). All of the functions within the stacks, with their own security needs, are necessary for providing the IoT and for managing the security of the IoT.

This management of interdependencies includes:

- *Business Technology Stack*: This is the technology stack integrated into the business for the provision of services. Blockchains with role-based authentication assist with information sharing between the layers as well as within a layer.
- *Shared Services Stack*: The common services that business areas use to deliver services. Each service has its own Business Technology Stack.

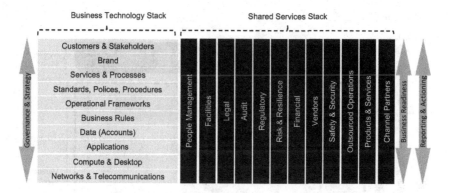

FIGURE 11.2   The Business Technology Stack and the Shared Services Stack in security of the IoT. (Reprinted with permission from [2]. © 2020 Palgrave Macmillan.)

Each of the services can use AI and blockchains for information management and security, including role-based access management.

- *Governance and Readiness*: This is the reporting, monitoring, frameworks, standards, policies, procedures, and protocols that are needed for assuring security and managing provision of security. AI is playing an increasing role with blockchains used for information sharing and assurance.

11.2.4   IoT Security from Layered Management

Whether it is in assisting with information exchange or the role-based permissions for information access, blockchains in security for the IoT across the different layers of the solution with nonalignment of points of failure to form an emergent security behavior (Figure 11.3).

Three broad layers are identified:

- *Identity Layer*: This is where the security issues are identified. Analogous to the three-layered defense seen in financial services:
  - *Operational Owned*: The areas of business are responsible for their own security and its operation and remediation. In this case, the areas within the Operational Layer have responsibilities within their areas, and the overall end-to-end security is the responsibility of the respective service owner(s) for the IoT.
  - *Internal Identified*: Internal audit, regulatory, and compliance also identify gaps in security and ensure the respective owner(s) address.
  - *External Identified*: External audit, regulatory, and compliance have also identified security issues with the respective owner(s) addressing.
- *Operational Layer*: These are the main layers that make up the provision of the IoT and have responsibility for services, including security, within the respective layer. Through the management of security in the

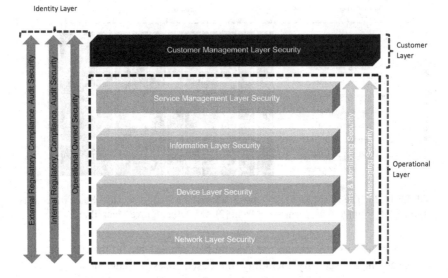

FIGURE 11.3   Layers for security in the Internet of Things.

Network Layer, the Device Layer, the Information Layer, and the Service Management Layer respectively, combined with Messaging and Alerts and Monitoring for an overall security for the IoT is achieved. While owner(s) exist for each layer with respective responsibilities, an overall owner for the IoT is required for assuring the end-to-end management. By addressing security at each layer, a breach in one layer may impact the service of the IoT in that layer, but there are still other layers operating with the service being maintained.

- *Customer Layer*: This is the management of the relationship with customers and stakeholders, including direct personal interaction as well as management through channel partners, device-to-device interactions, or provision of APIs. Through all of these channels, the security for the IoT is managed.

The emergent effect of the layered approach with proactive management is the resiliency and continuity of the service and the performance of the service to required levels. The focus is on the emergent behavior coming from the different layers and the components, elements, and functions within the IoT through the actions in specific areas. It is the emergent behavior that provides security to the IoT and is of value to the business.

11.3   OPERATIONS AND SERVICE MANAGEMENT IN SECURITY OF THE IOT

Like other ICT Operations and Services provided for other technologies, that for the IoT (with or without blockchain) requires the similar core business practices and principles, with a layered defense approach (multilayered solution with non-alignment of points of failure).

11.3.1 BUSINESS IN THE SECURITY OF THE IoT

The IoT is a service that is subjected to the normal factors of a business. The IoT is driven by customers, markets, costs, and regulatory bodies. The implementation and adoption of the IoT is subject to vested interests and the need to overcome incumbency (Figure 11.4).

Consequently, security for the IoT starts with the business and ends with the business having the required business priorities. From the business priorities, the components, elements, and functions identified previously can be applied within the layers and stacks necessary for service provision and management. The main business priorities needed for the establishment, operation, servicing, and consumption of a secure IoT by business are presented in Figure 11.5. Without these

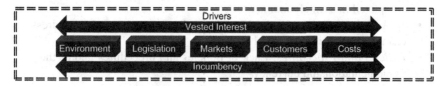

FIGURE 11.4   Drivers in business impact security in the Internet of Things .

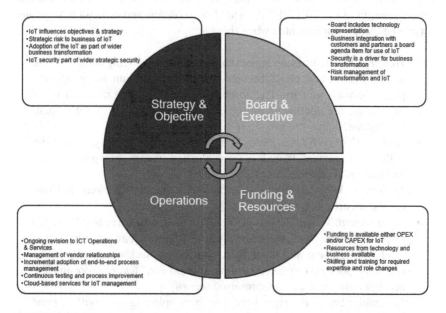

FIGURE 11.5   Business considerations in security for the Internet of Things.

main priorities, a secure IoT may be provided and consumed, but the benefit realization to the business are seldom achieved.

11.3.2   Business Management in Security of the IoT

11.3.2.1   Vulnerability Management

Security failings in the IoT, and in cybersecurity in general, are predominantly based in operational and service management-related issues (rather than technology failures per se), as reported by [8–11], among others, where vulnerabilities are seen in:

- *Passwords*: Default passwords remain in place.
- *Updates*: Updates are not implemented and/or configured to support the changes.
- *Expiry*: Users accounts are not expired and/or completed in a timely manner.
- *Roles*: Changes in roles and related permissions are not updated and/or completed in a timely manner.
- *IP Addresses*: Point-to-point IP address changes and group IP address changes are not implemented properly.
- *Policies and Procedures*: Lack of policies and procedures and/or their implementation and currency.

11.3.2.2   Ethics

The behaviors and outcomes of people that impact their motivations and actions impact security of the IoT [12]. Ethics are hard to define and difficult to legislate, but can be readily implemented when there is:

- *Training*: Where training is provided on ethics, including scenarios and actions to be taken; ethics become part of the operation and people are proactive with security. The absence of ethics training does not mean that people are unethical, but the ethical risk exists whether it is managed or not.
- *Accountability*: Where there is a lack of accountability, especially within management, there is a corresponding impact upon performance and increased risk for security.
- *Incentives*: With the correct incentives for the right actions and outcomes, security is stronger.
- *Environment*: Where the business environment conducive to ethical and responsible behaviors, the actions result. Conversely, an environment of weak management and leadership, poor ethics, lack of accountability, excepting poor performance, turning a "blind eye," fear of reporting and accountability leads to compromised security.
- *Behaviors*: Instill the right behaviors in people and they will respond accordingly. The behaviors are instilled through leadership, incentives,

environment, and personal responsibility. Conversely, where wrong behaviors are enabled and allowed to remain, weaker security for the IoT results.

11.3.2.3   Business Drivers in IoT Security

Business operations within an organization are concerned with the wider aspects of security within the IoT which are also drivers for security in the IoT, including:

- *Brand Management*: Breaches in security impact the brand of a business, especially those providing knowledge services. Managing the brand through a security breach in the IoT supporting business operations is becoming part of overall brand management. Damage to the brand and the need to protect the brand are drivers for security in the IoT.
- *Sales Management*: Where customers consider the services and/or products provided by a business are insecure, and/or their customer details are insecure, there is the corresponding impact upon the business (i.e., security impacts profitability through customer acquisition and retention).
- *Service Management*: The occurrence of security breaches impacts the delivery of the service. Not only are their risks of outages with customer impacts but the cost to manage customers and remediate the results of security breaches impact the business. The business driver for security in the IoT is strongest after a breach has occurred and/or when customers are lost.
- *Retention Management*: Similar to the sales management, breaches in security lead to customer dissatisfaction and churn. The management of a security breach and the customer impacts of a breach is part of customer retention with impacts on profitability, being a driver in security of the IoT.

11.3.2.4   Regulatory and Compliance in IoT Security

The management of reporting to regulators around security is often a shared function between the business and ICT Operations and Services. It is the business that is ultimately responsible to the regulator and for other compliance, which is in turn, the business driver for security in the IoT.

11.3.2.5   Audit in IoT Security

The audit function in a business is often a driver for improvements in processes, procedures, operations, and other aspects of the security of the IoT [13]. As the business becomes more complex, more knowledge worker dependent, and as operations become more automated, the audit function is increasingly a driver for security within the IoT. While privacy has been used alongside security within this chapter, audit of privacy is also driving security in the IoT. Like other business drivers, privacy is an agent that energizes security within the IoT. Similar considerations from the discussion of other layers apply to audit [14].

11.3.3 Service Management in Security of the IoT

The management of the service, including security [15] covers:

- *Service Support (Management)*: The day-to-day management of the service.
- *Application Support (Management)*: The day-to-day development and management of the IoT.
- *Service Improvement*: The ongoing revision and improvement of the service.
- *Service Development*: The development of the service and new products, growth of markets, strategy, and emergent technologies.

11.3.4 Technical Management in Security of the IoT

11.3.4.1 Network Layer

The security of the Network Layer (Figure 11.3) is changing around blockchains to enable self-healing networks with automated virtual devices alongside device redundancy and with automated routing and virtualized devices. AI is being used for assurance of the information used in security operations.

11.3.4.2 Device Layer

Managing security in the Device Layer (Figure 11.3) for the IoT has the environment, the physical, the device, and the virtual. Within the Device Layer, security has a dependency upon:

- *Asset Tracking*: In the consumer environment (e.g., TVs and fridges at home), there is less control of the assets, the ownership of the asset is less rigorous, and blockchains play a role in the asset accessing and validation.
- *Across Jurisdiction Management*: Management by multiple stakeholders, with different security requirements and standards applying, even within a business. In addition to complicating the management of the IoT and increasing the risk profile of the operation (as well as cost), this difference in standards creates opportunities for security vulnerabilities.
- *Uniformity and Standardization*: Given devices within the IoT maybe secure, but the interaction across the business silos often remains an area of vulnerability. Consistency in devices, versions, configurations, and levels of service are required.

11.3.4.3 Information Layer

The IoT generates a diverse range of information which is used in the operation and management of the IoT (e.g., asset details of the device in the IoT or people details for those accessing devices and information in the IoT or device information or device accessing information or storage information or sharing

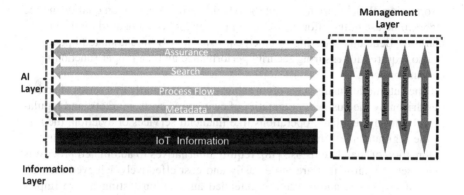

FIGURE 11.6 Automation within security of the Internet of Things. (Reprinted with permission from [2]. © 2020 Palgrave Macmillan.)

information, as well as a range of supporting and financial information) in which blockchains play a role.

11.3.5 AUTOMATION WITHIN SECURITY FOR IoT

Alongside technologies like blockchain, the use of AI, real-time decision-making, and machine-to-machine learning is automating the management and operation of the IoT, including its security. Consider the stylized management of the IoT (Figure 11.6) where:

- *Information Layer*: The information layer (discussed previously) is the different information from the IoT that are combined to provide the required information.
- *AI Layer*: The AI layer is used to combine the data sources to provide the value-added services. The AI layer is deployed on the information layer in the IoT and within the management layer.
- *Management Layer*: The management layer is used for managing the performance of the other layers. AI is used within the management as well.

AI combines the information to provide the value-added details [16]. Taking data from blockchains and combining with algorithms, the metadata (data about data), the process flow details, the search (whether by human or by machines), and the assurance of information quality is combined to provide security within the IoT. The AI is a combination of these component layers, each requiring its own service and operations that are combined to deliver operations. The use of AI is a risk-based approach, especially within assurance to manage delays other trade-offs in processing [17]. The AI is also about an emergent behavior and the use of benchmark testing to ensure the correct combination and aggregation of information. Benchmark testing of the AI addressing information quality and the algorithmic use of the information is required and AI is a tool used for the benchmarking. Benchmark testing of emergent behaviors from automated operations is also required for ensuring security performance and the proper functioning of the automation.

The automation using AI is often achieved incrementally as a series of activities driven by the business pragmatics of customers, costs, markets, and regulatory. In highly regulated operations like financial services or aviation, the role of the regulator is of increased importance. Regulators can restrict the use of automation or mandate its use [18], require alternatives to automated processes exist, set standards that are more easily and cost effectively achieved through manual processes, and/or require extended automation testing before implementation. From the deployment of driverless cars, to automated software control systems on aircraft, to automated assessments of loan approvals, regulatory impacts are seen on business and regulatory influence is increasing within the management and services from the IoT and in the management of automation within the IoT.

11.4 BLOCKCHAIN IN THE FUTURE TRENDS OF SECURITY IN THE IOT SECURITY

The role of blockchain within security of the IoT is emerging [19] with many areas still in research and development. The role of blockchain is increasing within business management, and the ICT Operations and Services, as well as in the network, data, and device layers of the IoT. Blockchains are also a component to the automation of the management and operations of the IoT. Some emerging opportunities for the use of blockchains in the security of the IoT are summarized in Table 11.1.

11.5 CONCLUSION

Blockchains are playing an increasing role within the security of the IoT, both directly through information exchange and role-based permissions at the device level, as well as within ICT Operations and Services and the wider business management and integration. Like other technologies, blockchains are enablers for the IoT and the security of the IoT, and are impacted by the wider business management issues including:

- *Drivers*: The business drivers of products and services, markets, regulatory, and costs.
- *Business Services*: The business services of access to capital, the skills of people, the supporting infrastructure, access to information, the required processes, and changes in the physical environment.

TABLE 11.1

Emerging Opportunities for Blockchains within the Security of the IoT

Trend	Description	Blockchains
Decision making	Use of AI in decision-making directly by machine to machine or by persons with the assistance of technology. Dependent upon the provision of information in workflow for real-time decision-making.	Enable direct machine-to-machine information sharing, while providing role-based permissions and metadata. The emerging area for blockchains is in sophisticated workflow (enhancement of emerging smart contract sophistication) to enable decision making. Similar trend for real-time decision-making by people.
Personality assistance	Adoption of personal profile information and related information to help assess expected responses in real time. Used for risk management and profiling of employees to assist in determining suitability for ICT Operations and Services in support of security.	Blockchains used for information sharing and permissions for real-time personality assistance.
Professional skills	Development of professional skills required for management of the IoT, including security.	Blockchains used across skilling for information sharing to support AI in curriculum development, analysis, profiling, suggested skilling, and assessments.
Benchmark testing	Benchmark setting and testing of automation to ensure performance and integrity with independent auditability of self-learning and AI. Used for assuring the security of the IoT and alerts and monitoring management.	Blockchains used in AI for self-testing and diagnosis, as well as playing a role in data validation and authentication for assessment. Blockchain information used for triggers for automated remediation.
Automation of operations	From machine-to-machine operations through to automated alerts and monitoring or manual intervention by exceptions, operations of the IoT are increasingly automated.	Blockchains enable the device-to-device information sharing with role-based authentication to support automated operations.

- *Business Management*: The business management including governance, leadership, the business environment, the behaviors of people, risk management, skilling and training, and the ability to improve.

Blockchain are also set to play an increasing role within the automation of the IoT and its security across the layers (network, device, information, service management) that make up the IoT as well as the management and operation of the service. The secure role-based access to information provided by blockchains is a key emerging area.

REFERENCES

1. J. Song (2018) Why Blockchain Is Hard. https://medium.com/@jimmysong/why-blockchain-is-hard-60416ea4c5c (last viewed March 2020).
2. K. Sherringham and B. Unhelkar (2020) *Crafting and Shaping Knowledge Worker Services in the Information Economy* (Palgrave Macmillan, Singapore, 570pp.).
3. P. McConnell (2017) ASX – Actually, a Failure of Block Chain. https://www.linkedin.com/pulse/asx-actually-failure-blockchain-patrick-mcconnell (last viewed March 2020).
4. P. McConnell (2019) Blockchain Examining the Technical Architecture (ITNOW 61 (1) 38–41). https://doi.org/10.1093/itnow/bwz016 (last viewed March 2020).
5. T. Ali Syed, A. Alzahrani, S. Jan, M. S. Siddiqui, A. Nadeem and T. Alghamdi (2019) "A Comparative Analysis of Blockchain Architecture and Its Applications: Problems and Recommendations," *IEEE Access*, vol. 7, pp. 176838–176869.
6. S. Wang, L. Ouyang, Y. Yuan, X. Ni, X. Han and F. Wang (2019) "Blockchain-Enabled Smart Contracts: Architecture, Applications, and Future Trends," *IEEE Transactions on Systems, Man, and Cybernetics: Systems*, vol. 49, no. 11, pp. 2266–2277.
7. T. T. A. Dinh, R. Liu, M. Zhang, G. Chen, B. C. Ooi and J. Wang (2018) "Untangling Blockchain: A Data Processing View of Blockchain Systems," *IEEE Transactions on Knowledge and Data Engineering*, vol. 30, no. 7, pp. 1366–1385.
8. G. Glover (2011) Top 10 Network Security Audit Fails. https://www.securitymetrics.com/blog/top-10-network-security-audit-fails (last viewed August 2019).
9. National Audit Office (2018) Investigation: WannaCry Cyber Attack and the NHS. https://www.nao.org.uk/report/investigation-wannacry-cyber-attack-and-the-nhs/ (last viewed March 2020).
10. H. Saini, B. Bhushan, A. Arora and A. Kaur (2019) "Security Vulnerabilities in Information Communication Technology: Blockchain to the Rescue (A Survey on Blockchain Technology)," In *2019 2nd International Conference on Intelligent Computing, Instrumentation and Control Technologies (ICICICT)*, Kannur, Kerala, India, 2019, pp. 1680–1684.
11. D. Arora, S. Gautham, H. Gupta and B. Bhushan (2019) "Blockchain-Based Security Solutions to Preserve Data Privacy and Integrity," In *2019 International Conference on Computing, Communication, and Intelligent Systems (ICCCIS)*, Greater Noida, India, 2019, pp. 468–472.
12. M. Abomhara and G. M. Køien (2015) "Cyber Security and the Internet of Things: Vulnerabilities, Threats, Intruders and Attacks," *Journal of Cyber Security and Mobility*, vol. 4, no. 1, pp. 65–88.
13. S. D. Gantz (2013), *The Basics of IT Audit: Purposes, Processes, and Practical Information* (Syngress Media, 144+pp.). ISBN-13 9780124171596 144pp
14. M. Chanson, A. Bogner, D. Bilgeri, E. Fleisch and F. Wortmann (2019) "Blockchain for the IoT: Privacy-Preserving Protection of Sensor Data," *Journal of the Association for Information Systems*. doi: 10.3929/ethz-b-000331556.
15. K. Sherringham and B. Unhelkar (2016) "Service Management in Big Data," In *Proceedings of the System Design and Process Science (SDPS2016) Conference*, 4–6 December 2016, Orlando, FL, USA, pp 135–143. https://sdpsnet.org/sdps/documents/sdps-2016/proceedings%20SDPS%202016.pdf

16. T. Gonsalves (2017), *Artificial Intelligence: A Non-Technical Introduction* (Sophia University Press, Tokyo).

17. P. Danzi, A. E. Kalør, Č. Stefanović and P. Popovski (2019) "Delay and Communication Tradeoffs for Blockchain Systems with Lightweight IoT Clients," in *IEEE Internet of Things Journal*, vol. 6, no. 2, pp. 2354–2365.

18. F. Provost and T. Fawcett. *Data Science for Business: What You Need to Know about Data Mining and Data-Analytic Thinking.* O'Reilly Media; July 27, 2013. Pages 340, 341

19. T. Sharma, S. Satija and B. Bhushan (2019) "Unifying Blockchain and IoT: Security Requirements, Challenges, Applications and Future Trends," In *2019 International Conference on Computing, Communication, and Intelligent Systems (ICCCIS),* Greater Noida, India, 2019, pp. 341–346.

12 IoT-Based Healthcare

Personalized Health Monitoring Using Machine Learning and Blockchain for Security of Data

A. Chauhan and Y. Hasija

Delhi Technological University

CONTENTS

12.1   INTRODUCTION

A breakthrough in technology came with the internet, which not only connected people globally but also provided access to information worldwide. While the use of a portmanteau of interconnected networks was initially limited to connecting computers and mobile phones, nowadays, every other device is linked by the Internet of Things (IoT), optimizing the performance of devices, enabling them to be controlled efficiently, and storing data. It has also brought about a revolution in the healthcare sector, where data collection is automated easily with devices such as fitness bands and glucose meter. The volume of data collected by smart healthcare devices using sensors comes in use for remote healthcare monitoring. The application of machine learning algorithms to process data stored by IoT-based healthcare devices using cloud computing (Hassanalieragh et al. 2015) is a new approach for making predictions with large amounts of data and minimizing human error.

The chapter is organized as follows – the second section deals with how IoT has been applied in the healthcare industry and as revolutionized the entire concept of personalized and remote healthcare. It also explains the various advantages and objectives of IoT application in healthcare. A basic methodological approach using the example of prediction model based on IoT has been discussed in the last subsection. The third section aims at reviewing the latest advances in the combination of IoT and machine learning used in the healthcare sector and how it has transformed personalized healthcare. With the ever-increasing data from IoT devices, the application of big data modeling systems has also been reviewed in this section. A core component in any IoT environment is the technology used for establishing communication between devices. The fourth section briefly discusses some of the technologies which are used and have the potential to be used in the healthcare sector and a comparison of the technologies, while the fifth section outlines the IoT communication models. No technology is free of challenges that need to be overcome for its successful application in any sector. The sixth section explores the various challenges faced while applying IoT in the healthcare sector, and how data security is one of the biggest blocks in the road to efficient and secure personalized healthcare systems. The section also deals with the implementation of blockchain technology in securing sensitive health data followed by conclusion in Section 12.7.

Healthcare industry has changed a lot over the years with detection of diseases being done at an earlier stage to facilitate treatment of the conditions whose diagnosis in later stages can prove to be fatal. Although IoT devices have been widely used in home as well as hospitals for automated control of devices, this chapter focuses mainly on how IoT application coupled with machine learning and cloud computing in the healthcare industry has led to the utilization of a plethora of health data made available by IoT devices has led to transformation of the sector. In addition, the security of sensitive health data with blockchain technology, which is still a less explored concept, has been discussed in detail.

12.2   IOT IN HEALTHCARE

As the life expectancy of people has risen across the globe, the population of older people has also expanded. As per the reports of the United Nations, 22% of the world population (i.e., 2 billion people) will be old by 2050 (United Nations 2014). Further, in surveys, it was found that about 89% of the older people are expected to live independently, with medical researches indicating that 80% of people above the age of 65 are affected by at least one chronic sickness (Atella et al. 2019). This has led to difficulties for the older people in looking after themselves, and a need has arisen for providing a decent solution to them for maintaining their quality of life. As the IT industry is rising rapidly, there has been an explosion of data available, and the modern healthcare solutions have been developing around it and showing promising results. Moreover, the monitoring of patients requiring regular checking has become more comfortable with automated health tracking technologies acting as an interface between the patient and the doctor.

With technological advancements reaching new heights in the 21st century, the IoT has become a paradigm of communication between devices. IoT environment is a system consisting of devices connected through the internet due to their ability to communicate and compute. These devices build a network because of microchips and transceivers placed in them (Rosadi and Sakti 2017). IoT is a sophisticated network of devices ("things"), each of which is addressable uniquely and is associated with a server responsible for providing suitable service. The interoperable objects in IoT are uniquely addressed by radio-frequency identification (RFID) technology (Zhong, Putnik, and Newman 2019). IoT has broadened the notion of Internet-enabling sensors fitted in health monitoring equipment to monitor the functions of the human body, and report any kind of life-critical activity inside the body remotely to healthcare experts or doctors. Incorporation of IoT in healthcare instruments enhances the quality and performance of medical services (Marques et al. 2019); thus, it has found broad applicability in the sector with researches carried out not only to improve the efficiency but also to secure the critical data collected. Cross-device connectivity achieved by IoT (Alulema, Criado, and Iribarne 2018) implements smart technology in the medical domain providing older population and patients requiring continuous supervision with on-point health services. Remote monitoring of the health statistics of people has become an attainable task by integrating hardware and software systems for health management.

Personalized healthcare systems are heavily dependent upon the collection of an individual's health data such as pulse rate, body temperature, glucose levels, and rate of respiration. IoT has facilitated the collection of such information with an integral part played by sensors. Physical devices that may also be wearable consist of several sensors (Mosenia et al. 2017) for detecting different parameters, thus sensors act as a collector of data. Second, the storage of this data is required for deriving meaningful results, which can be achieved easily by the use of microchips. The final step is the analysis of the collected information, which may be done manually.

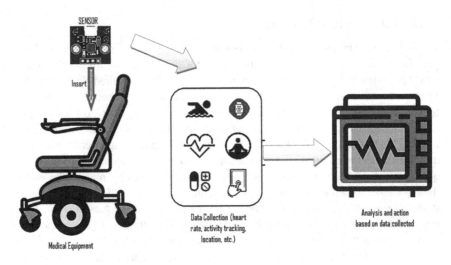

FIGURE 12.1 Basic workflow in IoT-based health management system.

But with the progress in artificial intelligence and machine learning, extracting useful conclusions and making prediction models from large datasets has become an attainable task. The use of IoT with artificial intelligence (Osifeko et al. 2020) has been an active area for research not only for the prediction of certain diseases but also for accomplishing the aim of personalized medication (see Figure 12.1).

A large number of medical instruments have been tagged with sensors such as heart rate monitoring devices, glucometers, wheelchairs, and fitness band, which provide patient data to the physicians time to time, thereby enabling them to track patient health stats remotely and make recommendations without requiring to visit physically. The transformation of the healthcare industry by IoT is discernible as it is enabling access to personalized health assistance at any point in time. IoT has redefined the association between health experts and the people requiring medical support or help by speeding the process of delivery of healthcare solutions. Web of things (or IoT) has aided patients and physicians alike. Majorly benefited are elderly patients who live independently and need timely medical supervision. For physicians, continuous monitoring is possible by IoT devices, which allows them to provide instant medical attention in case of emergency. In the case of careless patients, the doctors can track their adherence to the treatment plans. Further, the data collected using the sensor containing device can be utilized for preparing better personalized treatments for the patient, increasing the chances and speed of recovery. Furthermore, sensors can be incorporated in hospital equipment such as nebulizers, wheelchairs, and glucose drips for controlling them by machines and minimizing human error.

Remote health monitoring (Majumder, Mondal, and Deen 2017) which is achievable by application of IoT in healthcare has a formidable future as it will not only play a role in improving the quality of healthcare services but also reduce the cost of healthcare as it averts situations requiring hospitalization by identifying diseases and harmful health conditions at an earlier stage (Balandina et al. 2015; Taylor, Baron, and Schmidt 2015). The expenditure on healthcare services has risen nowadays as patients need to be hospitalized throughout the entire process of treatment and need to be kept under observation for preventing reoccurrence of disease or casualties that may occur during the healing period. IoT has a significant role to play in such a scenario as it will not only reduce the cost of the treatment phase but will also enable better utilization of hospital beds as they will become empty for cases that are more in need of medical attention. Furthermore, smart healthcare solutions that apply machine learning and artificial intelligence have led to the convergence of heterogeneous IoT architectures, facilitating complete imaging of a patient's physiological state. These IoT-based solutions are the premise of home healthcare services that are needed to surmount population aging-related issues (Konstantinidis et al. 2015). Several IoT frameworks are in place, such as seizure detection and fall detection, permitting prompt action from caregivers, and thereby averting a dangerous situation during seizures or critical body function failure. Another application of in-home healthcare service is the detection of sleep patterns in different age groups. This allows for the collection of sleep data that health professionals analyze to determine the effect of timing and duration of naps on a particular section of people. Sleep pattern differences can also be observed in the normal and disease state of a patient. If a patient has undergone treatment, an improvement in sleep pattern may indicate the effectiveness of the therapy as sleep patterns of patients tend to decline in disease state and while undergoing treatment. Effects of medication on sleep patterns can be studies by IoT-based devices, which is an ongoing research area.

The most widely used example of IoT is a fitness band (Madigan 2019), which is a mobile healthcare system for recording information such as steps walked, activities done, heart rate, and calories burned. Although the fitness band or health band was not considered an essential requirement less than a decade ago, it has now become imperative to the younger generation, which is growing more and more aware of the importance of good health. People have adopted a better lifestyle and can track their daily activities with a small band on their wrist. An improvement of fitness band integrating various types of wearable sensors can be used for monitoring patients' health even outside home. It can notify the location and condition of a person requiring emergency medical help to a healthcare facility and assist the person in reaching the nearest possible center, which can provide help. The instant access to data associated with a patient's health in real time can enhance the status of healthcare service, improve communication between patient and physician, and thus increase the satisfaction level of a patient and allow timely intervention when required.

12.2.1 Some of the Major Advantages of Applying IoT in Medical Sector

- *Reduction in Cost*: Constant supervision and immediate medical attention when required reduces the high cost incurred when emergencies occur due to negligence or delay in treatment. Moreover, regular visits to physicians are reduced, further saving cost and energy of both the patient and the doctor.

- *Earlier and Faster Diagnosis*: Consistent monitoring and availability of real-time statistics of the body operations allows detection of a disease at an early stage even when the symptoms have not been expressed externally.
- *Better Treatment*: It often happens that due to limited knowledge of the patient, the patient cannot express the problems or is not able to differentiate between normal and abnormal state of body functions; however, as the sensor regularly provides error-free data to the health professional, better decisions can be made based on evidence.
- *Proactive Medical Analysis*: As real-time monitoring is attainable by IoT devices, active medical treatment and management of emergency cases is not a farfetched goal.
- *Management of Medicines and Instruments in Healthcare Facilities*: Equipment fitted with sensors have the potential of being used in large medical centers for their real-time tracking. Moreover, automated control of pharmacy inventory allows improved management of drugs available in a medical facility.
- *Error Minimization*: Automated control of medical devices and the use of data collected by IoT devices can be used to minimize errors caused due to human heedlessness (see Figure 12.2).

12.2.2 OBJECTIVE OF IoT IN HEALTHCARE

The question that first comes to mind while studying the application of any technology in a field – what is the purpose of this technology, and how will it aid the services provided? So, for the application of IoT in the healthcare sector, the question of its requirement in the industry remains. While technology is considered the greatest boon to humans by some, it is also believed that technology is the cause of many disorders being faced by people nowadays. Integration of IoT in healthcare was a matter of debate, but with positive feedback, IoT has found its way in the medical industry with the following objectives which need to be fulfilled for the success of the IoT technology in healthcare:

- Wearable IoT devices have become a huge success owing to the drop in the prices of the Internet over the years, making it accessible to a large number of people. The final users of the technology, that is, the clients or customers of the intricate network of healthcare are human beings (either patients or informed individuals knowing the importance of health), therefore, it is essential to develop ambient intelligence.
- For the development of this ambient intelligence, it is required that there is consistent learning form the data acquired by the devices connected through IoT and execution of the desired action brought about by a known event. The incorporation of independent supervision, as well

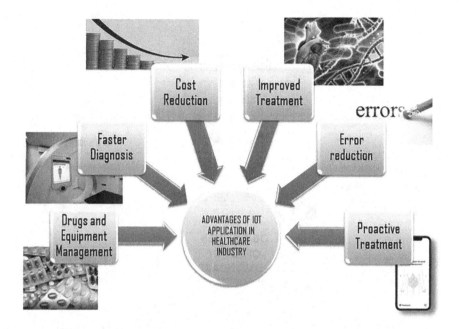

FIGURE 12.2   Advantages of the application IoT in the healthcare industry.

as human–computer interaction technology, can augment the ambient intelligence of IoT-assisted healthcare services.

- The flexible and wearable devices with IoT technology aid access to real-time information about human body functions.
- As computing technologies are becoming autonomous, IoT devices can also aid the preprocessing of data acquired about humans (if required).
- Early detection of chronic diseases by making use of data mining techniques and artificial intelligence helps in making informed decisions.
- Better accessibility to IoT healthcare solutions at any place and any time (see Figure 12.3).

Continual efforts are being put in to meet the goals and provide better healthcare facilities by uplifting the quality of services offered to substantiate IoT incorporation in healthcare. Recent studies focus on the integration of new-age technologies such as machine learning for facilitating on-point medical care, hence, improving the quality of life of people. Remote and real-time health monitoring are the products of IoT and considerable potential in promoting a healthy lifestyle (Martínez-Caro et al. 2018). As people are becoming health conscious, wearable IoT devices are a huge hit in the market, depicting the need for increased research in the area. The generation of statistical data associated with health conditions performed by a healthcare device is extensive and free of errors, which can be correlated with data from multiple IoT-linked devices and can be useful in

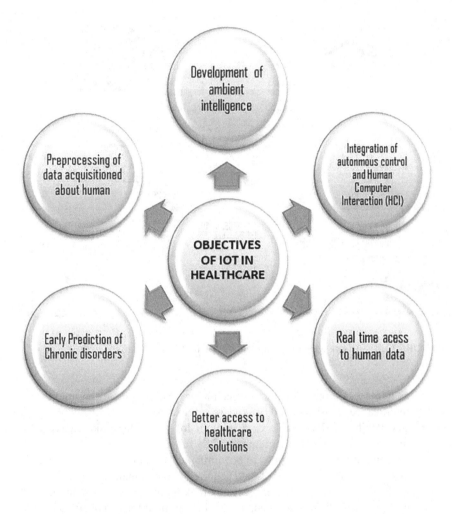

FIGURE 12.3   Current research focus on various objectives of IoT in healthcare.

deducing risk associated with a disease, surveillance of treatment procedures, and remote monitoring (Muthu et al. 2020).

12.2.3   METHODOLOGY FOR IMPLEMENTING IoT IN HEALTHCARE DEVICES

For the actualization of IoT in healthcare instruments, a proposed method is depicted in Figure 12.4. It shows how the interpretation of information and knowledge acquired from wearable health assistance technologies can be made for the prediction of severe health conditions or faster detection of chronic diseases. The first level is the collection of unprocessed data by healthcare systems/ health monitoring wearable gadgets fitted with sensors, which may include ECG

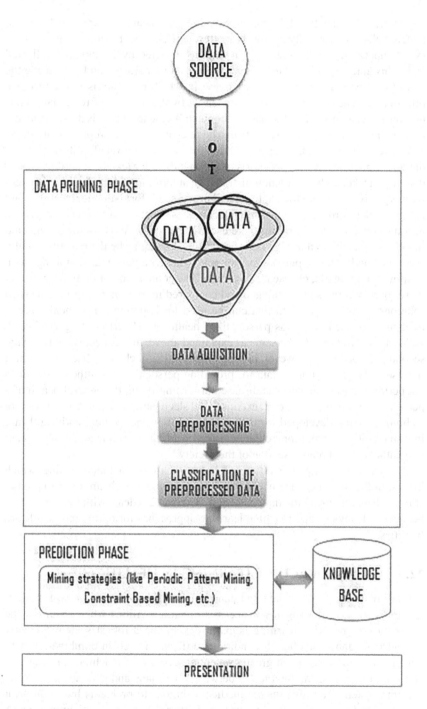

FIGURE 12.4   Inferring information and knowledge from healthcare data using prediction modeling (Darshan and Anandakumar 2016).

sensor (records electrical impulses generated by beating heart), EEG sensor (detects electrical activity of the brain through electrical impulses generated), skin temperature sensor, etc. The raw data is detected by sensors and collected and is instantly transferred to a server responsible for storing and processing the data. Level two is the data pruning phase in which raw data is filtered to give information. The data may also be condensed by the removal of redundant information and categorized on the basis of context. As the inference is drawn from the information received, level two can also be referred to as the preprocessing stage. The third level is the final phase involving the design of novel algorithms for prediction of poor health and prognosis of chronic disorder (can be a cardiovascular disorder, stroke, diabetes, mental disorder, an autoimmune disorder, etc.). A specified objective is set and the captured and processed information is structured and categorized accordingly. Data mining techniques (Li et al. 2019) (periodic pattern mining, constraint-based mining) are used for deriving valid conclusions that aid in informed decision-making. The conclusions drawn maybe the real-time analysis of the condition of a person, for example, whether a person is exhibiting ideal, healthy, symptomatic, or disease condition. The comparison of body functions in these four conditions are cataloged and compared in real time for giving learned solutions. Another approach that can be used for applying IoT in healthcare is by using AI-based device as personalized healthcare expert bot (Agarwal et al. 2020). First two tiers are the same as explained above, first collection of raw data, second preprocessing; however, instead of prediction of chronic disease, the third tier uses a large amount of data for providing personalized healthcare solutions not pertaining to some specific disease but for improving the general health of a person, which can relate to diet, exercise, and sleep pattern. There are many more techniques being developed for incorporating IoT in the medical and healthcare industry as it can revolutionize the industry, enabling economic and easier access to healthcare to a broader section of the society.

A key player in implementation IoT in healthcare is machine learning, which is used in the third tier of process. Machine learning models are used to predict chronic disorders using the data collected. Section 12.3 deals with the correlation between IoT devices and machine learning approaches for optimized healthcare facilities.

12.3   IOT AND MACHINE LEARNING IN HEALTHCARE

While IoT devices collect data and can alert healthcare experts in case of medical emergencies allowing faster access to health facilities when required, the application of machine learning helps in developing a robust system for early prognosis of dangerous health conditions. IoT, when used in combination with machine learning, has much greater potential. The use of machine learning also helps in minimizing the interaction between physicians and patients as machine learning systems can give instant medical solutions to problems based on prior information, thereby speeding up the recovery process and diminishing costs. A continuous learning system can also be used, which improves as more and more data is fed into it. The IoT sensors are responsible for gathering error-free data and sometimes preprocessing it (removal of redundant information and filtering), while the machine learning algorithms analyze the data and derive meaningful results. The three-tiered network technology incorporating machine learning in the third tier has come into use widely across various fields, with healthcare being one of them. Earlier, the three-tiered model consisted of only interconnected systems delivering information to the caretakers when required for averting potential dangers. Some of the examples of the traditional methodology (not involving big data) are listed in Table 12.1.

The traditional systems did not use big data technologies for the storage and processing of the voluminous clinical data collected by IoT devices. However, with machine learning gaining momentum in the past few years and the development of distributed system frameworks to manage and process massive volumes of statistical data, approaches have been designed for incorporating machine learning models (Durga, Nag, and Daniel 2019) and training it on the collected large amount of data stored in a distributed manner. The machine learning algorithm makes the system more efficient as it analyzes the data before sending it to a medical professional in contrast to the earlier system, which used to send raw data to the doctor, making the procedure slower and sometimes susceptible to error.

Figure 12.5 shows the framework and data flow of one such method, which uses big data and machine learning for improving the performance of IoT-based healthcare systems. The proposed structure for the system consists of three blocks, with the first tier being data collection. Wearable IoT sensing devices are responsible for the constant collection of patient data of clinical importance. When the value of a clinical variable exceeds the range observed in the ordinary state of body, an alert message is relayed to the healthcare professional along with the value of a clinical variable. The alert message, as well as the clinical measure, is stored in a database every time there is an abnormal fluctuation in the value. Apache HBase is the distribution system used for storage and processing of the huge data generated continuously from IoT wearable devices (Zhang et al. 2018). The large volume of data is stored in a scalable manner using Hadoop distributed file system. Transfer of data occurs between Amazon S3 to Apache HBase using Apache Pig. The final tier is data analytics, which employs a logistic regression algorithm of Apache Mahout Libraries for machine learning. The prediction algorithm is used for the early detection of heart diseases in the given framework.

12.4 COMMUNICATION TECHNOLOGY USED IN IOT IN HEALTHCARE

Healthcare devices are interconnected and have the power to communicate with each other in an IoT environment, which is achieved by different communication technologies. The following IoT technologies have the potential of being used in the healthcare sector.

TABLE 12.1

Three-Tier Methodology Projects (Not Incorporating Big Data or Machine Learning) of IoT-Based Healthcare Systems

Project/System Name	Organization/ University	Details
CodeBlue	Harvard University	It monitors individuals' health parameters such as ECG and EKG, which are continually recorded on various computing devices like PCs, PDAs, and laptops enabling doctors and healthcare professionals to take appropriate actions if the patient's health deteriorates. The architecture used in the CodeBlue project is widely published, and it helps in the constant supervision of patient's health statistics (Chacko and Hayajneh 2018; Varatharajan et al. 2018; Patel, Singh, and Kazi 2017).
Alarm-net	University of Virginia	It is also a three-tiered system with the first tier involving attachment of IoT devices with body function sensors such as ECG and SpO$_2$ which sense the physiological statistics of a patient. It also includes environmental sensors for recording parameters such as heat, humidity, motion, and light (Patra 2017). The second tier transfers clinical data from the second to the third tier through the shortest-path-first routing protocol. The third tier uses IP-based connectivity for achieving wireless communication between the source and destination. The system predicts emergencies based on data available beforehand in a patient's record.
MobiCare (Varatharajan et al. 2018)	Chakravorty et al.	This project uses HTTP POST protocol for transfer of accurately sensed physiological data efficiently to healthcare personnel through a PDA or mobile phone. MobiCare architecture consists of a body sensor network. BSN is composed of multiple sensors connected by IoT such as ECG and SpO$_2$, or placed in the body to track body functions. This application has found extensive usage in the health monitoring of large facilities. It uses fog or cloud computing for sending physiological variables of a patient being observed scrupulously to caretaker or physician.
PAM Project	Blum et al.	IoT devices containing sensors for mental health monitoring are used to observe the mental health conditions of a patient (Malasinghe, Ramzan, and Dahal 2019). Identification of bipolar disorder is one of the essential aims of this project. PAM-I and PAM-A blocks are developed using infrastructure and architecture-based technologies, respectively. Electronic devices such as mobile phones, PDAs, and wearable bands are part of PAM-I-based systems. The connection between multiple IoT sensors in the health monitoring system is achieved through Bluetooth and 5G networks in mobiles.

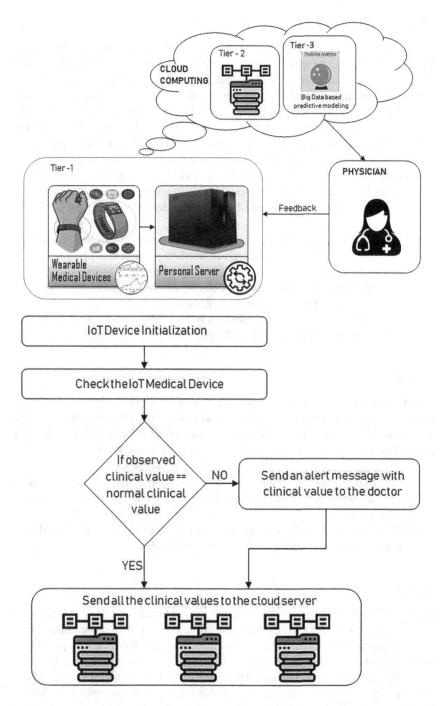

FIGURE 12.5 (a) An example of framework for IoT-based health monitoring system. (b) Data flow diagram for the framework given in (a) (Kumar and Devi Gandhi 2018).

12.4.1 RFID AND NFC

Unambiguous recognition of an entity is possible by assignment of Unique Identifier (UID) in correspondence to a single entity or device; this allows retrieval of data pertaining to a uniquely identified device. Every resource in a healthcare setting such as an emergency center, nurses, and hospitals are assigned a digital UID (Yin et al. 2016). The technology of RFID is used for short-range communication between entities tagged with electronic product code (EPC), which contains data for the unique identification of objects. The cost-effectiveness and reliability of RFID tags make it a convenient technology for use in IoT devices.

NFC is another technology suitable for short-range communication with a mechanism for secure and straightforward authentication among the objects. The devices in which NFC is incorporated can be operated in three modes:

a. *Reader or Writer Mode*: In this mode, the information can be accessed, or action can be triggered through contactless reading or writing operation of the system.
b. *Peer-to-Peer Mode*: The channel of communication is open both ways in this mode; thus, two-way communication can occur.
c. *Card Emulation Mode*: The devices behave like a smart card in this mode (Sethia, Gupta, and Saran 2019).

12.4.2 LR-WPAN

Low-rate wireless personal area network (LR-WPAN) is a subtype of WPAN, which is more efficient in terms of cost and power utilization, but offers a reliable protocol for data transfer (Dahiya 2017; Bockstael, Clercq, and Botteldooren 2017). Two types of devices are capable of operating in LR-WPAN:

a. *Full-Function Device (FFD)*: It is a flexible system that can function as a PAN coordinator or general coordinator or even as a normal device in a network.
b. *Reduced Function Device (RFD)*: It is a device that is used when the transfer of a huge amount of data is not required.

12.4.3 BLUETOOTH

Bluetooth is a short-range wireless communication technology that can be used to connect one or more devices at a time consuming very minimal power. Encryption and authentication before the establishment of connection are the two protection methods utilized by this technology (Cvitić, Vujić, and Husnjak 2015). About 79 radio-frequency channels having a bandwidth of 1 MHz on the 2.4 GHz band are used by this technology with a range of 100 m and a connection speed of 3 Mbps. The low power solution provided by Bluetooth is advisable for IoT applications in telemonitoring (Gentili, Sannino, and Petracca 2016).

12.4.4   ZigBee

This technology has found wide applicability in the IoT environment as it served as the foundation for IoT by allowing entities to operate while maintaining communication with each other. It operates upon end nodes, routers (for communication), a coordinator for maintaining contact, and a processing unit, which is the center for data accumulation and processing (Deese and Daum 2018). Features such as flexibility of network, interoperability, and minimum consumption of power make it preferable for use in IoT implementations. Its use of mesh network model adds another advantage of continued operation and communication between nodes even after failure of one of the devices connected to an end node (Ndih and Cherkaoui 2016).

12.4.5   Wireless-Fidelity (Wi-Fi)

Wi-Fi has become the master of communication technologies because of its wide availability and speed. Even though it is expensive than some of the technologies, its wider range and ease of use make it convenient for use in wearable IoT healthcare devices. WLAN (Wireless local area network) is also considered in the Wi-Fi category if it meets IEEE 802.11 standards (Saleem et al. 2019). LANs based on Wi-Fi have been applied in many hospitals.

12.4.6   Worldwide Interoperability for Microwave Access (WiMAX)

WiMAX is a constituent of WMAN (Wireless metropolitan area network), 802.16 series standards. WiMAX standard IEEE 802.16a having frequency band spectrum range of 2–11 GHz has the capability of execution in licensed and unlicensed frequency bands was the first standard developed (Kanno et al. 2016). WiMAX standard IEEE 802.16b, on the other hand, has frequency band spectrum range of 5–6 GHz and provides customers with high-quality voice and data services in real time. WiMAX IEEE 802.16c standard with operating frequency band range of 10–66 GHz provides interoperable communication between the devices or gadgets in which it is incorporated.

12.4.7   Mobile Communications

The mobile communications systems have developed with new generations overcoming the limitations of previous ones. Analog systems based on the first generation allowed voice transmission over the network in real time. The progression into second generation (2G) based upon digital infrastructure supported text messaging. Third-generation network systems (3G) developed for providing online exchange of information for the creation of global infrastructure supporting multiple services efficiently (Parashar 2019). Optimization of 3G is possible through the separation of equipment for data access, facilities of transport, and user interface. Concept of fourth generation (4G) with services of the same quality to the fixed internet was introduced to overcome limitations of 3G improving quality while decreasing resource costs. Fifth-generation (5G) networks offer higher efficiency in terms of system capacity and power consumption, thereby bringing about perfect wireless connectivity. The concept of sixth-generation (6G) network provides wide area coverage by unifying satellites (Blanco et al. 2017).

12.4.8 WIRELESS SENSOR NETWORKS (WSN)

WSN come in frequent use in the healthcare sector due to their ability to connect heterogeneous sensor networks. WSNs provide the following benefits that make it suitable for monitoring patients at home and home automation:

    a. Wider coverage
    b. Lower cost of installation
    c. Collection of data real-time

WSN plays a role in monitoring the physiological conditions of the patient, drug monitoring, and management of emergency situations and control of equipment in hospitals, thus playing an integral role in healthcare monitoring systems. WSN systems can either consist of a wearable device or implanted devices for sensing physical world conditions (Chen et al. 2013; Dressler et al. 2016) (see Table 12.2).

12.5 IOT COMMUNICATION MODELS

Construction of IoT environment involves connection of connected and communicating devices to the internet, which can be achieved via different ways of deployment. Internet architecture board (IAB) gave guidelines according to which four deployment models are permissible – device-to-device, device-to-cloud, device-to-gateway, and backend data sharing model.

12.5.1 DEVICE-TO-DEVICE COMMUNICATION MODEL

The basis of this communication model is IP networks in which multiple connected entities can communicate. Technologies such as Bluetooth or ZigBee are applied to create stable communication for the exchange of information to achieve proper functioning (Rose, Eldridge, and Chapin 2015).

12.5.2 DEVICE-TO-CLOUD COMMUNICATIONS

TCP/IP network or Wi-Fi connections are used to establish connectivity between IoT-based devices and cloud storage services. Exchange of data between connected devices and service providers is achieved by the shortest route protocol (Rose, Eldridge, and Chapin 2015).

TABLE 12.2
Comparison of IoT Communication Technologies

Parameters	Wi-Fi	WiMAX	LR-WPAN	Mobile Communication	Bluetooth	RFID	ZigBee	WSN
Standard	IEEE 802.11 a/c/b/d/g/n	IEEE 802.16	IEEE 802.15.4	2G-GSM, CDMA 3G-UMTS CDMA2000 4G-LTE	IEEE 802.15.1	ISO/IEC 15, 693	802.15.4	IEEE 802.15.4
Frequency band	5–60 GHz	2–66 GHz	868/915 MHz 2.4 GHz	865 MHz, 2.4 GHz	2.4 GHz	860–960 MHz and 2.4 GHz	868/915 MHz 2.4 GHz	902–928 MHz
Data rate	1 Mb/s–6.75 Gb/s	1 Mb/s–1 Gb/s (fixed) 50–100 Mb/s (mobile)	40–250 Kb/s	2G: 50–100 Kb/s 3G: 200 Kb/s 4G: 0.1–1 Gb/s	1–24 Mb/s	106 k–424 Kb/s	20 k–250 Kb/s	20–250 Kb/s
Transmission range	20–100 m	<50 km	10–20 m	Entire cellular area	8–10 m	Up to 100 m (325 feet)	10–75 m	20–100 m
Energy consumption	High	Medium	Low	Medium	Bluetooth: Medium BLE: very low	Medium	Medium	High
Cost	High	High	Low	Medium	Low	Low	Medium	High

Source: Ahmadi et al. (2019).

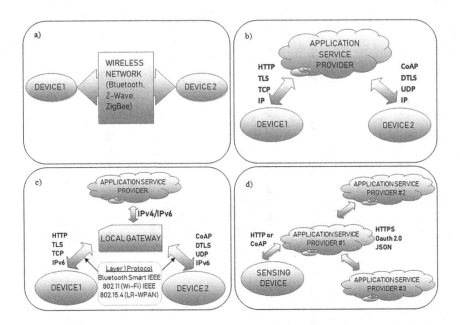

FIGURE 12.6 Internet of Things communications models – (a) device-to-device communications model, (b) device-to-cloud communications model, (c) device-to-gateway model, (d) backend data sharing model (Rose, Eldridge, and Chapin 2015).

12.5.3 DEVICE-TO-GATEWAY MODEL

The information is passed on to cloud services from the interconnected devices via a software app that acts as a gateway for transmission of data between IoT devices and cloud services. This model provides interoperability to the integrated smart devices and legacy computing structures (Rose, Eldridge, and Chapin 2015).

12.5.4 BACKEND DATA SHARING MODEL

This model integrates cloud application into the service allowing interoperability within the cloud environment. It controls access to information by authorization techniques. Secure transmission of accumulated data is possible using this framework (Rose, Eldridge, and Chapin 2015) (see Figure 12.6).

12.6 CHALLENGES OF IOT APPLICATION IN HEALTHCARE

- Even though the technology is very advanced, IoT faces the following challenges in operation in healthcare systems: There is a requirement of robust and efficient infrastructure for processing of voluminous data acquired at high speed.

- A competent analytics platform is a prerequisite for high-efficiency and cost-efficient data mining of physiological data aggregated.
- Scalability of the technology used for connection of a diverse set of devices should be kept in mind before applying it to a wider scale as the aim is to incorporate more and more devices into the network, and the number of devices is bound to escalate and diverge with the expansion of the system.
- The range of bandwidth should be large for efficient access to raw data being generated by a huge number of interconnected devices.
- As large amount of data is aggregated, data security becomes a big challenge because of the risk of hacking (Balas 2020). A secure authentication system needs to be established for the security of sensitive data. Data security issues will be discussed in detail in Section 12.6.1.
- Filtering, removal of redundant data, and data pruning become challenging as the amount of data gathered increases (see Figure 12.7).

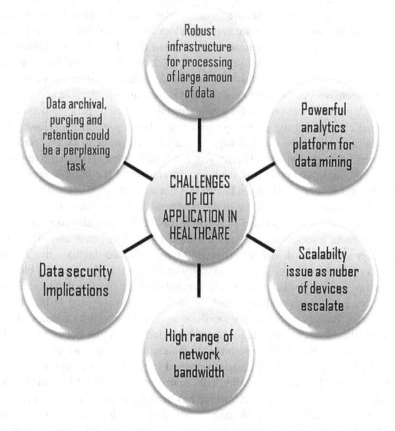

FIGURE 12.7 Despite being one of the most powerful technologies, IoT is not spared by shortcomings; the figure depicts few of the challenges faced in application of IoT in healthcare industry.

12.6.1 Data Security in IoT Applications in Healthcare

As IoT has opened new frontiers in the healthcare sector, it has also exposed it to the bigger challenge of data security (Farahani et al. 2018). The healthcare devices generate a large amount of sensitive data, unauthorized access to which can have dire consequences. Thus, secure gateway-based architectures are required for implementing IoT in healthcare equipment with proper authentication frameworks. Data privacy is the right of a customer, which should not be compromised at any cost, be it banking or healthcare (Williams and McCauley 2017). There have been many attempts to build secure IoT networks between healthcare equipment and cloud services; some of the algorithms and frameworks used are listed in Table 12.3.

12.6.2 Blockchain to the Rescue in Healthcare Data Security

After successful application in cryptocurrencies, blockchain has found its way into the healthcare sector for secure storage and transmission of sensitive medical data. Big healthcare industries are applying blockchain technology for the protection of massive amounts of patient data collected by IoT devices (Nikolaevskiy, Korzun, and Gurtov 2014; Ali Syed et al. 2019). Novel blockchain-based models are being proposed for IoT in smart healthcare devices as blockchain is computationally expensive. As the number of patients is increasing, wearable and implantable IoT devices are proving to be a boon for taking care of patients outside the traditional clinical setting allowing remote healthcare monitoring. However, if conventional blockchain methodology is used, it will prove to be computationally expensive, reducing the efficiency of the healthcare system (Ali Syed et al. 2019). Thus, improved protocols are being developed for the feasible application of blockchain security systems in health monitoring systems (Figure 12.8).

As IoT systems are developing in the healthcare industry, the data is transmitted not only from patient to healthcare professionals but also shared between the systems; hence, a robust security system should be put in place, and blockchain technology provides a solution to that. Being a distributed database that records new sets of data as blocks (each block consists of a link to the previous block as well as timestamp), blockchain not only manages to store the data dynamically but also transmit it securely and completely as each block is encrypted and the blocks form a chain. Based on the *proof of work* concept, the blockchain validates a transaction on the basis of computational work carried out by the nodes responsible for authorization. New blocks are added to the chain by miners which are consistently solving cryptographic puzzles (presented in the form of hash computation) for the creation of enciphered blocks. The identification of a block is made by hash present in the header of the block, which has been generated by the Secure Hash Algorithm (SHA-256). As the blockchain-based system does not allow meddling with the blocks, that is, deletion or changing data is not possible; it is a promising solution for the security of data in healthcare and medical IoT structures.

TABLE 12.3

Security Frameworks and Algorithms Applied in IoT System in Healthcare Environment for Securing Communication and Protecting Patient Data from Unwanted Intervention

Framework/Algorithm	Founder	Description
Continua Health Alliance (CHA) framework (Darwish et al. 2019)	Zhang et al.	This framework aims to resolve the issues of interoperability and data security. Security issues are catered to by making use of Secure Sockets Layer (SSL), individual verification, and scrutiny. Remote Method Invoking (RMI) is combined with SSL for achieving an optimized system, in which SSL encrypting the distributed and encapsulated objects is generated by RMI, which can be accessed via key exchange. Authentication is achieved by the MD5 algorithm, while auditing of security system is done through activity logs for identifying security issues and stopping dangerous activities.
Lightweight sensor authentication scheme (Lee, Sung, and Park 2016)	Lee et al.	The security technology applied to counter the damage that can result from intrusion in IoT devices. In this scheme, is the LEACH routing protocol, which is a lightweight protocol consuming low power. Security is improved by incorporation of the following parameters – mutual authentication, security against relay attacks, secure against counterfeit message attacks, only encrypted phrase upon snooping, network secure from snooping and side-channel attacks, error detection, and forward security.
Security for Mobile Health Systems applying Host Identity Protocol (HIP) (Nikolaevskiy, Korzun, and Gurtov 2014)	Nikolaevskiy et al.	To prevent malicious adversaries from gaining access to the patient's sensitive medical data, different key exchange schemes (for authentication) are applied depending upon the communication channel: Channel 1: *Sensors to gateway communication* – preshared keys are used, which are installed by the medical personnel while configuring the device. Channel 2: *Gateway to healthcare system* – The key exchange system is based on HIP. Channel 3: *Sensors to PMT* – Key exchange system based on HIP DEX (HIP Diet EXchange) used for secure communication while lightweight key exchange used for mutual authentication. Channel 4: *Gateway to PMT* – Same key exchange used as Channel 4 for simple and light security solution.

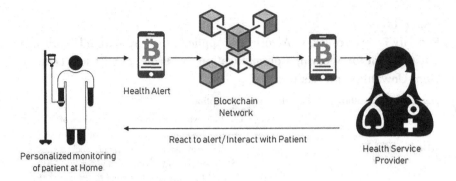

Health Alert

Blockchain
Network

React to alert/Interact with Patient

Personalized monitoring
of patient at Home

Health Service
Provider

FIGURE 12.8   Basic framework for remote healthcare monitoring applying blockchain technology (Dwivedi et al. 2019).

12.7 CONCLUSION

IoT has proved to be a strong contender in achieving the connectivity of health-care devices and has great potential if the challenges are overcome. Security is a pressing issue when it comes to data transmission by IoT equipment, and the current frameworks are not well suited for use in the healthcare industry. Blockchain technology provides a promising solution for solving one of the biggest problems associated with IoT devices in healthcare, that is, the security of storage and transmission of sensitive health data of the user or patient, which is susceptible to interception by malicious adversaries who can cause harm by using or tampering with the data. Appropriate architectures for blockchain-based systems, like the one provided by Dwivedi et al. (2019), are essential for providing a solution to the privacy and security threats faced by IoT devices in the healthcare sector.

REFERENCES

Agarwal, Yashasvi, Mahima Jain, Shuchi Sinha, and Sanjay Dhir. 2020. "Delivering High-Tech, AI-Based Health Care at Apollo Hospitals." *Global Business and Organizational Excellence* 39 (2): 20–30. https://doi.org/10.1002/joe.21981.

Ahmadi, Hossein, Goli Arji, Leila Shahmoradi, Reza Safdari, Mehrbakhsh Nilashi, and Mojtaba Alizadeh. 2019. *The Application of Internet of Things in Healthcare: A Systematic Literature Review and Classification.* Universal Access in the Information Society. Vol. 18. Springer, Berlin; Heidelberg. https://doi.org/10.1007/s10209-018-0618-4.

Ali Syed, Toqeer, Ali Alzahrani, Salman Jan, Muhammad Shoaib Siddiqui, Adnan Nadeem, and Turki Alghamdi. 2019. "A Comparative Analysis of Blockchain Architecture and Its Applications: Problems and Recommendations." *IEEE Access* 7: 176838–69. https://doi.org/10.1109/ACCESS.2019.2957660.

Alulema, Darwin, Javier Criado, and Luis Iribarne. 2018. "A Cross-Device Architecture for Modelling Authentication Features in IoT Applications." *Journal of Universal Computer Science* 24 (12): 1758–75.

Atella, Vincenzo, Andrea Piano Mortari, Joanna Kopinska, Federico Belotti, Francesco Lapi, Claudio Cricelli, and Luigi Fontana. 2019. "Trends in Age-Related Disease Burden and Healthcare Utilization." *Aging Cell* 18 (1): 1–8. https://doi. org/10.1111/acel.12861.

Balandina, Ekaterina, Sergey Balandin, Yevgeni Koucheryavy, and Dmitry Mouromtsev. 2015. "IoT Use Cases in Healthcare and Tourism." In *Proceedings – 17th IEEE Conference on Business Informatics, CBI 2015* 2: 37–44. https://doi.org/10.1109/CBI.2015.16.

Balas, Valentina E. 2020. *A Handbook of Internet of Things in Biomedical and Cyber Physical System.* Vol. 165. Springer International Publishing. https://doi. org/10.1007/978-3-030-23983-1.

Blanco, Bego, Jose Oscar Fajardo, Ioannis Giannoulakis, Emmanouil Kafetzakis, Shuping Peng, Jordi Pérez-Romero, Irena Trajkovska, et al. 2017. "Technology Pillars in the Architecture of Future 5G Mobile Networks: NFV, MEC and SDN." *Computer Standards and Interfaces* 54: 216–28. https://doi.org/10.1016/j.csi.2016.12.007.

Bockstael, Annelies, Lies De Clercq, and Dick Botteldooren. 2017. "Communication with Low-Cost Hearing Protectors : Hear, See and Believe." https://biblio.ugent. be/publication/8537540/file/8537543

Chacko, Anil, and Thaier Hayajneh. 2018. "Security and Privacy Issues with IoT in Healthcare." *EAI Endorsed Transactions on Pervasive Health and Technology* 4 (14): 1–8. https://doi.org/10.4108/eai.13-7-2018.155079.

Chen, Dan, Zhixin Liu, Lizhe Wang, Minggang Dou, Jingying Chen, and Hui Li. 2013. "Natural Disaster Monitoring with Wireless Sensor Networks: A Case Study of Data-Intensive Applications upon Low-Cost Scalable Systems." *Mobile Networks and Applications* 18 (5): 651–63. https://doi.org/10.1007/s11036-013-0456-9.

Cvitić, Ivan, Miroslav Vujić, and Siniša Husnjak. 2015. "Classification of Security Risks in the Iot Environment." In *Annals of DAAAM and Proceedings of the International DAAAM Symposium 2015*, January 2016: 731–40. https://doi.org/10.2507/26th. daaam.proceedings.102.

Dahiya, Menal. 2017. "A Relative Study of Heterogeneous Wireless Protocols." *Advances in Computational Sciences and Technology* 10 (2): 159–64.

Darshan, K. R., and K. R. Anandakumar. 2016. "A Comprehensive Review on Usage of Internet of Things (IoT) in Healthcare System." In *2015 International Conference on Emerging Research in Electronics, Computer Science and Technology, ICERECT 2015*, 132–36. https://doi.org/10.1109/ERECT.2015.7499001.

Darwish, Ashraf, Aboul Ella Hassanien, Mohamed Elhoseny, Arun Kumar Sangaiah, and Khan Muhammad. 2019. "The Impact of the Hybrid Platform of Internet of Things and Cloud Computing on Healthcare Systems: Opportunities, Challenges, and Open Problems." *Journal of Ambient Intelligence and Humanized Computing* 10 (10): 4151–66. https://doi.org/10.1007/s12652-017-0659-1.

Deese, Anthony S., and Julian Daum. 2018. "Application of ZigBee-Based Internet of Things Technology to Demand Response in Smart Grids." *IFAC-PapersOnLine* 51 (28): 43–48. https://doi.org/10.1016/j.ifacol.2018.11.675.

Dressler, Falko, Margit Mutschlechner, Bijun Li, Rüdiger Kapitza, Simon Ripperger, Christopher Eibel, Benedict Herzog, Timo Hönig, and Wolfgang Schröder-Preikschat. 2016. "Monitoring Bats in the Wild: On Using Erasure Codes for Energy-Efficient Wireless Sensor Networks." *ACM Transactions on Sensor Networks* 12 (1): 1–29. https://doi.org/10.1145/2875426.

Durga, S., Rishabh Nag, and Esther Daniel. 2019. "Survey on Machine Learning and Deep Learning Algorithms Used in Internet of Things (IoT) Healthcare." In *Proceedings of the 3rd International Conference on Computing Methodologies and Communication, ICCMC 2019*, no. Iccmc: 1018–22. https://doi.org/10.1109/ICCMC.2019.8819806.

Dwivedi, Ashutosh Dhar, Gautam Srivastava, Shalini Dhar, and Rajani Singh. 2019. "A Decentralized Privacy-Preserving Healthcare Blockchain for IoT." *Sensors* 19 (2): 1–17. https://doi.org/10.3390/s19020326.

Farahani, Bahar, Farshad Firouzi, Victor Chang, Mustafa Badaroglu, Nicholas Constant, and Kunal Mankodiya. 2018. "Towards Fog-Driven IoT EHealth: Promises and Challenges of IoT in Medicine and Healthcare." *Future Generation Computer Systems* 78: 659–76. https://doi.org/10.1016/j.future.2017.04.036.

Gentili, M., R. Sannino, and M. Petracca. 2016. "BlueVoice: Voice Communications over Bluetooth Low Energy in the Internet of Things Scenario." *Computer Communications* 89–90: 51–9. https://doi.org/10.1016/j.comcom.2016.03.004.

Hassanalieragh, Moeen, Alex Page, Tolga Soyata, Gaurav Sharma, Mehmet Aktas, Gonzalo Mateos, Burak Kantarci, and Silvana Andreescu. 2015. "Health Monitoring and Management Using Internet-of-Things (IoT) Sensing with Cloud-Based Processing: Opportunities and Challenges." In *Proceedings – 2015 IEEE International Conference on Services Computing, SCC 2015*, 285–92. https://doi.org/10.1109/SCC.2015.47.

Kanno, A, P T Dat, N Yamamoto, and T Kawanishi. 2016. "Millimeter-Wave Radio-over-Fiber System for High-Speed Railway Communication." In *2016 Progress in Electromagnetic Research Symposium (PIERS)*, 3911–3915. IEEE. https://doi.org/10.1109/PIERS.2016.7735471.

Konstantinidis, Evdokimos I., Giorgos Bamparopoulos, Antonis Billis, and Panagiotis D. Bamidis. 2015. "Internet of Things for an Age-Friendly Healthcare." *Studies in Health Technology and Informatics* 210: 587–91. https://doi.org/10.3233/978-1-61499-512-8-587.

Kumar, Priyan Malarvizhi, and Usha Devi Gandhi. 2018. "A Novel Three-Tier Internet of Things Architecture with Machine Learning Algorithm for Early Detection of Heart Diseases." *Computers and Electrical Engineering* 65: 222–35. https://doi.org/10.1016/j.compeleceng.2017.09.001.

Lee, Jaeseung, Yunsick Sung, and Jong Hyuk Park. 2016. "Lightweight Sensor Authentication Scheme for Energy Efficiency in Ubiquitous Computing Environments." *Sensors* 16 (12). https://doi.org/10.3390/s16122044.

Li, Rumei, Chuantao Yin, Xiaoyan Zhang, and Bertrand David. 2019. *Recent Developments in Intelligent Computing, Communication and Devices*. Vol. 752. Springer, Singapore. https://doi.org/10.1007/978-981-10-8944-2.

Madigan, Elizabeth A. 2019. "Fitness Band Accuracy in Older Community Dwelling Adults." *Health Informatics Journal* 25 (3): 676–82. https://doi.org/10.1177/1460458217720399.

Majumder, Sumit, Tapas Mondal, and M. Jamal Deen. 2017. "Wearable Sensors for Remote Health Monitoring." *Sensors* 17 (1). https://doi.org/10.3390/s17010130.

Malasinghe, Lakmini P., Naeem Ramzan, and Keshav Dahal. 2019. "Remote Patient Monitoring: A Comprehensive Study." *Journal of Ambient Intelligence and Humanized Computing* 10 (1): 57–76. https://doi.org/10.1007/s12652-017-0598-x.

Marques, Gonçalo, Rui Pitarma, Nuno M. Garcia, and Nuno Pombo. 2019. "Internet of Things Architectures, Technologies, Applications, Challenges, and Future Directions for Enhanced Living Environments and Healthcare Systems: A Review." *Electronics* 8 (10): 1–27. https://doi.org/10.3390/electronics8101081.

Martínez-Caro, Eva, Juan Gabriel Cegarra-Navarro, Alexeis García-Pérez, and Monica Fait. 2018. "Healthcare Service Evolution towards the Internet of Things: An End-User Perspective." *Technological Forecasting and Social Change* 136: 268–76. https://doi.org/10.1016/j.techfore.2018.03.025.

Mosenia, Arsalan, Susmita Sur-Kolay, Anand Raghunathan, and Niraj K. Jha. 2017. "Wearable Medical Sensor-Based System Design: A Survey." *IEEE*

Transactions on Multi-Scale Computing Systems 3 (2): 124–38. https://doi. org/10.1109/TMSCS.2017.2675888.

Muthu, Bala Anand, C. B. Sivaparthipan, Gunasekaran Manogaran, Revathi Sundarasekar, Seifedine Kadry, A. Shanthini, and Antony Dasel. 2020. "IOT Based Wearable Sensor for Diseases Prediction and Symptom Analysis in Healthcare Sector." *Peer-to-Peer Networking and Applications*. https://doi.org/10.1007/s12083-019-00823-2.

Ndih, Eugene David Ngangue, and Soumaya Cherkaoui. 2016. "On Enhancing Technology Coexistence in the IoT Era: ZigBee and 802.11 Case." *IEEE Access* 4: 1835–44. https://doi.org/10.1109/ACCESS.2016.2553150.

Nikolaevskiy, Ilya, Dmitry Korzun, and Andrei Gurtov. 2014. "Security for Medical Sensor Networks in Mobile Health Systems." In *Proceeding of IEEE International Symposium on a World of Wireless, Mobile and Multimedia Networks 2014, WoWMoM 2014*. https://doi.org/10.1109/WoWMoM.2014.6918926.

Osifeko, Martins O, Gerhard P Hancke, and Adnan M. Abu-Mahfouz. 2020. "Artificial Intelligence Techniques for Cognitive Sensing in Future IoT: State-of-the-Art, Potentials, and Challenges." *Journal of Sensor and Actuator Networks* 9 (2): 21. https://doi.org/10.3390/jsan9020021.

Parashar, Ruchi. 2019. "Survey: Internet of Things." *AKGEC International Journal of Technology*, 9 (1). https://www.akgec.ac.in/wp-content/uploads/2019/06/7-Survey_ Internet-of-Things-Ruchi-Parashar.pdf.

Patel, Ankit, N. M. Singh, and Faruk Kazi. 2017. "Internet of Things and Big Data Technologies for Next Generation Healthcare." 23 (October). https://doi.org/10.1007 /978-3-319-49736-5.

Patra, Manoj Kumar. 2017. "An Architecture Model for Smart City Using Cognitive Internet of Things (CIoT)." In *Proceedings of the 2017 2nd IEEE International Conference on Electrical, Computer and Communication Technologies, ICECCT 2017*, 492–96. https://doi.org/10.1109/ICECCT.2017.8117893.

Rosadi, Imron, and Setyawan P. Sakti. 2017. "Low-Cost Wireless Sensor Network for Small Area in a Building." In *Proceedings – 2017 International Seminar on Sensor, Instrumentation, Measurement and Metrology: Innovation for the Advancement and Competitiveness of the Nation, ISSIMM 2017* January 2017s: 115–18. https://doi. org/10.1109/ISSIMM.2017.8124273.

Rose, K., S. Eldridge, and L. Chapin. 2015. "The Internet of Things: An Overview. The Internet Society." no. October: 1–50. https://www.internetsociety.org/wp-content/ uploads/2017/08/ISOC-IoT-Overview-20151221-en.pdf.

Saleem, Yasir, Noel Crespi, Mubashir Husain Rehmani, and Rebecca Copeland. 2019. "Internet of Things-Aided Smart Grid: Technologies, Architectures, Applications, Prototypes, and Future Research Directions." *IEEE Access* 7: 62962–3. https://doi. org/10.1109/ACCESS.2019.2913984.

Sethia, Divyashikha, Daya Gupta, and Huzur Saran. 2019. "Smart Health Record Management with Secure NFC-Enabled Mobile Devices." *Smart Health* 13. https:// doi.org/10.1016/j.smhl.2018.11.001.

Taylor, Robin, David Baron, and Daniel Schmidt. 2015. "The World in 2025 – Predictions for the Next Ten Years." In *2015 10th International Microsystems, Packaging, Assembly and Circuits Technology Conference, IMPACT 2015 – Proceedings*, 192–5. https://doi.org/10.1109/IMPACT.2015.7365193.

United Nations. 2014. "World Population Ageing, 2014." *Department of Economic & Social Affairs Population Division*, 73. https://doi.org/10.2307/1524882.

Varatharajan, R., Gunasekaran Manogaran, M. K. Priyan, Valentina E. Balaş, and Cornel Barna. 2018. "Visual Analysis of Geospatial Habitat Suitability Model Based on

Inverse Distance Weighting with Paired Comparison Analysis." *Multimedia Tools and Applications* 77 (14): 17573–93. https://doi.org/10.1007/s11042-017-4768-9.

Williams, Patricia A.H., and Vincent McCauley. 2017. "Always Connected: The Security Challenges of the Healthcare Internet of Things." In *2016 IEEE 3rd World Forum on Internet of Things, WF-IoT 2016*, 30–35. https://doi.org/10.1109/WF-IoT.2016.7845455.

Yin, Yuehong, Yan Zeng, Xing Chen, and Yuanjie Fan. 2016. "The Internet of Things in Healthcare: An Overview." *Journal of Industrial Information Integration* 1: 3–13. https://doi.org/10.1016/j.jii.2016.03.004.

Zhang, Lu, Qi Li, Ye Li, and Yunpeng Cai. 2018. "A Distributed Storage Model for Healthcare Big Data Designed on HBase." In *Proceedings of the Annual International Conference of the IEEE Engineering in Medicine and Biology Society, EMBS* July 2018: 4101–5. https://doi.org/10.1109/EMBC.2018.8513400.

Zhong, Ray Y., Goran D. Putnik, and Stephen T. Newman. 2019. "A Heterogeneous Data Analytics Framework for RFID-Enabled Factories." *IEEE Transactions on Systems, Man, and Cybernetics: Systems*: 1–10. https://doi.org/10.1109/tsmc.2019.2956201.

13 Blockchain-Based Regenerative E-Voting System

S. Vishnuvardhan, R. Aswath Srimari, S. Sridevi,
B. Vinoth Kumar, and G. R. Karpagam
PSG College of Technology

CONTENTS

13.1 INTRODUCTION

Voting is an integral part of any organization. It plays a significant role in democracy and the people's freedom to make choices. Paper ballot system is considered to be a traditional yet powerful medium of voting. Such election process is tedious to conduct and can lead to manual errors. Electronic voting system is an ongoing research directed at mitigating the cost of running an election while assuring the trustworthiness of voting with compliance of protection and privacy. Optical scanning machines used to scan the paper votes are often interrogated for manufacturing defects and the privacy of votes. Electronic voting system played a profound role in the evolution of the voting domain after the paper ballot system [1–3]. Any modification to the system must adhere to the integrity, credibility, and anonymity of the voters [4,5]. Direct recording-electronic (DRE) voting system registers votes through a ballot screen operated by the voter using touchscreens or buttons and captures the voting information in the local memory component. Elections in Virginia that used DRE system reported manipulation and incorrect recording of votes leading to vote hack and thefts. Because Internet is an essential part of our lives, during recent years, interest in Internet voting has picked up momentum [6,7]. Transparency and security continue to lack in Internet voting as voters cannot audit or review their vote. Online voting in a centralized storage is prone to hack and other vulnerable attacks. Blockchain is an emerging technology assisted by distributed network connected by various nodes with a strong cryptography foundation qualifying an election system to leverage these abilities to achieve a secure solution. Blockchain is system consisting of blocks and chains where transactions are recorded in a block. Blocks are connected to one other by a digital chain system.

The objective of the chapter is to discuss an easy process of voting with reliable information and security. The major contributions involve the creation of separate nodes to each users involved. Smart contracts are well defined for the application that does not overlap the proceedings of each use cases, thereby executing a clear flow of transaction that have credibility and avoid false transactions that violate smart contracts, which are nothing but a set of rules to be followed. Different phases of voting take place in accordance, thereby maintaining the chain of events required for the voting process.

Blockchain technology is still being explored by people around the world and many have succeeded to deploy it in unique applications. Research, publications, and various applications developed are discussed in Section 13.2 The usage of blockchain application in the voting system is elaborated in Section 13.3. Section 13.3.1 deals with the types of blockchain and consensus. Section 13.3.2 involves registering the users in the blockchain network and discusses how users' credentials are stored in a database. Section 13.3.3 illustrates the procedures before the election process. Section 13.3.4 verifies the eligibility of voters with face recognition. Section 13.3.5 explains how the voter casts vote and the transactions involved in the blockchain network. Section 13.3.6 notes how the process concludes and the changes in the blockchain once the vote is cast. This process involves the role of various designated persons who play a major role in mining the blockchain who form the miners. Use cases of the application are mentioned in Section 13.4. Section 13.5 discusses the possible attacks on blockchain and previous attacks that highlight the potential exploitations in blockchain. Section 13.6 presents the results of our application. The future trends and the final conclusion arrived from the results are mentioned in Section 13.7. The references are included in Section 13.8.

13.2 LITERATURE SURVEY

In the last few years, blockchain technology innovations have gained a huge impetus. Blockchains are distributed ledgers that enable parties who do not trust each other entirely to manage a set of global states. The parties agree on the values, existence, and histories of the states. Given the rapid growth of the technology landscape, it is necessary to comprehend the importance and challenges to have a firm grasp of what the core technologies have to offer, especially with regard to their data processing capabilities. The working of the blockchain system is the same as the database system, but with enhanced security using cryptography hash and easy access using a decentralized property. If we want to develop a blockchain platform, at the beginning, it is necessary to postulate a benchmarking framework to signify quantitative evaluation and comparison among the different blockchain platforms, as well as the existing database system in terms of both in-production and research systems. It postulates the performance of various blockchain platforms and traditional data processing workloads [8]. Blockchain technology was first introduced for the basis known as the cryptocurrency – Bitcoin – by Nakamoto et al. Blockchain is a distributed, decentralized, transparent ledger data structure which records all transactions in a chronological order same as a Merkle tree [9]. Blockchain technology stores sensitive and invariant data in a secured manner and in an encrypted format to ensure the integrity of a transaction, thereby proving a safe and trustable facility for users. Although cryptocurrency like Bitcoin is the most popular face, blockchain has recently gained massive attention due to its properties of easy and secure incorporation in various application domains such as Internet of Things (IoT), healthcare, and logistics [10]. The hash function in the block structure guarantees the security of the stored data and the trusted mechanism, which synchronizes the changes over transactional data to create a tamper-proof digital network platform. The incorporation of blockchain and the Internet diversifies the interactive systems between peer-to-peer (P2P) networks. The Bitcoin blockchain platform has been used in a broad variety of applications such as real estate transactions, and national tax collection [11]. Although blockchain technology is considered as the future of data protection and integrity, there are some potential research issues that exist in a distributed network, privacy preservation among permissioned and permission-less platforms, the consensus mechanism is still open and needs to be tackled to preserve privacy while using blockchain [12]. The rapid development of quantum computing in the near future has opened the possibility of carrying out attacks based on the algorithms of Grover and Shor. Both public cryptography and hash functions can be challenged by these algorithms. This forces blockchain cryptography (i.e., its public-key encryption algorithms and hash functions) to be revamped to withstand quantum attacks, thereby creating quantum-proof, quantum-safe post-quantum, or quantum-resistant cryptosystems. It is also important to review the most applicable post-quantum schemes to examine their application to the blockchain along with the key challenges [13]. By considering all, the potential secured blockchain technology platform should be created and incorporated in various application domains.

In the current era, the IoT is considered as one of the most exciting information technology innovations of the century. The IoT facilitates ubiquitous data collection and network connectivity together to fetch significant and vital data to provide intelligent ease of knowledge in everyday life as well as in industrial operations. At present, the IoT is expanding to make our day to days smart via smart devices. Although providing protection for IoT devices is still an immense problem, blockchain has the solution to ensure the security and efficiency of IoT devices. Blockchain is a combined technology that incorporates a creative and well-constructed eyesight to provide wide-ranging securities to IoT devices [14]. However, many resource-constrained IoT devices cannot bear the overhead of communication, traditional storage, and authentication protocols. Resource-rich devices need to be updated with their protocols as a lightweight version to interoperate with these devices so that all devices can communicate and function in a modern and unified manner. These lightweight communication protocols and schemes, however, face challenges in their reliability. Hence, when a large number of devices are connected with poor security, communication privacy, software malfunction, poor system management, and poor data storage platform guarantee to hold and transmit personal and sensitive data. For the aforementioned challenges in IoT, blockchain technology hopes to overcome these challenges [15]. Accordingly, blockchain empowers IoT devices to make IoT networks more secure and transparent. According to IDC, 20% of all IoT implementations will require blockchain-based solutions by 2019. Blockchain affords a decentralized and scalable environment for IoT devices. Financial firms and banks like HSBC, ING, and Deutsche Bank conduct PoC to validate the potential attainment of incorporation of blockchain technology in IoT. On the other hand, the IoT opens up endless possibilities for companies to conduct smart operations. Every system around us now features sensors that send data to the cloud. The combination of these two technologies, therefore, will make the systems effective [16]. One of the applications of using IoT is secured e-voting system.

The state-of-the-art study of existing e-voting systems and the rules of governing the election and protection of results encourage research into an alternative method for voting in a secured manner based on the latest innovative technologies. Voting is a way to express an opinion and to make a collective decision between groups of people. Election is conducted in different instances, for example, for selecting a leader of a country, selecting a head in shareholders markets, selecting a head for college and school communities, etc. In general, voting results in debates and campaigns. To cast vote, people need privacy. To reduce conflicts and irregularities arising during organizing and running an election, election campaigns are created. By trust development of Internet along with information technologies, e-voting system is introduced. Here, voters can submit their vote electronically without divulging their location. This system saves time and effort of election organizations efficiently. Kiayias et al. proposed a new approach for an E2E verifiable e-voting system without any assumption structure [17]. Chen et al. proposed a methodology for an e-voting system based on certificate-less deniably authenticated encryption method [18]. Kshetri and Voas et al. in 2018 delineated the blockchain technology for voting process, which ensures recording of votes in a transparent, secure, accurate, and permanent manner [19]. However, the scientific research for achieving an e-voting system is still far from replacing the traditional voting system completely. The replacement policy needs a thorough investigation, and high supremacy implementation is essential for developing solutions to overcome some critical issues, such as voters secrecy. Khan et al. proposed e-voting on multichain platform emphasizing on fundamental requirements such as integrity preservation and anonymity of vote [20]. In 2018, Agora Technologies developed a new Ethereum public blockchain-grounded digital democracy voting platform that protects voter's privacy. In this model, they used ELGamal cryptography method for protecting casted votes and voter's anonymity, and the developed system was called Neff shuffling [21]. Moura et al. develop an e-voting system to provide great transparency, while protecting voter's privacy and make easy auditability for interested citizens. Most of above systems deal only with transactional data [22]. Yet, no importance is given for multimedia data. For that, Bhowmik et al. proposed a novel watermarking-based multimedia blockchain system framework by generating a separate hash for image which preserves original media content [23]. Voters can have only limited amount of time to cast their vote. Somehow, if voters want to recast or change their decision and wish to vote for candidates, e-voting protocol makes a provisions for this. An e-voting protocol is designed that follows fundamental properties of the voting system and provides choice for voters to change their vote within a permissible time [24,25]. The distribution and integration of management procedures in the phases of election have been articulated by Francesco [26]. Real-time implementations of e-voting in blockchain can be seen in FollwMyVote, BitCongress, and FIVI.

From the previous works, it was observed that the current system lacks the unique identification of a voter. With the proposed system, the unique identification can be done at the voting booth by the authorized election officials using image recognition. Hence, this system is better than the existing systems that use both image recognition and blockchain for voting.

13.3   BLOCKCHAIN-BASED REGENERATIVE E-VOTING SYSTEM

Blockchain is a distributed ledger technology connecting various nodes to form a decentralized network. It contains a chain of blocks with information. Each block includes a timestamp, transaction data, and previous block's cryptographic hash. The participants of the blockchain are miners. Transactions are initiated by a source node. The requested transactions are broadcasted to a P2P network consisting of computers called nodes. Using the established algorithm, the network of nodes verifies the transaction. A confirmed transaction can include cryptocurrency, contracts, records, or any other information. Once the transaction is checked, it is appended with other transactions to produce a block of data to be added into the ledger, which is unalterable and permanent. The transaction is now complete.

13.3.1   TYPES OF BLOCKCHAIN AND CONSENSUS PROTOCOLS

There are three types of blockchain:

- Public blockchain,
- Private blockchain, and
- Hybrid/Consortium blockchain

In public blockchain, anyone can participate in the mining process to validate the transaction (e.g., Bitcoin, Arc Block, Symbiont, etc.). In private blockchain, only members inside of the consortium can participate in the mining process (HyperLedger Fabric, Ripple, Quorum, etc.). In consortium blockchain, anyone after meeting certain predefined criteria can participate in the mining process (e.g., Ethereum, Stellar, Aion, etc.). Initially, Ethereum platform is created only for public blockchain for its properties like smart contract; recently, it has been developed for private blockchain to support efficient business processes.

The legitimacy and validation of blockchain transactions is maintained by consensus mechanism. The consensus mechanism is accomplished by miners. Miners are special nodes who wish to participate in the mining process. Mining is a method of verifying, validating, authenticating, checking the legitimacy of the transaction, and adding the transaction as blocks by generating hash values. Different types of consensus mechanism are used to validate the transaction. Some of the different consensus mechanisms are:

Proof of Work (PoW): Any miner can randomly select and validate the transaction. Subsequently, miners spend computational power and time to generate the hash values to add the transaction as a block in the block-chain. Generally, the generated hash value must be equal to or smaller than the specific explicit values.

Proof of Stake (PoS): Miners will be randomized with the perception of coinage. Coinage is derived by the product of belonging number of coins by a miner and the number of days that coins impounded by a miner. If a coin has been unspent at least for 30 days might begin to participate for mine next block. Initially, coin starts over with zero age. This proce-dure secures the blockchain network from the domination of miners and reduces the significant consumption of computational power.

Proof of Burn (PoB): It postulates the principle of destroying and burning of grand virtual currency tokens to miners to buy virtual mining rig for mining the blocks. If more coins are burned resulting in more virtual mining rig, there is a chance to gain more rewards.

Proof of Elapsed Time (PoET): It was hosted by Intel in 2016. It deploys a Trusted Execution Environment (TEE) to provide security and random-ness for the leaders election process. It needs Software Guard Extension (SGX) processor to provide a security guarantee of TEE.

Proof of Deposit (PoD): Miners who wish to participate in the mining pro-cess should deposit some security amount before they mine and propose the blocks. This procedure is applied in Tendermint blockchain.

Proof of Importance (PoI): PoI not only relies on prolonged stakes holding by miners in the system but monitors the usage and spending of tokens to achieve the level of trust and importance, for example, NEM coin Blockchain.

Proof of Capacity (POC): POC was first introduced by Burstcoin crypto-currency. It utilizes hard disk space as a resource for the mining process. This process is also called hard disk mining.

Proof of Storage (PoS): PoS allows the utilization of storage capacity. It works on the concept of a certain piece of data would probably be stored by peers, which induces peers to participate in the mining process. In the proposed e-voting system, we used Ethereum private blockchain platform and PoS consensus mechanism for mining process.

The overall architecture is illustrated in Figure 13.1. The main features of blockchain are as follows:

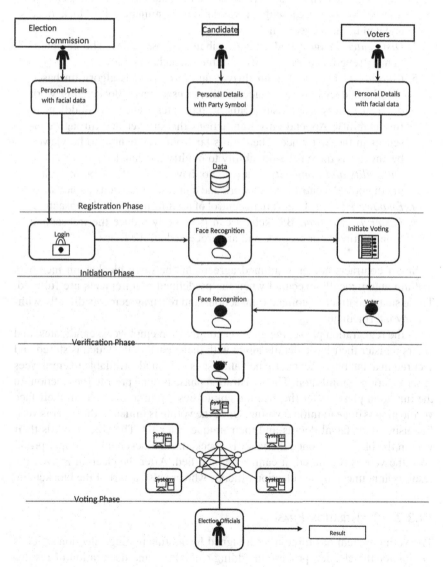

FIGURE 13.1 Overall architecture for blockchain e-voting system.

1. *Corruption-Proof*: Digital ledger is available for all nodes in the block-chain. When a new transaction is initiated, all the miners (the participants of the blockchain) need to verify the transaction. Only after the verification, the transaction forms a block in the ledger. This promotes transparency and is corruption-proof.
2. *Decentralized Technology*: The network is decentralized as a result there is no single governing head. No single node can enjoy ownership of the data.
3. *Enhanced Security*: As there is no central authority, no data can be inserted or modified without validation by miners. The block once updated in the ledger cannot be altered.
4. *Distributed Ledger*: Ledger is distributed across nodes. The computational strength is divided among all the blockchain nodes.
5. *Consensus*: The blockchain thrives due to consensus algorithm based on which decisions are made in the transactions. Blockchains work on a trust-less consensus without the interference of middlemen. Immutability: Append only transactions that are verified through consensus in the first place. The data is far from confidential to be viewed by anyone as there is no possibility to modify the data later.
6. *Durability and Longevity*: They do not have a centralized point of failure due to a decentralized network and can resist malicious assaults.
7. *Empowered User*: Users are in control of all information and transaction.
8. *Faster Transaction*: Blockchain can potentially reduce the transaction time to minutes or seconds and are processed all the time.

Smart contracts are programmed agreements between blockchain members that are automatically executed when the predefined requirements are fulfilled. These smart contracts automate transactions and reach agreements directly without any intermediate.

In the registration phase, the election commission employees, candidates, and voters register their basic details along with their facial data, which is stored and secured in a database. Access to the database is only made available to employees of the election commission. The registration phase is done prior to the election. In the initiation phase, after the election employees log in to the system, their face verification is done to initiate voting. Once the voting is initiated, the voter is verified using their facial details and their unique voter ID. The voter records their vote in the blockchain once the voter is checked. As blockchain is tamper-proof, once the vote is registered, it cannot be modified. After the election is over, the result is announced by the election officials who are also a part of the blockchain.

13.3.2 REGISTRATION PHASE

The voting system has three roles assigned for its functioning. Election officials to monitor the election process in adding candidates and declaration of results, election commission employees to initiate the voting, and voters who vote for the

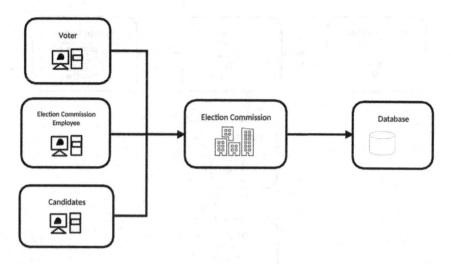

FIGURE 13.2   Registration phase.

candidates. Facial recognition scanner scans the face print of the participants in the blockchain and stores it in the database along with their unique ID. The details of election commission employees are added. The registered officials will now be able to add voters to the system. Database of voters contains the unique voter ID along with the faceprint. Officials of the election commission will add the candidates to the blockchain who are contesting the election and assign them an ID. A voter's role in the blockchain workbench is allocated to the registered voter. The blockchain creates three different nodes each for the voter, election commission officials, and election commission employees. The volume of nodes relies on the registration count for each role. Now the blockchain generates a chain of nodes. The registration process is illustrated in Figure 13.2.

13.3.3 INITIATION PHASE

Election commission employees are given the privilege to oversee the voting procedure. Election commission employees will log in to the system with their electronic mail ID with a secure password. The voting system requires the facial recognition of the employees to allow access to the database. Facial recognition is one of the biometric software that maps individual facial characteristics mathematically and stores it in database as a faceprint. The individual is identified by matching the facial features captured using the camera with the faceprint stored in a database.

Once the face recognized is verified, the employee initiates the voting. Initializing the voting process is a transaction that requires verification by all the miners to form a foundation block in the ledger. The transaction is sent to all miners or stored in a pool of unconfirmed transactions.

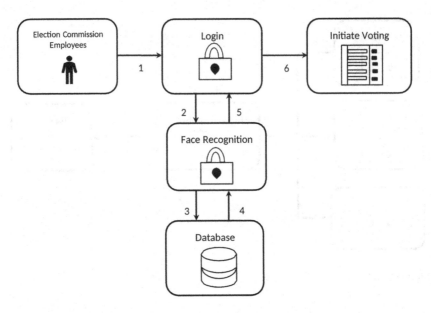

FIGURE 13.3   Initiation phase.

The miners pick the transaction from the pool and validate it based on the contract. The new transaction gets added as a block only after it is validated by all the miners. All miners are responsible to add the transactions or messages in the block. The transactions in the block cannot be modified or updated after being added to the blockchain. All the people involved in the voting system are now aware that the voting process is initiated and this transaction is updated as a new block. The voters can now initiate voting. The initiation process is exemplified in Figure 13.3.

13.3.4   VERIFICATION PHASE

Once the voter registration is done, the voter becomes valid to participate in the election. The registered election commission official will handle the verification of the voters during the election. The voter must carry voter card with an identification number on it for verification. The voter's faceprint is verified with the faceprint stored in the database corresponding to the identity code brought by the voter in person. This is done to check if the eligible holder of the voter card is casting a vote. If the verification fails, then the voter is not eligible to vote. The database can be accessed only by the registered election officials who have their faceprint stored in the database. The overall verification phase is shown in Figure 13.4.

13.3.5   VOTING PHASE

Voting process is now initiated and a single block with one transaction data is available with all the miners. Each block has a PoW called a signature. The block

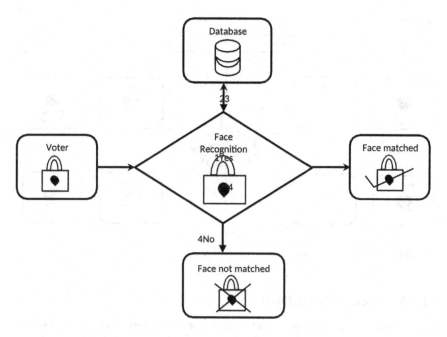

FIGURE 13.4 Verification phase.

created will execute a mathematical algorithm for cryptographic hash function to generate a hash output from a string input. The hash input which is a string of data is generated based on the transaction data. The hash input is converted to a hash output called signature of the block using a mathematical hash function that is unique to each block. A single change in the data of the block will cause it to get a completely different signature. The signature connects blocks to form a blockchain. This method is called mining to generate hash values for the blocks. The miner broadcasts the block to other miners included in its signature. Other miners will verify the legitimacy of the signature by using a string of date of the broadcasted block to generate a hash output and verify it with the included signature. The blocks are inserted to the blockchain if the signature was successfully verified. If it doesn't pass the verification, it indicates modification of the data in the block and it cannot be added in the chain. A miner cannot generate a transaction for another miner unless they are aware of the earlier block's digital signature.

A voter can now vote for a specific candidate and will create a fresh transaction. All miners validate the transaction and a fresh block is only generated with the signature of the earlier block and produces a fresh signature for the present block which should match the previous signature in the sequence of zeros. The random nonce which is the data part of the block is modified indefinitely until the output meets the signature requirements. The voting phase is illustrated in Figure 13.5.

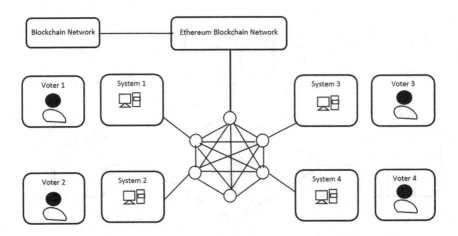

FIGURE 13.5   Voting phase.

13.3.6   ELECTION RESULT PHASE

Voters vote for their desired candidate. When the voters make their vote for Candidate "A" then the count of candidate "A" gets updated. The count value is visible only to the election commission officials only after the results get published. Smart contracts are developed to accomplish each vote count of election candidates and data is stored in the block. These contracts are the behavioral rule-set for all the participants of the blockchain. The results are published by the election commission officials. Once they initiate this transaction, all the voters will not be able to cast their vote. The candidate with maximum count gets displayed in the interface which is fetched from the ledger of the election commission officials. The election result phase is illustrated in Figure 13.6.

13.4   USE CASES

The following use cases are considered to validate our system.

Use Case 1 – Verified Voter: The voter details are registered in the database. Before voting, his/her face is identified and he/she is allowed to vote. He/she votes successfully and his/her vote is registered in the blockchain.

Use Case 2 – Unregistered Voter: The voter details are not registered in the database. When he/she tries to vote, his/her face will not be matched with the database. He/she will not be allowed to vote and the necessary actions will be taken.

Use Case 3 – Double Voting: The voter has already voted. He/she tries to vote again. The system will not accept a voter voting more than once. Hence, he/she will not be allowed to vote again and the necessary actions will be taken.

Use Case 4 – Location Issues: The voter's home address is one location "A". He/she currently works in location "B." He can vote in any of the voting booths available in location "B."

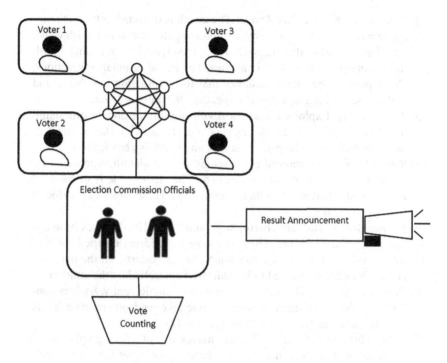

FIGURE 13.6 Election result phase.

The current voting system lags in the following areas: location independency, security, transparency, and privacy. Because blockchain is secure and transparent, it is used to implement the voting system. This is called regenerative voting system because it does not completely change the voting process but only the way in which we vote. The voting process is made completely transparent and decentralized. This system prevents anyone from tampering the votes which are casted, thereby giving actual results.

13.5 POSSIBLE ATTACKS ON BLOCKCHAIN NETWORK

Attacks on blockchain network, especially bitcoins, make way for hackers with cryptocurrencies. Recent attacks such as Ransomware WannaCry was massive due to extreme exploitation of blockchain security by hackers. Some of the possible attacks include.

1. *Distributed denial of service (DDoS)*: Though DDoS are hard to execute on blockchain network, blockchain technology is susceptible to DDoS attacks. Hackers intend to bring down the server by consuming all its processing resources, such as mining pools, e-wallets, and cryptoexchanges with numerous requests.

2. *Transaction Malleability Attack*: The attack is intended to trick the victim to pay the amount twice by manipulating the transaction hash value. Attackers can alter this transaction ID, broadcast it with changed hash to the network, and have it confirmed before the original transactions. Consequently, the victim assumes that the transaction has failed and makes a second attempt, thereby spending the amount twice.

3. *Time Jacking*: Exploits a theoretical vulnerability in Bitcoin timestamp handling. During this attack, hackers alter the network time counter of the node and forces the node to accept an alternative blockchain.

4. *Sybil Attack*: It is arranged by assigning several identifiers to the same node. Hackers control many nodes here. The victim node is surrounded by fake nodes that close all their transactions. These attacks are difficult to detect and prevent.

5. *Eclipse Attack*: Hackers control large number of IP addresses or maintain a distributed botnet. They overwrite the address on tried table of victim node and waits for it to restart. Once it restarts, all the outgoing connections are redirected to IP addressed controlled by the attacker.

6. *Phishing*: Attacks take place on user wallets too. Recently, hackers conducted a phishing campaign with a online fake seed generator on IOTA wallets and collected logs and secret seeds [27].

7. *Vulnerable Signatures*: Blockchain makes use of cryptographic algorithm to create user signatures, but these technologies have many vulnerabilities. For example, bitcoin uses ECDSA cryptographic algorithm to automatically generate private keys, which resulted in same random value in more than one signature due to its insufficient entropy [28].

8. *Smart Contract Attacks*: Vulnerabilities in smart contracts have risk to parties the sign the contracts. One such is Reentrancy attacks that make contract A call function from contract B that has an undefined behavior.

9. *Double Spending*: All transactions need to be verified by all the users which takes some time. This delay is exploited by hackers to trick the system into using the same coins or tokens in different transactions.

10. *Finney Attack*: One transaction is premined into a block and an identical transaction is created before the premined block is released to the network.

11. *51% or Majority Attack*: Hackers control 51% of the network hash rate and collect enough hashing power of the network. It is impossible in large networks, therefore, small networks are prone to this attack as it is easy for hackers to rent the computing power to create a majority of share of the small networks.

13.6   RESULTS AND DISCUSSIONS

The unique blockchain-based e-voting system makes maximum utilization of smart contracts to enable secure and cost-effective solution while guaranteeing voter privacy. It offers a new possibility to the democratic country to follow more

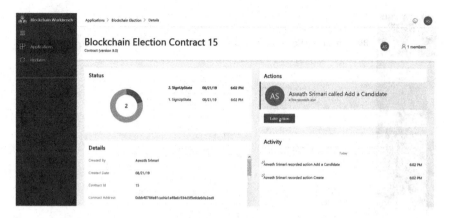

FIGURE 13.7 Blockchain application deployed in Microsoft Azure blockchain workbench.

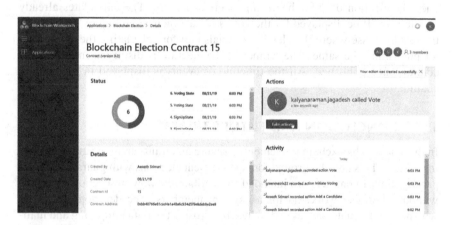

FIGURE 13.8 Voting phase: voters casting vote in the blockchain application.

time-efficient electoral process while increasing the transparency from the view of every voter. For a country with a large population, such a system allows greater throughput of transactions per second. The details of the voters, candidates, and the election officials are stored in Microsoft SQL Database. The image recognition is implemented in Python programming language. The smart contracts in written in solidity language and is deployed in Microsoft Azure Blockchain Workbench. Figure 13.7 shows how the blockchain applications are deployed in the Microsoft Azure Blockchain Workbench. The candidates contesting the election are added. The application will ensure its stability even when multiple voters vote simultaneously. The election process is location-independent and there is no necessity for the voter to have internet connection as it is provided by the voting booth. Figure 13.8 depicts the voting phase. Voters can begin to cast vote

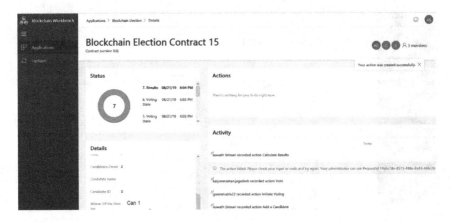

FIGURE 13.9   Election result phase: publication of the winning candidate.

once the election officials begin the election process. The candidates already registered will be displayed for the voters to cast their vote. Figure 13.9 depicts the results phase where the election officials call for calculating the results and for publishing the same. The winner of the election is displayed in the details column once published. All the timeline of events is displayed in the activity column.

13.7   CONCLUSION AND FUTURE TRENDS

In this chapter, blockchain implements a secure and transparent means of the voting system. This lays the ground for cost-efficient elections with guaranteed voter privacy. Smart contracts are intended to facilitate the flow of election processes without additional workload for all system members. The decisions are made with no intermediate parties and reduce the prospect of data tampering and malicious attacks. Voting is more open and voters can independently audit their votes. The concept of a decentralized network makes voting location-independent, and, overall, it will lead us to the following future trend.

IoT market will change its focus on security as more safety challenges are about to occur on connecting devices across the world. The main vulnerability of interconnected devices is the centralized architecture; therefore, blockchain offers a ray of hope for IoT security. Blockchain in voting can integrate IoT that aids the analysis of the voting pattern and prediction of voting trends by testing various data. This involves interconnection of all voting machines with the Internet. As years pass by, the turnout at polling booth shows a drastic downfall. Blockchain can come in aid of voters to cast votes at the leisure of their homes. Voters can use decision tokens to cast votes from a mobile phone or PC, which are then logged into a immutable blockchain and used to verify the outcome of the election. There can be no manipulation, recording errors, or tampering.

REFERENCES

1. Timothy B. Lee, "A Top Voting machine firm calls for paper ballots", Accessed on February 10, 2020 from https://www.wired.com/story/a-top-voting-machine-company-is-finally-ditching-paperless-voting/, 2019.

2. W. D. Eggers, *"Government 2.0: Using technology to improve education, cut red tape, reduce gridlock, and enhance democracy"*, Lanham, MD: Rowman & Littlefield, 2007.

3. Michal Chatlas, "Pros and cons of blockchain: Do I even need one for my business?", 2018, Accessed on February 10, 2020 from https://espeoblockchain.com/blog/pros-cons-blockchain-advisor.

4. K. Dalia, R. Ben, Y. A. Peter, and H. Feng, "A fair and robust voting system by broadcast", In *Fifth International Conference on E-Voting*, Bregenz: Gesellschaft fur Informatik (GI), 2012.

5. S. Bell, J. Benaloh, M. D. Byrne, D. Debeauvoir, B. Eakin, P. Kortum, and M. Winn, "Star-vote: A secure, transparent, auditable, and reliable voting system", In *2013 Electronic Voting Technology Workshop/Workshop on Trustworthy Elections (EVT/WOTE 13)*, Washington, DC: USENIX Association.

6. Pavel Tarasov, and Hitesh Tewari, "The future of e-voting", *IADIS International Journal on Computer Science and Information Systems*, 2017, 12(2): 148–165.

7. U. Madise, and T. Martens, "E-voting in Estonia 2005: The first practice of country-wide binding Internet voting in the world", In *Conference: Electronic Voting 2006: 2nd International Workshop, Co-organized by Council of Europe, ESF TED, IFIP WG 8.6 and E-Voting.CC*, August 2–4, 2006, Castle Hofen, Bregenz, Austria.

8. Tien Tuan Anh Dinh, Rui Liu, and Meihui Zhang, "Untangling blockchain: A data processing view of blockchain systems", *IEEE Transactions on Knowledge and Data Engineering*, doi: 10.1109/TKDE.2017.2781227.

9. Satoshi Nakamoto, "Bitcoin: A peer-to-peer electronic cash system", Accessed on February 15, 2020 from https://bitcoin.org/bitcoin.pdf, 2008.

10. Tejsi Sharma, Shivangi Satija, and Bharat Bhushan, "Unifying blockchian and IoT: Security requirements, challenges, applications and future trends", In *2nd International Conference on Intelligent Computing, Instrumentation and Control Technologies (ICICICT)*, Vimal Jyothi Engineering College, Kannur, Kerala, India jointly with IEEE India Council, 2019.

11. Sumit Soni, and Bharat Bhushan, "A comprehensive survey on blockchain: Working, security analysis, privacy threats and potential applications", In *2nd International Conference on Intelligent Computing, Instrumentation and Control Technologies (ICICICT)*, Vimal Jyothi Engineering College, Kannur, Kerala, India jointly with IEEE India Council, 2019.

12. Himanshu Saini, Bharat Bhushan, Aman Arora, and Anureet Kaur, "Security vulnerabilities in information communication technology: Blockchain to the rescue (A survey on blockchain technology)", In *2nd International Conference on Intelligent Computing, Instrumentation and Control Technologies (ICICICT)*, Vimal Jyothi Engineering College, Kannur, Kerala, India jointly with IEEE India Council, 2019.

13. Tiago M. Fernández-Caramés, and D. Paula Fraga-Lamas, "Towards post-quantum blockchain: A review on blockchain cryptography resistant to quantum computing attacks", *IEEE Access*, doi: 10.1109/ACCESS.2020.2968985.

14. Ayasha Malik, Siddharth Gautam, Shafiqul Abidin, and Bharat Bhushan, "Blockchain technology-future of IoT: Including structure, limitations and various possible attacks", In *2nd International Conference on Intelligent Computing,*

Instrumentation and Control Technologies (ICICICT), Vimal Jyothi Engineering College, Kannur, Kerala, India jointly with IEEE India Council, 2019.

15. Pinchen Cui, Ujjwal Guin, Skjellum Anthony, and David Umphress, "Blockchain in IoT: Current trends, challenges, and future roadmap", *Journal of Hardware and Systems Security*, doi: 10.1007/s41635-019-00079-5.

16. LeewayHertz, "Blockchain IoT use cases", Accessed on February 20, 2020 from https://www.leewayhertz.com/blockchain-iot-use-cases-real-world-products/, 2019.

17. Kiayias Aggelos, and Helger Lipmaa, "E-voting and identity", In *Third International Conference, VoteID 2011*, Tallinn, Estonia, September 28–30, 2011.

18. Guanhua Chen, Jianyang Zhao, Ying Jin, Quanyin Zhu, Chunhua Jin, Jinsong Shan, and Hui Zong, "Certificateless deniable authenticated encryption for location-based privacy protection", *IEEE Access*, doi: 10.1109/ACCESS.2019.2931056.

19. Nir Kshetri, and Jeffrey Voas, "Blockchain-enabled e-voting", *IEEE Software Digital Library*, 2018, 35: 95–99, doi: 10.1109/MS.2018.2801546.

20. K. M. Khan, J. Arshad, and M. M. Khan, "Secure digital voting system based on blockchain technology", *International Journal of Electronic Government Research (IJEGR)*, 2018, 14(1): 53–62.

21. C. Andrew Neff, "A verifiable secret shuffle and its application to e-voting", January 2001, Accessed on March 8, 2020 from http://web.cs.elte.hu/~rfid/p116-neff.pdf.

22. Teogenes Moura, "Blockchain voting and its effects on election transparency and voter confidence", In *Conference: The 18th Annual International Conference*, doi: 10.1145/3085228.3085263.

23. Deepayan Bhowmik, and Tian Feng, "The multimedia blockchain: A distributed and tamper-proof media transaction framework", In *Conference: International Conference on Digital Signal Processing*, doi: 10.1109/ICDSP.2017.8096051.

24. Yi Liu and Qi Wang, "An e-voting protocol based on blockchain", *IACR Cryptology*, ePrint Archive 2017, Accessed on March 18, 2020 from https://www.semantic scholar.org/paper/An-E-voting-Protocol-Based-on-Blockchain-Liu-Wang/5b6a0b0 ff2c574d9bb8bad9e191b22f44c92add7.

25. Freya Sheer Hardwick, Apostolos Gioulis, Raja Naeem Akram, and Konstantinos Markantonakis, "E-voting with blockchain: An e-voting protocol with decentral-isation and voter privacy", arXiv:1805.10258v2 [cs.CR] Jul 3, 2018, Accessed on February 25, 2020 from https://arxiv.org/pdf/1805.10258.pdf.

26. Francesco Fusco, Maria Ilaria Lunesu, Filippo Eros Pani, and Andrea Pinna, "Crypto-voting, a blockchain based e-Voting System", In *10th International Conference on Knowledge Management and Information Sharing*, doi: 10.5220/0006962102230227.

27. A. A. Andryukhin, "Phishing attacks and preventions in blockchain based projects", In *2019 International Conference on Engineering Technologies and Computer Science (EnT)*, supported by the Russian Academy of Sciences, Ministry of Science and Higher Education of the Russian Federation (grant 19-07-20005), IEEE, Moscow, Russia, 2019, pp. 15–19.

28. Joachim Breitner and Nadia Heninger, "Biased nonce sense: Lattice attacks against weak ECDSA signatures in cryptocurrencies", Accessed on March 15, 2020 from https://eprint.iacr.org/2019/023.pdf.

14 Blockchain Application of IoT for Water Industry and Its Security

Wu Linjing
Donghua University

Liu Xinyue
Shanghai National Engineering Research
Center of Urban Water Resources Co. Ltd

Shu Shihu
Donghua University

CONTENTS

14.1 INTRODUCTION

Under the trend of smart cities and smart water services, as an important part of the public service system, will usher in rapid development. The emerging security issues and privacy concerns caused by the increasing number of sensors/actuators are becoming prominent. Hence, the security threat that these services are likely attacked by hackers requires efficacious solutions. Over the past decades, with further construction of smart water, many practical cases around the world have proven that intelligent water based on blockchain technology has the potential to address such problems. Therefore, blockchain combined with the Internet of Things (IoT) will definitely be instrumental in advancing the automation and intelligence of water management in the future.

This chapter consists of five sections. First, possible issues and security challenges of IoT are summarized in Section 14.2, including general challenges, technical challenges, issues caused by centralized servers, and security attacks. Second, the issues and challenges in water management integrated with IoT are listed in Section 14.3, which includes monitoring, a tremendous number of wells, water supply system, and statistics. Furthermore, this part elaborates the salient futures of blockchain to meet quality-of-service (QoS) and intelligent water applications which is conducive to revealing latent issues and unforeseen challenges and providing favorable engineering experience. Third, the abstract concept of blockchain is introduced in Section 14.5. The presentation incorporates synopsis, structure, working mechanism, types, and development of blockchain. The different technologies in blockchain over time, particularly the reformed technologies, are compared in this chapter. Then, because future waterworks requires more service functions for approaching water crisis and urgent need of smart water based on blockchain, the security issues and challenges can be recapitulated as three points. In addition, four targeted factors should be considered in blockchain application in water industry to reinforce the technical performance. Finally, some countries, especially Australia, have already attached importance to efficient and environmental model of water industry. In Section 14.6, five practical applications based on blockchain in water industry are presented to briefly describe development status. Finally, three recommendations according to the current situation are proposed in the end.

14.2 ISSUES AND CHALLENGES IN IOT

14.2.1 BRIEF INTRODUCTION TO IoT

The notion of IoT was originally proposed by Kevin Ashton in 1999; according to him, IoT is "A network concept that creates a connection between material objects and online network for information interchange through several sensing equipment e.g. Radio Frequency Identification (RFID), Infrared Sensor etc., which can realize intelligent identification, positioning, tracking, monitoring, and management." In 2005, ITU Internet reports 2005—the Internet of Things reported by the International Telecommunication Union (ITU) formally defined the concept of the IoT, which is not just based on RFID but more technologies to achieve interconnection at any time, any place, and any object. In general, it is possible to achieve the goal that everything would be monitored or controlled by calculations or engines because the IoT can have access to add math logic to the things (objects) in the network.

The concrete idea of IoT can be expressed as a simple formula based on the latest conceptual framework [1].

$$IoT = Service + Data + Networks + Sensors \qquad (14.1)$$

14.2.2 SECURITY ISSUES AND CHALLENGES OF IoT

14.2.2.1 General Challenges

On September 20, 2016, a malware, the famous latterly software named Mirai, launched massive DdoS attacks on Krebsonsecurity.com all over the world [2]. The security vulnerabilities are becoming more destructive along with the emergence and development of Industrial Internet of Things (IIoT). The security would be vulnerable due to the possible issues of data leakage and unnecessary data damages when the terminals send instantaneous data to users or intelligent equipment. This is referred to as privacy concern and security vulnerability (see Tables 14.1–14.5), and there are many hidden risks [3].

14.2.2.2 Centralized Sever

Traditional centralized servers are based on third-party entities to guarantee privacy and security and to meet data handling requirements. The entities that process information proved the feasibility in abusing or worse situation in spite of an effective supervision, such challenges generally classified into four aspects as shown in Table 14.1.

TABLE 14.1
Issues Caused by a Centralized Sever [1]

NO	Issues
1	Any single failure of a central point may paralyze the entire network.
2	Users have no access to control over their transaction records or sensitive information in centralized server, and privacy concerns cannot be dismissed.
3	The data centrally saved in a server can be attacked or accidentally struck out due to the lack of guaranteed traceable and accountable features.
4	A centralized server may severely restrict the growth of IoT further, resulting in inefficient processing of a large amount of end-to-end communications.

TABLE 14.2
Technical Challenges Slowing Down Global IoT Adoption

NO	Problems	Content	Ref.
1	Cybersecurity	Eavesdropping, message spoofing/alteration, traffic analysis, distributed denial of service, password security attacks, Sybil attack, etc.	[4]
2	Privacy	The barriers to develop schemes that protect privacy by reasonable design are difficult to tackle, particularly detailed information.	[5]
3	Enormous data management	As huge amounts of data are generated by IoT devices, it can be difficult to manage in terms of refinement, communication/transmission, and storage.	[6]
4	Lack of standardization and interoperability	Although the solutions over standards in the field of IoT have been supported by IEEE, ETSI, IETF, W3C, OneM2M, ITU-T, OASIS IEC, OMG, etc., the proliferation and fragmentation of standards would be further aggravated by the lack of standardization.	[7,8]
5	Lack of skills	In the light of the complexity and heterogeneity of the techniques in IoT, which claims designative skills to design, implement, and deploy the solution.	[9]

14.2.2.3 Technical Challenges

In the past two years, the IoT platforms are developing from universal to be more specific accommodate certain needs. Such transformation has caused the emerging numerous types of licensing models from various platforms, either proprietary or open source, resulting in slowing down the existing global IoT applications [2] (see Table 14.2).

14.2.2.4 Security Attacks Types

Broadly, the security attacks are numerous in IoT, which can be summarized into four domains [3], as shown in Table 14.3.

TABLE 14.3
Four Broad Domains of Security Attacks

NO	Domains	Security Attacks	Ref.
1	Physical attacks	Tampering, fake node injection, malicious code injection, RF interference/jamming, permanent denial of service (PDoS), sleep denial attack, side channel attack	[10]
2	Network attacks	Traffic analysis attack, RFID spoofing, RFID unauthorized access, routing information attacks, selective forwarding, sinkhole attack, Sybil attack, man in the middle attack (MiTM), replay attack, wormhole attack, distributed denial of service (DoS/DDoS) attacks	[10–12]
3	Software attacks	The Mirai botnet, the jeep hack	[11,12]
4	Data attacks	Data inconsistency, unauthorized access, data breach	[13]

TABLE 14.4
Design Issues of Blockchain-Driven IoT Services [14]

NO	Issues	Content
1	Locations of nodes	In the blockchain network, the functionalities of nodes (also called endpoints) are responsible for tracking data and verifying transaction. In a PoS network, the nodes participate in a process described as "mining" (generating a new block). The layout of nodes arranged authentically plays a significant role on bandwidth, computation, and space.
2	Deployment of business logic and data	Smart contracts (stored scripts in a blockchain) provide a useful approach to implement the business logic. Some design concerns of smart contract in blockchain is that only a part of logic and data can be appropriately used in deploying.
3	How to realize cyberphysical integration	Cyberphysical integration in IoT services is a significant question to manage. For example, it is assumed that if every intelligent object has a pathway to link to a smart contract, is it possible to pay by delivering transactions to the contract address? Another issue is the way to make the service provider find and control objects?

14.2.2.5　Design Issues

The IoT and blockchain are both emerging concepts followed by the inevitably existing design faults of blockchain-driven IoT (B-IoT) services. Three unique design issues of B-IoT services are presented and discussed in Table 14.4.

14.2.2.6　Blockchain Challenges with IoT

Challenges in blockchain technology integrated with IoT are shown in Table 14.5.

TABLE 14.5
Challenges in BC Technology Integrated with IoT

NO	Challenges	Content	Ref.
1	Limitation with storage capability	Compare to the DAO-based blockchain technology, a single central server store facility is much more efficient than others (i.e., CAO) as each ledger must be stored at their own nodes.	[15]
2	Lack of skills in the field	As a new technology, for exploiting application in the field of water management, many solutions to fix practical difficulties need to researched. So far, the number is extremely less when it is integrated with the IoT, not to mention smart water.	[16]
3	Lack of Legal	This is one of the most challenging issues as the new technology does not have any legal codes to regulate it.	[17]
4	Complexity in computing ability	Not just in smart city system. When the blockchain technology is integrated with the vast network connected with diverse IoT system, the computing ability of algorithm for running the encryption becomes much more complex.	[18]
5	Processing time	The problem of reducing the necessary processing time for computing is still a challenge.	[19,20]
6	Scalability	The advantages would decline if continuous development leads to centralization.	[21]

14.3 ISSUES AND CHALLENGES IN INTELLIGENT WATER INDUSTRY INTEGRATED WITH IOT

14.3.1 CURRENT ISSUES IN WATER MANAGEMENT

"Urban disease" is becoming increasingly serious. As an important part of the public service system, under the global trend, smart water has gone through rapid progress in solving urban water issues and achieving sustainable development.

14.3.1.1 Monitoring Limitations

The traditional water environment monitoring technology mainly depends on laboratory analysis. Limitations such as inaccurate result, long detection cycle, high manpower input, large cross-section severely restrict its application to large watershed range.

14.3.1.2 Number of Wells

There is a large number of working wells, inspection wells, valve wells, and instrument wells in the municipal water networks, which needs large labor input for regular and irregular inspections, as well as involvement of citizens. Therefore, the management mode is complicated and challenging with heavy workload and long response time.

14.3.1.3 Design Issues

Water supply problems such as serious leaks, irregular water supply services, and water quality risks still exist.

14.3.1.4 Statistics

Relative statistics have shown that the number of secondary water supply tanks in Shanghai exceeds 200,000, accounting for over one-fifth of the water supply [22] and resulting in management difficulties.

14.3.2 Is Blockchain the Rescue?

The increasing massive intelligent devices and corresponding data terms have already hindered the space of meeting QoS [23]. The way of storing data and conducting the anonymous transactions for decentralized system is supported by blockchain, which is different from traditional centralized processing. It is conducive to further eliminate the malfunctioning points, and accordingly a advanced and resilient ecosystem for long-term operation of equipment is being created [17]. In addition, it is of great benefit to intensify private consumer data safeguards because of the cryptographic algorithms of blockchain [24]. The most important features of blockchain [25] are decentralization, immutability, auditability, and fault tolerance.

14.3.2.1 Decentralization

In fact, the data interchanges (i.e., the water quality) would be smoothly stored in a centralized network only after being authorized by central third-party entities. Therefore, there is a fee for centralized server maintenance, and it is hard to strike the optimal balance between performance and cost [26]. However, in a decentralized system, such as blockchain-based infrastructures, two nodes can engage in mutual transactions (peer-to-peer network) despite the third party. Decentralization cannot rarely cut down the cost and eradicate multiple-to-single traffic workflows to preclude delays and failures caused by a single point [27].

14.3.2.2 Immutability

Unlike tamper, the records are immutable as blockchain is a censorship-resistant system. For the changes would be easily detected in this system, attackers can rarely alter the data unless the major nodes in the water pipe networks are attacked [27].

14.3.2.3 Auditability

The transparency of blockchain provides the authority for peers to inspect and verify specific addresses, because the pseudo-anonymity refers to the addresses in blockchain paralleled with real identities.

14.3.2.4 Immutability

Especially in water monitoring, although record faults and leaks are rarely caused by transmission, detection limit, omission, etc., it requires significant efforts because of such a large number of sensors or other equipment. In blockchain, data errors can be identified and the replicas (peers will copy the ledger records) stored in peers can be proven to be effective in mitigating the concern of data leakages.

14.3.3 OTHER SALIENT FEATURES

14.3.3.1 Anonymity

As a result of the independent method of encryption (public and private keys) in blockchain network, people can protect their privacy in anonymous state by enabling third parties for verification.

14.3.3.2 Veracity

Real-time records of all nodes will be stored and replicated as the same copy of the ledger. Moreover, the consensus level forces each record to be verified by consensus mechanism, even stimulating the emergence of wrong bogus entries. While it is not a fretting trouble because such cases can be eliminated with efficiency by consensus successfully.

14.3.3.3 Transparency

In the past, transactions were not transparent or broadcasted to the public, which is a necessary part to end a process, but only in a public blockchain.

14.3.3.4 Trust

When the blockchain network adds data, the trust is built on a mechanism that major (more than half) participants are in agreement. Hence, intermediation can be avoided, and the degree of trust increases significantly.

14.3.3.5 Turing-Complete

Turing-complete aims to construct a distributed application by taking advantages of the abundance of programing functions (programmable) [27,28]. Therefore, the overall blockchain networks, for example, Ethereum, Hyperledger Fabric, and EOS, adhere to Turing Completeness and render full support to the stateful contracts besides Bitcoin.

14.3.4 IoT APPLICATION FOR SMART WATER MANAGEMENT

To solve above problems, the application of IoT technology can significantly improve the management efficiency and level of waterworks construction and the security of water systems. It would be introduced into four levels, terminal services (sensor), network, platform, and client (application) (shown in Figure 14.1), which have been widely constructed around the world for water monitoring, purification, leaks detection, pipes corrosion, distribution, management, and cost analysis. Nowadays, there are more and more applications around the world (see Table 14.6).

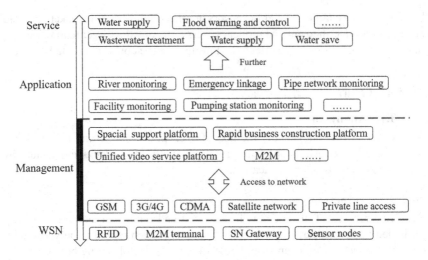

FIGURE 14.1   Current architecture model of smart water.

TABLE 14.6
Examples of Waterworks Integrated with IoT

NO	Nation	Applications	Ref.
1	Hong Kong	WATERIG has conducted a pilot study laying rainwater collection points and connecting these collection points with water treatment systems to enter vertical agriculture and urban greenhouse projects.	[29]
2	Australia	Power Ledger cooperates with governments to use blockchain technology for operating a solar photovoltaic power plant and district water treatment facility to verify the reliability of blockchain technology in water resources and energy distribution systems.	[30]
3	The United States	IBM has built a comprehensive real-time sensing network and a real-time online monitoring system for the Hudson River.	[31]
4	Netherlands	IBM has worked on a project of monitoring the condition of flood inundation.	[31]
5	Australia	The SEQ smart water network.	[32]
6	Singapore	The recycling water project.	[32]
7	China	According to the status of their water conservancy informatization and the most urgent water problem, a model of smart water construction has been proposed.	[33]

14.4   ISSUES AND CHALLENGES IN IOT

14.4.1   SYNOPSIS OF BLOCKCHAIN TECHNOLOGY

Blockchain originated from the article *Bitcoin: A Point-to-Point Electronic Cash System*, in which a novel and reforming framework, distributed infrastructure, and computing paradigm based on an asymmetric encryption algorithm was presented and described [34,35]. The blockchain that is different from tradition follows the two top trending technologies, namely, "big data" and "artificial intelligence (AI)" and the positive prospect of blockchain combined with IoT is revealing. This concept terms from 2009 to 2013 blockchain 1.0; now, blockchain has developed to the era of programmable and provides decentralized and distributed methods for IoT applications [36].

Four restrictive factors of current blockchain are security, efficiency, resources, and game. Toqeer et al. proposed a comparative analysis of blockchain architecture and its application [37].

14.4.2 THE CONSTRUCTION OF BLOCKCHAIN [38]

14.4.2.1 Nodes

Each node plays an indispensable role. Once the old hash from a block has been validated by the next block successfully, the next block would be authorized to build a link with last one. Even one node has the right to start a transaction and broadcast to all the network nodes.

14.4.2.2 Components

A complete block consists of two parts: block header and block body (list of transaction). A general blockchain consists of six layers (see Figure 14.2a), and every layer in an IoT application structure should be integrated with blockchain to improve the security of responding system (e.g., smart water).

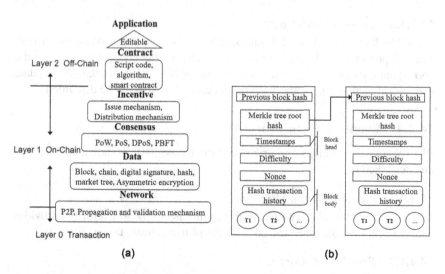

(a) (b)

FIGURE 14.2 Structure of blockchain technology. (a) Structure of popular blockchains. (b) Organization of a single blockchain.

14.4.2.3 Mining Formula

The block header includes previous harsh, Merkle tree root harsh, mining statistics, and so on, and the following formula is used to explain the concept of "mining."

$$H_k = \text{Hash}(H_k - 1|T||\text{Nonce}|) \qquad (14.2)$$

Broadly, all the transactions, described as block body part, in Merkle tree play a significant role in organizing, which is called "Merkle Tree Root" with excellent performance in verification procedures. This kind of structure can guarantee the digital invariability, but follows the block hash falsified by attackers.

14.4.3 How Blockchain Works? [39]

14.4.3.1 Hash Encryption [38]

The former block will receive a series of hash numbers related with corresponding transactions from the later block, which creates a connection between them. The exception is the first block of the chain (not in Figure 14.2b), called genesis, created by the first man who developed blockchain (Satoshi Nakamoto in Bitcoin).

14.4.3.2 Electronic Signature

Private key used to encrypt was produced by a set of random numbers while public key used to decrypt is from a private key which can produce an address. Such a process is irreversible. Hash is a tool to encrypt transaction information, and using public key and password, is broadcasted to decrypt a result. Then, the validated transactions within a promised time interval would be sequentially packaged as a matched candidate block with timestamp using a set of judging criteria.

14.4.3.3 Correct Mechanism

If one block (M) contains valid transactions needs to be checked and confirmed, then the previous blocks (N) would be corrected by the concept of harsh on the corresponding chain. They add the right case of block to their chain and update their world view in time, otherwise the checking block would be discarded.

14.4.4 Various Types of Blockchain

Blockchain is frequently summarized into three types [40], as shown in Table 14.7 and described below.

14.4.4.1 Public (Bitcoin, Ethereum)

The ledger of public blockchain is visible to Internet users, which further provides endorsement to verification and adds blocks of transactions to the chain [41].

14.4.4.2 Private (R3, Corda)

This kind of private blockchains only permits specific people with verification authority and block addition in the organization [42].

TABLE 14.7

Types, Features, and Application of Blockchains

Types	Public	Private	Consortium
Consensus	PoW/PoS/DPoS	Distributed consensus algorithm	Distributed consensus algorithm
Ledger	All	Only specific people	Only a group of organization
Degree of centralization	Decentralized	Polycentric	Polycentric
Features	Self-establishment of credit	Transparent and traceable	Efficiency and cost priority
Representative application	Virtual currency, etc.	Audit, issuance, etc.	Payment, settlement, etc.

14.4.4.3 Consortium (Monax, Multichain)

The ledger of organization (such as waterworks Ltd.) on the Internet is only allowed to be in the progress of verification and transaction addition, in addition, who can view the information is restricted to the selected group.

14.4.5 DEVELOPMENT OF BLOCKCHAIN TECHNOLOGY

14.4.5.1 Development Path

The development of blockchain technology is described in Figure 14.3 and Table 14.8. More details from blockchain 1.0 to 3.0 is introduced below.

14.4.5.2 Typical Applications

Blockchain technology has been and will be widely used in IoT and IIoT due to several innovative features. With the development of blockchain, some typical applications are summarized in the following subsections.

    a. Electric vehicle clouds and edge (EVCE) [44–47];
    b. Mobile commerce [48–50];
    c. Trace food source [51,52];
    d. Cloud storage [53–55];
    e. Authentication and access management [55,56];
    f. Big data [57,58];
    g. Smart home [59–61];
    h. Smart cities [62,63];

14.4.5.3 Blockchain 1.0—Cryptocurrencies

Blockchain based on cryptography was first developed in the domain of cryptocurrencies, which set up a safer financial model [64] for the existing advantages. In addition, it is reported that some cryptocurrencies (e.g., Dogecoin, Ethereum)

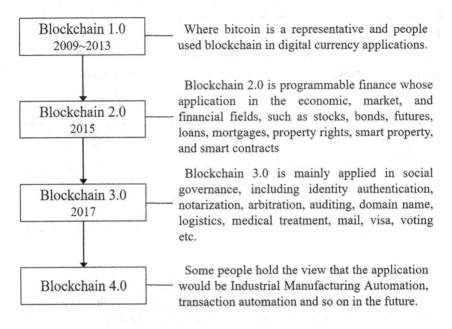

FIGURE 14.3   Development of blockchain technology [43].

TABLE 14.8

Comparison between Blockchain 1.0 and 2.0

Layer	Blockchain 1.0	Blockchain 2.0
Application	Transfer and keep accounts	*EVM, DAPP*
Incentive	Distribution and issue mechanism	Distribution and issue mechanism (Reducing to 16 s)
Consensus	Proof of work (PoW)	*PoW, PoS, DPoS, PBFT*
Network	P2P, propagation, and validation mechanism	P2P, propagation, and validation mechanism
Data	Block, chain, digital signature, hash, market tree, asymmetric encryption	Block, chain, digital signature, hash, market tree, asymmetric encryption (*supporting to input data and variable*)

The italic stands for new additives.

are regarded as more suitable for applications in the financial market. So far, the applications of blockchain have been various and far beyond Bitcoin.

14.4.5.4   Blockchain 2.0

Based on blockchain 1.0, the updated versions are available in economic, market, and decentralization of financial markets (stokes, bonds, smart contact, etc.) [65]. In addition, the emergence of blockchain 2.0 due to the development of smart contracts through turning-complete and universal computer program languages [64]. The merits of smart contact, defined as a lightweight decentralized application (DApp), are the following:

a. *DAO*: Smart contract can run autonomously without any participation of third-party entities.
b. *Strong Stability*: The hacker or any other attack would be possible only when the majority of the whole nodes fail; in such a scenario, wrong execution can be efficiently avoided to maintain stability.
c. *Traceable and Transparent*: The DApp always runs on the same block, and all the valid records and outcome can be traced.
d. The smart contract runs on blockchain, and the quality and advantages are applicable to the blockchain through smart contracts [66].

14.4.5.5 Blockchain 3.0

Blockchain 3.0 can be used in applications beyond the previous version such as healthcare and culture [67]. Some cases have been conceptualized and summarized in Table 14.9.

14.4.6 THE COMPARISON OF TECHNOLOGY

14.4.6.1 Smart Contracts in Blockchain [80]

As mentioned, although the nodes do not trust each other, the transactions can be maintained and trusted. To distribute to the "trustless network," a key feature is that the entire network is combined with cryptography. Smart contract is a self-executing script on blockchain. In addition, there are several benefits of a smart contract, notably among them can be classified into automated transactions, excellent accuracy, trustless nodes in network (with the help of third-party mechanism, there is no need to trust a centralized organization), and lower costs. This has attracted huge attention of programmers due to its proper, distributed, heavily automated workflows [39].

14.4.6.2 Distributed Autonomous Organization (DAO) and Centralized Autonomous Organization (CAO)

In IoT or IIoT network, the comparison of architecture CAO and DAO (shown in Figure 14.4) and the benefits of decentralizing IIoT or IoT are clearly superior to traditional centralized systems, as shown in Tables 14.10 and 14.11:

In the application of IIoT (i.e., smart city with cloud), the different architectures are shown in Figure 14.5.

14.4.6.3 Ethereum in Blockchain Application with IoT

The blockchain technology is originated from bitcoin that is regarded as the blockchain 1.0 and exists several engineering limitations in application integrated with IoT. With the need for further functionalities, Ethereum has recently become

TABLE 14.9

Several Aspects of Application of Blockchain 3.0

NO	Domains	Cases	Ref.
1	Finance	The largest banks in the world have collaborated to develop cryptocurrency (UBS of Switzerland) and have made rapid progress in the financial markets; a stock exchange system supporting blockchain is planning to launch in the near future by Australian Stock Exchange in 2020.	[68,69]
2	Identity safety	SelfKey is an identity wallet that helps users (individuals/companies) to manage their own digital identity; ShoBadge is a secure enterprise identification authentication tool using which users can authenticate using own identity recognition.	[70,71]
3	Foreign assistance	Usizo is a crowdfunding platform that is planning to launch in South Africa. The Bankymoon meter of the platform based on blockchain can pay the bill directly to meters due to the function of token; United Nations assists the refugees in Jordan by providing a few shops (merchants) for nearby refugees, who have the right to receive funds from selected shop after identity verification.	[72]
4	Voting	In multiparty states, online voting instead of other ways is becoming an alternative at the country level and politicians to avoid two issues, low voter and voter fraud. Horizon State is a kind of online digital voting box, which is more credible and inspiring.	[73]
5	Transportation	"Arcade City Token" as the cryptocurrency for payment was proposed by Arcade City (a platform of matching passengers and drivers), which aims to pay the daily transportation fee of users by the same cryptocurrency.	[74]
6	Agriculture	National Taiwan University has developed a decentralized system based on the GCOIN technology for managing irrigation information (pesticide use, crop costs, climate changes, energy use, soil quality, etc.) collected by deployed neighborhood sensors.	[75]
7	Food	Walmart started to develop a food tracking system in cooperation with IBMl Data61 of CSIRO in Australia to introduce blockchain from preventing food fraud fare.	[76,77]
8	Healthcare	The Media Lab in MIT is developing an electronic medical records (EMRs) system, the MedRec, to adapt diverse systems in solving the interoperability; OmniPHR, a personal health record (PHR) management system, provides the construction of semantic interoperability and integration of different health standards.	[78,79]

a symbol for the onset of blockchain 2.0. Bitcoin only realized built-in cryptocurrency and transaction, while the functions extended by Ethereum can support the maximum extent of blockchain platform. It involves the majority of decentralized applications (financial products and service, ownership traceability, digital identification, voting, etc.) and makes them come true.

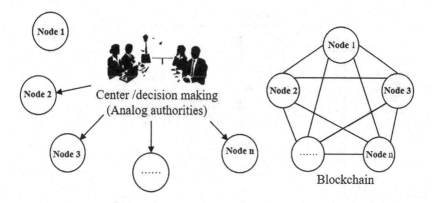

FIGURE 14.4 CAO and DAO (based on blockchain) architecture.

TABLE 14.10

The Advantages of DAO (Compare to CAO) [81]

NO	Drawbacks	Content
1	Scalability	As the amount of devices that operates in a centralized system and generate data are increasing, more storage facility is required to maintain the running and updating of the system.
2	Communication overhead and synchronization	The distributed and immutable nature of blockchain leads to high communication overhead, and the cost of replica (all peers) should further affiliate to the overhead.
3	Efficiency	If a transaction needs to be approved (n peers totally), the one must be verified by $(n-1)$ by the other peers. This demonstrates reducing efficiency in great measure.
4	Energy wastage	When a novel BC technology is applied practically (to consensus) by using proof of work (PoW), of the cost of calculation increases.

TABLE 14.11

Drawbacks of the Traditional Blockchains (Ethereum and Bitcoin)

NO	Advantages	Content
1	Enhance confidence and security	The single point of vulnerability from attackers in IoT system is inevitable. Hence, the method of unforgeable signatures cryptographically was employed in the data collection layer (node-to-node transaction). It is considered immutable and resistant from attacks.
2	More robust	Eliminating intermediaries produced by CAO through decentralizing IoT, which will probably reduce related cost.
3	Autonomy	As Section 14.3.2 (3) auditability part described
4	Data provenance	When the ledger has recorded all transmission procession and the devices/entities have signed the corresponding data, people have access to data provenance.
5	Fairness	A relative fair way is to incentive efficient parties by native cryptocurrency.

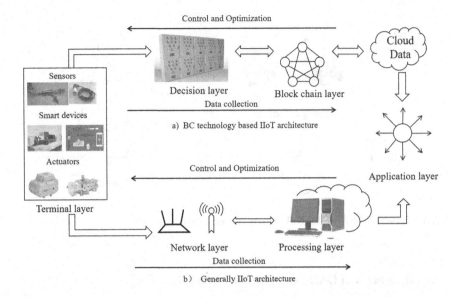

FIGURE 14.5   BC technology-based IIoT architecture and general IIoT architecture.

Ethereum, first proposed by Vitalik Buterin, is an open-source public block-chain platform with smart contract. This concept was recognized as "the next generation of cryptocurrency and decentralized application platform." It provides a decentralized Ethereum Virtual Machine (EVM) to handle peer-to-peer con-tracts through its dedicated cryptocurrency (Ether). The structure of Ethereum is shown in Figure 14.6.

Nicola et al. [82] conducted a survey of attacks on Ethereum smart contracts (SoK). Dongchao et al. explicitly summarized the graphical structure and statistical proper-ties of Ethereum transaction relationships [83]. Matevž presented some approaches to front-end IoT application development for the Ethereum blockchain [84]

14.5   BLOCKCHAIN APPLICATION FOR
WATER INDUSTRY SECURITY

14.5.1   WHY BLOCKCHAIN HAS GOOD PROSPECTS IN APPLICATION?

14.5.1.1   Water Crisis Is Approaching [85]
In 2017, Officials in Cape Town have found that the city's water reserve can only sustain for 6 months. Fortunately, they avoided running out of water through col-lective water storage operations and some unusual precipitation.

Tokyo, London, Istanbul, Barcelona, and Mexico City have been facing the same situation wherein their water reserves will run out in the coming decades. The fact is that human beings always have conflicts over unfair resource distribution.

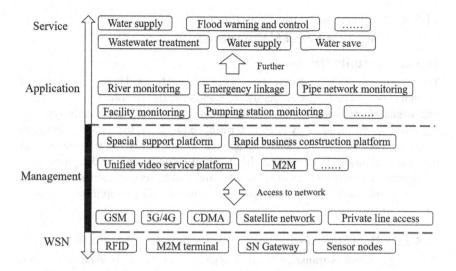

FIGURE 14.6 Structure of general Ethereum.

The dispute between Karnataka and Tamil Nadu in India for water resources has existed for a long time. Because the climate has become drier in the past few years, the Bolivian government advocated privatization of water resources in 2000, which directly triggered the "Water Battle in Cochabamba." Goldman Sachs predicts that water resources will become "oil" of the 21st century, so they have already taken some actions to obtain vested interests.

14.5.1.2 Smart Water Based on Blockchain Technology Is Urgent [85]

In 2019, lawmakers in Colorado hoped that the U.S. government will study the potential of blockchain technology in water rights management. Republican Senator Jack Tate, Democratic Rep. Jeni James Arndt, and Republican Rep. Marc Catlin submitted Bill 184 to the Senate, proposing to authorize the Colorado Water Institute to study how blockchain technology can help improve its operations [86]. On October 24, 2019, President Xi Jinping declared to accelerate the development of blockchain technology and enhanced the industrial innovation. Shenzhen Water Group released Shenzhen Water Strategy 2035 on November 16, 2019 [87], encouraging the practice and application of new informationized technology in the water industry, emphasizing the innovation and development of water technology, and supporting Shenzhen to build a city of water technology innovation engines. The blockchain technology has certain applicable potential in water industry management and industrial development due to the multiple attributes of water. Especially for refined water resources management and water capital management in smart water affairs. Specific ideas need to be further combined with research and development at technology maturity, system construction, and management level.

14.5.2 Security Issues and Challenges of Intelligent Water Management

14.5.2.1 Security Standards

The diverse sensors from different companies would be used in the terminal layer of smart water. The hardware, hardware standards, information source, content, and format of captured information originated from services are not always identical. To address the security issue of terminal equipment, the foremost approach is to establish industrial standards, as well as standards for admission, testing, and evaluation. Furthermore, the standard formulation and design must be implemented to improve security of vital and public security services, such as trusted architecture, secure development processes, and preapplication security assessment.

14.5.2.2 Data Transmission

Short-range wireless transmission is frequently used in the IIoT (RFID, WIFI, ZigBee, Bluetooth etc.). Two factors have considerably increased the chance of data being stolen during data transmission. (1) The interface data of various wireless sensors generally utilize a custom format, and data encryption uses a simple one-by-one encryption. (2) To ensure compatibility among components because there are too many ports in the open network. The commonly used solutions are intrusion detection systems, public key infrastructure, and physical unclonable function.

14.5.2.3 Protection System [37]

The Internet is always the key to the entire IIoT technology, which would deliver real-time and accurate information from objects through various wired and wireless networks. This illustrates that the information security risks of IIoT mainly exist in two aspects: (1) It comes from the natural structural characteristics of the IIoT. (2) It comes from the risks caused from external network integrated with the Internet. More intelligent devices are deployed to the edge of the network, even disorderly and randomly. Consequently, network devices become invisible or unidentified just like an "island effect" – the central security agency cannot monitor all devices and networks. Then, the risk of causing vulnerabilities is increasing; although security measures are taken into account for critical business data nodes, ignoring certain noncore business in the IoT would risk the security of the entire system.

14.5.3 Factors Need to be Considered for Blockchain Applications in Water Industry [88]

14.5.3.1 Record

Water sectors need to access lots of information including market share, consumption and transaction of consumers (management of bills), consumption partners, etc. Blockchain technology has the facility to balance and settle real-time data more efficiently at lower cost.

14.5.3.2 Compliance Reporting/Audit

Considering the natural characters of blockchain regarding immutability, immediacy, and transparency, the real-time data (instead of post-hoc records) provided would properly keep the output creating audit trails for regulators and streamlines compliance. This method can not only drastically improve data accuracy and confidence but also reduce time and effort. Therefore, a more powerful system would offer scalability across multiple networks, segregation of data for industrial privacy, and flexible redundancy across related nodes.

14.5.3.3 Data Reconciliation

The blockchain network creates a distributed data repository by computers, and each computer (trustless node) has a duplicate of the ledger. Under this structure, the data will be saved independently, which can be scheduled by the processing center due the connected network. In addition, it can enable regulators to monitor and control the real-time status of water allocation balance, waterworks licenses, the state of transactions, and water management information, instead of a centralized regulatory reporting process.

14.5.3.4 Bond Issuance on Blockchain

As the water credit project, bond leasing provides an incentive mechanism to improve water efficiency, resilience, and industrial innovation. Blockchain smart contracts can automatically execute bond issuance, which can simplify green bond issuance processes. The completion of agreement signing and certification on the blockchain has the potential to replace the current offline paper agreement printing process and improve the efficiency of agreement signing. In addition, the bond based on blockchain will provide a shared space for various agencies.

14.6 APPLICATION SUMMARY

Blockchain technology is progressively entering the smart water industry, with many applications in areas such as water data transactions, waterworks service contracts, and sharing. There are five practical engineering cases in water industry as follows [89]:

14.6.1 ORIGIN CLEAR

An American water treatment technology supplier has been working on a blueprint for a blockchain protocol called WaterChain [89]. The agreement aims to increase the transparency and efficiency of the water treatment industry through the use of smart contracts and cryptocurrencies.

14.6.2　Civic Ledger

"Water Ledger" was developed after Australian Blockchain Company received a $80,000 federal government grant. Civic Ledger has signed a contract to explore the potential of developing a "blockchain-based rainwater trading platform for rainwater trading" in partnership with South East Water, a utility company, in a new residential development in Melbourne, Australia.

14.6.3　AQUAOSO

According to Christopher Peacock, CEO and founder of AQUAOSO, a water rights management platform and water transaction market, AQUAOSO will integrate smart contracts with its existing platform through blockchain with water flow monitoring sensors supporting IoT. It will act as a "gateway" to confirm delivery to clients and automatically issue payments to sellers.

14.6.4　Newater Technology Works

In China, a company called "NW Blockchain Ltd." intends to use blockchain technology to raise funds for industrial wastewater projects by selling asset-backed tokens, similar to private equity funds.

14.6.5　A Water Management Partner [88]

For example, the Water Credit project presented by Smart4Tech, a monetary incentive program aimed to improve water efficiency, resilience, and advance public (urban water users) and private sectors (industrial water users). Blockchain is regarded as a separate mechanism of verification and confident source of shared information.

14.6.6　Recommendation

1. The prominent advantages of blockchain e.g. distributed storing, big data technology, P2P network, consensus mechanism, etc. should be taken to improve the security and privacy. In addition, the communication protocol of its network interface is highly versatile, which can connect with many external advanced technology systems to improve the expansibility in physical applications.
2. Blockchain will fundamentally change the pattern of arrangement and traceability of water resources. The most critical blockchain based technology in water source management are: (1) Transparency helps consumers to make decisions regarding when to store or use water, and to prevent corruption where local authorities tamper with or detain water quality data (e.g., Power Ledger). (2) Decentralized solutions can play a key role in these environments to complement traditional methods

and expand the reach of safe water and sanitation services (e.g., Scott Bryan) [90].

3. Australia is recognized as a leader in intelligent water industry throughout the world, which includes water source management, municipal engineering, agriculture use, environmental protection, etc. We are in the fastest developing and most potential period as well as the beginning of B-based smart water industry. These cases from Australia or other nations are valuable for future applications in China.

14.7 CONCLUSION AND FUTURE WORKS

With the development of IoT/IIoT and blockchain technology, B-based IoT services have become popular with their notable characteristics. Some conceptual and abstract technologies including IoT/IIoT, blockchain, smart contract, Ethereum, DAO, smart water, and others were summarized in this chapter. Furthermore, we demonstrated the advantages/salient features of B-based IoT applications and compared it with the traditional centralized sever integrated with IoT, as well as listed a series of IoT-based smart water applications in Tables 14.1–14.5. The mechanism, types, and development of blockchain were explicitly described in Section 14.4, in which the latter was introduced from two routes. One of them is described in order of time, the other is in order of emerging technologies according DAO/CAO, smart contract, and Ethereum. Blockchain applications for B-based water industry security have a good prospect due to the existing water crisis and the urgent need for decentralized and trustless organization. Finally, an overview of practical blockchain-based smart water applications integrated with IoT all over the world was proposed in Section 14.6.

Security issues and privacy concerns in intelligent water industry are taken into consideration, especially security standards, data transmission, protection system. It is necessary to pay attention to the characteristics of the IoT itself while constructing smart water, for Industrial Control System (ICS) security protection, to create a perfect defense mechanism. Four factors (record, compliance reporting/audit, data reconciliation, bond issuance on blockchain) should be considered for future blockchain applications in water industry. The current cases from Australia or other countries are valuable experiences for future applications in China, and the most critical blockchain-based solutions in water source management are transparency and decentralized solutions. The applications of blockchain-based water solutions are not yet put into practical use. The combination of blockchain technology and artificial intelligence will produce greater progress in automation and smart water as artificial intelligence matures.

ACKNOWLEDGEMENTS

Part of this work was supported by key research projects of Shanghai Science and Technology Commission (19DZ1204400) and Open Project of State Key Laboratory of Urban Water Resource and Environment, Harbin Institute of

Technology (QA201612). The authors would like to thank all the people who contributed to the study.

REFERENCES

1. I. Antonio Luigi, and M. Giacomo. Understanding the Internet of Things: definition, potentials, and societal role of a fast evolving paradigm. *Ad Hoc Networks*, 2017, 56: 122–140.
2. M. S. Ali, M. Vecchio, M. Pincheira, K. Dolui, F. Antonelli, and M. H. Rehmani. Applications of blockchains in the Internet of Things: a comprehensive survey. *IEEE Communications Surveys & Tutorials*, 2019, 2(21): 1676–1717.
3. S. Jayasree, R. Sushmita, and D. B. Sipra. A comprehensive survey on attacks, security issues and blockchain solutions for IoT and IIoT. *Journal of Network and Computer Applications*, 2020, 149: 1–20.
4. B. Amira, B. Abdelmadjid, and G. Saïd. M2M security – challenges and solutions. *IEEE Communications Surveys & Tutorials*, 2016, 18(2): 1241–1254.
5. J. C. Zhou, Z. F. Dong, L. Xiao et al. Security and privacy for cloud-based IoT: challenges. *IEEE Communications Magazine*, 2017, 55(1): 26–33.
6. M. Mohammadi, and A. Al-Fuqaha. Enabling cognitive smart cities using big data and machine learning – approaches and challenges]. *IEEE Communications Magazine*, 2018, 56(2): 94–101.
7. H. Saini, B. Bhushan, A. Arora, and A. Kaur. Security vulnerabilities in information communication technology – blockchain to the rescue (A survey on Blockchain Technology). In *2019 2nd International Conference on Intelligent Computing, Instrumentation and Control Technologies (ICICICT)*, Kannur, Kerala, India, 2019, pp. 1680–1684.
8. V. Gazis. A survey of standards for machine-to-machine and the Internet of Things. *IEEE Communications Surveys & Tutorials*, 2017, 1(19): 482–511.
9. D. S. Nunes, P. Zhang, and J. Sa Silva. A survey on human-in-theloop applications towards an internet of all. *IEEE Communications Surveys & Tutorials*, 2015, 17(2): 944–965.
10. I. Andrea, C. Chrysostomou, and G. Hadjichristofi. Internet of Things – security vulnerabilities and challenges. In *2015 IEEE Symposium on Computers and Communication (ISCC)*, Larnaca, 2015, pp. 180–187.
11. M. M. Ahemd, M. A. Shah, and A. Wahid. IoT security—a layered approach for attacks & defenses. In *2017 International Conference on Communication Technologies (ComTech)*, Rawalpindi, 2017, pp. 104–110.
12. P. Varga, S. Plosz, G. Soos, and C. Hegedus. Security threats and issues in automation IoT. In *2017 IEEE 13th International Workshop on Factory Communication Systems (WFCS)*, Trondheim, Norway, 2017, pp. 1–6.
13. M. Chan. "Why Cloud Computing Is the Foundation of the Internet of Things" [Online]. Available: https://www.thorntech.com/2017/02/cloud-computing-foundation-internet-things/ (Accessed 22 March 2020).
14. C. Liao, S. Bao, C. Cheng, and K. Chen. On design issues and architectural styles for blockchain-driven IoT services. In *2017 IEEE International Conference on Consumer Electronics – Taiwan (ICCE-TW)*, Taipei, 2017, pp. 351–352.
15. M. Mohammadi, and A. Al-Fuqaha. Enabling cognitive smart cities using big data and machine learning – approaches and challenges. *IEEE Communications Magazine*, 2018, 2(56): 94–101.

16. D. S. Nunes, P. Zhang, and J. Sa Silva. A survey on human-in-the loop applications towards an internet of all. *IEEE Communications Surveys & Tutorials*, 2015, 2(17): 944–965.

17. M. S. Ali, M. Vecchio, M. Pincheira, K. Dolui, F. Antonelli, and M. H. Rehmani. Applications of blockchains in the Internet of Things – a comprehensive survey. *IEEE Communications Surveys & Tutorials*, 2019, 2(21): 1676–1717.

18. A. Al-Fuqaha, M. Guizani, M. Mohammadi, M. Aledhari, and M. Ayyash. Internet of things – a survey on enabling technologies, protocols, and applications. *IEEE Communications Surveys & Tutorials*, 2015, 4(17): 2347–2376.

19. F. Tschorsch, and B. Scheuermann. Bitcoin and beyond – a technical survey on decentralized digital currencies. *IEEE Communications Surveys & Tutorials*, 2016, 3(18): 2084–2123.

20. H. M. A. Aljassas, and S. Sasi. Performance evaluation of proof-of-work and Collatz conjecture consensus algorithms. In *2019 2nd International Conference on Computer Applications & Information Security (ICCAIS)*, Riyadh, Saudi Arabia, 2019, pp. 1–6.

21. J. Kwon. "Tendermint – Consensus without Mining" [Online]. Available: http://tendermint.com/docs/tendermint.pdf (Accessed 22 March 2020).

22. X. Y. Liu, S. H. Shu, K. Yang, et al. Intelligent management of secondary water supply systems in Downtown Shanghai. *Procedia Computer Science*, 2019, 154: 206–209.

23. F. Mohamed, D. Makhlouf, and M. Mithun. Blockchain technologies for the Internet of Things – research issues and challenges. *IEEE Internet of Things Journal*, 2019, 6(2): 2188–2204.

24. Y. Zhang, and J. Wen. The IoT electric business model – using blockchain technology for the internet of things. *Peer-to-Peer Networking and Applications*, 2016, 10(4): 983–994.

25. P. Srinath, N. Samudaya, M. N. N. Rodrigo, et al. Blockchain technology – is it hype or real in the construction industry? *Journal of Industrial Information Integration*, 2020, 17: 1–20.

26. GOV.UK. "Distributed Ledger Technology – Beyond Block Chain" [Online]. Available: https://www.gov.uk/government/news/distributed-ledger-technology-beyond-block-chain (Accessed 22 March 2020).

27. D. Ali, S. K. Salil, and J. Raja. "Blockchain in Internet of Things – Challenges and Solutions" [Online]. Available: https://arxiv.org/ ftp/arxiv/papers/1608/1608.05187.pdf (Accessed 22 March 2020).

28. K. Christidis, and A. M. Devetsikiotis. Blockchains and smart contracts for the Internet of Things. *IEEE Access*, 2016, 4: 2292–2303.

29. Oliver Russell. "Water Crisis" [Online]. Available: https://medium.com/@oliver.russellcw/we-have-a-water-crisis-can-blockchain-technology-save-us-b3941bb5625d (Accessed 22 March 2020).

30. Power Ledger. [Online]. Available: https://mp.ofweek.com/blockchain/a745683826266 (Accessed 22 March 2020).

31. D. H. Kim, K. H. Park, G. W. Choi, et al. A study on the factors that affect the adoption of Smart Water Grid. *Journal of Computer Virology and Hacking Techniques*, 2014, 10: 119–128.

32. J. H. Kim. Case study of Smart Water Grid development Australia and Singapore. *Water Future*, 2011, 44(8): 19–24.

33. T. A. Yu, Y. Z. Jiang, and M. X. Yang. Foundation and development strategy for wise water affair management. *China Water Resource*, 2014, 20: 14–15.

34. M. Iansiti, and R. L. Karim. The truth about blockchain. *Harvard Business Review*, 2017, 1(95): 118–127.

35. S. John. "Blockchain Technology in Finance – Five Use Cases Beyond Cryptocurrencies that You Need to Know" [Online]. Available: https://www.linkedin.com/pulse/blockchain-technology-finance-five-use-cases-beyond-you-john-soldatos (Accessed 22 March 2020).

36. S. Charlebois. "How Blockchain Technology Could Transform the Food Industry" [Online]. Available: https://theconversation.com/how-Blockchain-technology-could-transform-the-food-industry-89348 (Accessed 22 March 2020).

37. T. T. A. Dinh, R. Liu, M. Zhang, G. Chen, B. C. Ooi, and J. Wang. Untangling blockchain – A data processing view of blockchain systems. *IEEE Transactions on Knowledge and Data Engineering*, 2018, 7(30): 1366–1385.

38. Yu, Y., Li, Y., Tian, J., and Liu, J. Blockchain-based solutions to security and privacy issues in the Internet of Things. *IEEE Wireless Communications*, 2018, 25(6), 12–18.

39. K. Christidis, and A M. Devetsikiotis. Blockchains and smart contracts for the Internet of Things. *IEEE Access*, 2016, 4: 2292–2303.

40. T. Sharma, S. Satija, and B. Bhushan. Unifying blockchian and IoT – security requirements, challenges, applications and future trends. In *2019 International Conference on Computing, Communication, and Intelligent Systems (ICCCIS)*, Greater Noida, India, 2019, pp. 341–346.

41. X. Lei, S. Nolan, C. Lin, et al. Enabling the sharing economy – privacy respecting contract based on public blockchain. *Proceedings of the ACM Workshop on Blockchain, Cryptocurrencies and Contracts*, 2017, 5(51): 15–21.

42. D. Tien Tuan, W. Ji, C. Gang, et al. Blockbench – a framework for analyzing private blockchains. In *Proceedings of the 2017 ACM International Conference on Management of Data*, 2017. Chicago, USA, pp. 1085–1110.

43. N. Dong, and X. T. Zhu. Blockchain technology and its prospect of industrial application. *Journal of Information Security Reasearch*, 2017, 3(3): 200–210.

44. S. K. Datta, J. Haerri, C. Bonnet, et al. Vehicles connected resources – opportunities and challenges for the future. *IEEE Vehicular Technology Magazine*, 2017, 12(2): 26–35.

45. L. Hong, Z. Yan, and Y. Tao. Blockchain-enabled security in electric vehicles cloud and edge computing. *IEEE Networks*, 2018, 32(3): 78–83.

46. M. Conoscenti, A. Vetrò, and J. C. De Martin. Blockchain for the Internet of Things – a systematic literature review. In *2016 IEEE/ACS 13th International Conference of Computer Systems and Applications (AICCSA)*, Agadir, 2016, pp. 1–6.

47. J. Kang, R. Yu, X. Huang, et al. Enabling localized peer-to-peer electricity trading among plug-in hybrid electric vehicles using consortium blockchains. *IEEE Transactions on Industrial Informatics*, 2017, 13(6): 3154–3164.

48. Z. Li, J. Kang, R. Yu, et al. Consortium blockchain for secure energy trading in industrial internet of things. *IEEE Transactions on Industrial Informatics*, 2018, 14 (8): 3690–3700.

49. K. Suankaewmanee, D. T. Hoang, D. Niyato, et al. Performance analysis and application of mobile blockchain. In *Proceedings of the 2018 International Conference on Computing, Networking and Communications (ICNC)*, 2018, chongqing, China, pp. 642–646.

50. K. Christidis, and M. Devetsikiotis. Blockchains and smart contracts for the internet of things. *IEEE Access*, 2016, 4: 2292–2303.

51. S. Ramamurthy. "Leveraging Blockchain to Improve Food Supply Chain Traceability" [Online]. Available: https://www.ibm.com/blogs/Blockchain/2016/11/leveragingBlockchain-improve-food-supply-chain-traceability/ (Accessed 22 March 2020).

52. L. Dong. "What's the Future of Blockchain in China?" [Online]. Available: https://www.weforum.org/agenda/2018/01/what-s-the-future-of-Blockchain-in-china/?from=timeline (Accessed 22 March 2020).

53. A. Dorri, S. S. Kanhere, and R. Jurdak. "Blockchain in Internet of Things – Challenges and Solutions" [Online]. Available: https://www.researchgate.net/publication/314949919 (Accessed 22 March 2020).

54. N.Joshi."DistributedCloudStoragewithBlockchainTechnology"[Online].Available: https://www.allerin.com/blog/distributed-cloud-storage-with-Blockchain-technology (Accessed 22 March 2020)

55. N. Kshetri. Can blockchain strengthen the internet of things? *IT Professional*, 2017, 4(19): 68–72.

56. S. Ravindra. "The Role of Blockchain in Cybersecurity" [Online]. Available: https://www.infosecurity-magazine.com/next-gen-infosec/Blockchain-cybersecurity/ (Accessed 22 March 2020).

57. A. Venkat. "Introduction to Blockchains & What It Means to Big Data" [Online]. Available: https://www.kdnuggets.com/2017/09/introduction-blockchain-big-data.html (Accessed 22 March 2020).

58. T. Ahram, A. Sargolzaei, S. Sargolzaei, J. Daniels, and B. Amaba. Blockchain technology innovations. In *2017 IEEE Technology & Engineering Management Conference (TEMSCON)*, Santa Clara, CA, 2017, pp. 137–141.

59. Dragos Mocrii, Yuxiang Chen, and Petr Musilek. IoT-based smart homes – a review of system architecture, software, communications, privacy and security. *Internet of Things*, 2018, 1: 81–98.

60. A. Dorri, S. S. Kanhere, R. Jurdak, and P. Gauravaram. Blockchain for IoT security and privacy – the case study of a smart home. In *2017 IEEE International Conference on Pervasive Computing and Communications Workshops (PerCom Workshops)*, Kona, HI, 2017, pp. 618–623.

61. M. A. Ferrag, M. Derdour, and M. Mukherjee. Blockchain technologies for the internet of things research issues and challenges. *IEEE Internet of Things Journal*, 2018, 12(18): 2077–2080.

62. D. Bruneo, S. Distefano, M. Giacobbe, et al. An IoT service ecosystem for smart cities – The smartME project. *Internet of Things*, 2019, 5: 12–33.

63. S. N. O. Scekic, and S. Dustdar. Blockchain-supported smart city platform for social value co-creation and exchange. *IEEE Internet Computing*, 2019, 23(1): 19–28.

64. X. Li, P. Jiang, T. Chen, X. Luo, and Q. Wen. A survey on the security of blockchain systems. *Future Generation Computer Systems*, 2017, 107: 841–853.

65. S. Melanie. *Blockchain – Blueprint for a New Economy*. Newton, MA: O'Reilly Media, Inc, 2015: 152.

66. C. Kinnaird, M. Geipel, and M. Bew. *Blockchain Technology*. London: Arup, 2017: 71.

67. P. Otte, M. de Vos, and J. Pouwelse. TrustChain – a Sybil-resistant scalable blockchain. *Future Generation Computer Systems*, 2020, 107: 770–780.

68. A. Martin. "Six Global Banks Join Forces to Create Digital Currency" [Online]. Available: http://www.gata.org/node/17605 (Accessed 22 March 2020).

69. H. Hossein. "Research Methods in Computer Science – The Challenges and Issues" [Online]. Available: https://www.researchgate.net/publication/314949919 (Accessed 22 March 2020).

70. Selfkey. [Online]. Available: https://selfkey.org/ (Accessed 22 March 2020).

71. ShoCard. "ShoBadge Secure Enterprise Identity Authentication Built Using the Blockchain" [Online]. Available: https://apps.apple.com/cn/app/shobadge/id1243205791?l=en (Accessed 22 March 2020).

72. Bankymoon. [Online]. Available: http://bankymoon.co.zasocial-projects (Accessed 22 March 2020).

73. "Online Voting" [Online]. Available: http://shorizonstate.com/ (Accessed 22 March 2020).

74. E. B. Hamida, K. L. Brousmiche, H. Levard, et al. "Blockchain for Enterprise: Overview, Opportunities and Challenges" [Online]. Available: https://hal.archives-ouvertes.fr/hal-01591859/ (Accessed 22 March 2020).

75. M. F. Xie. "Taiwan National University Financial Blockchain" [Online]. Available: https://fintech.csie.ntu.edu.tw/?s=GCOIN (Accessed 22 March 2020).

76. P. Zhang, D. C. Schmidt, J. White, and G. Lenz. Blockchain technology use cases in healthcare. *Advances in Computers*, 2018, 111: 1–41.

77. F. Scott, and C. Berkeley. "Putting Australia on the Blockchain Map – Standards Australia Releases Roadmap for Blockchain Standards" [Online]. Available: https://www.kwm.com/en/au/knowledge/insights/standards-australia-blockchain-roadmap-issues-use-cases-dlt-applications-20170302 (Accessed 22 March 2020).

78. MIT Lab. [Online]. Available: https://www.media.mit.edu/publications/medrec/ (Accessed 22 March 2020).

79. N. K. B. Maged, et al. "Geospatial Blockchain: Promises, Challenges, and Scenarios in Health and Healthcare" [Online]. Available: https://www.researchgate.net/publication/325870739 (Accessed 22 March 2020).

80. S. Wang, L. Ouyang, Y. Yuan, X. Ni, X. Han, and F. Wang. Blockchain-enabled smart contracts: architecture, applications, and future trends. *IEEE Transactions on Systems, Man, and Cybernetics: Systems*, 2019, 11(49): 2266–2277.

81. S. Madhusudan, and S. H. Kim. Blockchain technology for decentralized autonomous organizations. *Advances in Computers*, 2019, 115: 115–139.

82. B. Massimo Nicola, and C. Tiziana. A survey of attacks on Ethereum smart contracts (SoK). In M. Maffei and M. Ryan (Eds.), *Principles of Security and Trust.* Lecture Notes in Computer Science, vol. 10204. Berlin, Heidelberg: Springer, 2017, pp. 164–186.

83. D. Guo, J. Dong, and K. Wang. Graph structure and statistical properties of Ethereum transaction relationships. *Information Sciences*, 2019, 492: 58–71.

84. M. Pustišek, and A. Kos. Approaches to front-end IoT application development for the Ethereum blockchain. *Procedia Computer Science*, 2018, 129: 410–419.

85. United Nations. [Online]. Available: https://www.unwater.org/water-facts/water-food-and-energy/ (Accessed 22 March 2020).

86. X. Qian. "Colorado Lawmaker Looks at Water Rights Management with Blockchain" [Online]. Available: https://www.jinse.com/news/bitcoin/331585.html (Accessed 23 March 2020).

87. Shenzhen Water Authority. "Shenzhen Water Authority's Feedback on Publicly Seeking 'Shenzhen Water Strategy 2035' (Draft for Comments)" [Online]. Available: http://swj.sz.gov.cn/gzhd/zjfk/201912/t20191216_18935378.htm (Accessed 23 March 2020).

88. A. Poberezhna. *Transforming Climate Finance and Green Investment with Blockchains – Addressing Water Sustainability with Blockchain Technology and Green Finance.* New York: Academic Press, 2018: 189–196.

89. Originclear. [Online]. Available: https://www.originclear.com/ (Accessed 22 March 2020).

90. S. John. "Blockchain Technology in Finance – Five Use Cases Beyond Cryptocurrencies that You Need to Know" [Online]. Available: https://www.linkedin.com/pulse/blockchain-technology-finance-five-use-cases-beyond-you-john-soldatos (Accessed 22 March 2020).

Index

Taylor & Francis
Taylor & Francis Group
http://taylorandfrancis.com

Printed in the United States
By Bookmasters